Lecture Notes in Economics and Mathematical Systems

397

Jaap Wessels   Andrzej P. Wierzbicki (Eds.)

User-Oriented Methodology and Techniques of Decision Analysis and Support

Proceedings of the International IIASA Workshop
Held in Serock, Poland, September 9-13, 1991

Springer-Verlag Berlin Heidelberg GmbH

Editors

Prof. Dr. Jaap Wessels
Technical University Eindhoven
Department of Mathematics and Computing Science
P.O. Box 5/3
5600 MB Eindhoven, The Netherlands

Prof. Dr. Andrzej Piotr Wierzbicki
Institute of Automatic Control
Warsaw University of Technology
Nowowiejska 15/19
00-665 Warsaw, Poland

ISBN 978-3-540-56382-2   ISBN 978-3-662-22587-5 (eBook)
DOI 10.1007/978-3-662-22587-5

© Springer-Verlag Berlin Heidelberg 1993
Originally published by Springer-Verlag Berlin Heidelberg New York in 1993

Typesetting: Camera ready by author/editor
42/3140-543210 - Printed on acid-free paper

Preface:
User-Oriented Decision Support

Over the last 10 years great progress has been made in integrating different types of decision aids in decision support systems (DSS). In this development the International Institute for Applied Systems Analysis (IIASA) played an important role with its own research in the System and Decision Sciences program, and particularly, by organizing a series of meetings on this topic in close cooperation with research institutes in countries where there is a National Member Organization. A broader public has been informed about the progress made via the publication of all the proceedings of previous meetings by Springer-Verlag in the series of Lecture Notes on Economics and Mathematical Systems. These previous mettings took place in Austria (1983, Lecture Notes volume 229), Hungary (1984, volume 248), Germany (1985, volume 273), Japan (1986, volume 286), Bulgaria (1987, volume 337), USSR (1988, volume 351), and Finland (1989, volume 356).

The present volume reports on the proceedings of the next meeting in this sequence which took place in Serock near Warsaw, Poland, on September 9-13, 1991. For the organization, IIASA cooperated with the Institute of Automatic Control of Warsaw University of Technology and with the Systems Research Institute of the Polish Academy of Sciences. The meeting was co-sponsored by the Committee of Automatics and Robotics of the Polish Academy of Sciences. A grant from the Polish Ministry of Education helped substantially in making the meeting possible.

An important reason for organizing this meeting in Poland was the fact that the group of Polish researchers working on the theory, methodology, and tools for interactive decision analysis and support has always been very active. The Polish group contributed substantially to all previous meetings and, over the years, the group has cooperated intensively with IIASA. A separate volume in this Lecture Notes Series (volume 331) documents the Polish activities in decision analysis and support. The meeting in Serock was also an occasion for reviewing the results of a cooperative research agreement of IIASA with several Polish research institutes working in this field. Several of these results are presented in this volume.

The title of the volume (and of the workshop) stresses an important tendency in the development of decision support tools: the final users are placed more and more in the center of the deliberations. There might be several types of users of decision support systems: analysts, policy advisers, operational officers, modellers encoding their expertise in substantive models, groups of negotiators, and even sometimes actual decision makers. For all of them it is important that the particular decision support system is customized to their needs and this in turn requires that the decision support system is developed with the active participation of the different groups of future users. As a result, compromises can be made between the demand for flexibility, on the one hand, and easiness of use, on the other. The resulting decision support system should stress the stages of the decision process in which the users are most interested and should include in its graphic interface those symbols and representations with which the user is accustomed to work and which stimulate his or her creativity. In this respect it is important to remember that decision support systems have various tasks: not only to help in choosing sensible decisions and in comparing the effects of different decisions but also to help the user in sharpening his/her intuition and in searching for new options and approaches.

The above-mentioned postulates are accepted guidelines for development rather than accomplished goals, which can be seen from the papers from the Workshop; however, the title of the Workshop was selected to present a challenge and future guidelines. Some papers presented responded to this challenge, and several fine contributions to this theme – on constructing user interfaces, on organizing intelligent DSS, on modifying theory and tools in response to user needs – are included in this volume.

Another major trend indicated by some papers in this volume is an emerging connection between tools and software environments for modelling and for decision support. A DSS, to be useful, must contain knowledge about substantive aspects of decision situations encoded in the form of either logical models (as in expert systems) or analytical models; the latter are more diversified and there exists a highly developed methodology and art of modelling. However, in order to easily incorporate such models in modern decision support systems, new collections of tools and more flexible, modular software environments must be developed. This trend was strongly indicated in our papers (the editors of this volume), as well as several other papers. In order to make the DSS friendly for the final user, we must also make the model base friendly for modellers and analysts. This is important since it is the expertise and knowledge of various users (experts, modellers, analysts, as well as the final user), encoded ultimately in the DSS, which will determine on the final usefulness of the system.

There were also some other interesting directions indicated by the papers from the Workshop. This volume contains 26 papers selected from Workshop presentations. The papers are subdivided in two major parts. The first part is devoted to the methodology of decision analysis and support and related theoretical developments; the second part reports on the development of tools – algorithms, software packages – for decision support as well as on their applications.

The first part starts with the paper by Wessels (one of the editors) on the relations between the problem specification and further mathematical analysis in decision support systems, concentrating on an example of a new approach for specifying goods flow situations for decision support. This is followed by five papers on various aspects of DSS methodology – group and bargaining decision situations, utility-based models in intelligent decision support, combining rule-based and analytical approaches, special min/max graphs useful in the organization of a decision process, and graphical user interfaces.

The next group of papers represents contributions to established areas of decision analysis: multiple criteria optimization, and decision theory. Three of the papers contribute to such important topics in multi-criteria optimization as parametric approximations of the efficient frontier, dynamic multi-criteria optimization, and quantitative analysis of properly Pareto-optimal solutions. The next five present contributions to the study of preference structures in decision theory: the problem of linear approximations of preference structures, smooth relations in such structures, pairwise comparisons, interval specification of tradeoffs, and a consistent ordering of the importance of criteria.

The second part begins with a paper by Wierzbicki (one of the editors), reflecting on the issues of classification of decision support systems, reviewing Polish contributions to this field, and representing a methodological perspective on the requirements for modern analytical modelling for decision support. This is followed by two more theoretical papers on DSS tools: analytic centers in nondifferentiable optimization algorithms for DSS and methods of classification by rough set theory.

The next five papers concentrate on software tools for decision support. They start with a difficult problem of representing two-dimensional irregular shapes in a graphic user interface. This paper is followed by tutorials on two well-developed and useful software packages: DINAS for interactive multi-criteria network analysis and HYBRID for multi-objective linear and quadratic dynamic modelling and optimization. Two other software packages are also represented: RZtools for building interactive systems and a collection of algorithms for multi-objective optimization. The last paper in this group presents the ICCS package for supporting interactive multi-attribute choice.

The last group of four papers on applications of decision support tools starts with a presentation of engineering applications of interactive multi-objective programming. This is followed by a paper on modelling social resource allocations for decision support. Another one reports in detail on an application of decision support and multi-objective optimization in the management of meat-processing production. Finally, an important area of applications is indicated by a report on decision support in competitive selection of R&D projects.

We would like to thank the authors for their contributions and all participants in the Workshop in Serock for their part in the discussions. We are grateful for the support of the organizing and sponsoring institutes. Finally, we hope that this volume not only will find interest in the growing community of researchers working on decision analysis and support but also will stimulate new research in this field.

Jaap Wessels

Andrzej P. Wierzbicki

Table of Contents

A2. Multiple Criteria Optimization and Decision Theory

B2. Applications

Part A:
Methodology of Decision Analysis and Support

Decision Systems:
The Relation Between Problem
Specification and Mathematical Analysis

Jaap Wessels

International Institute for Applied Systems Analysis
Laxenburg, Austria

Technical University Eindhoven
Eindhoven, The Netherlands

Abstract

In this paper it is demonstrated that automated support for decision making of a tactical or strategic nature requires a solver-independent medium for describing decision situations. Such a medium may be specific for one environment, but it is also possible to develop media for certain types of environments. By using such a medium one obtains a decoupling of problem formulation and method of analysis. This makes it possible to use (parts of) the problem formulation as input for different types of models. Such problem formulations may provide mathematical models themselves, although they might also contain some less formal features.

The decoupling makes it possible to choose problem formulations which are much closer to the original decision situation than would otherwise be possible with formulations in terms of a preselected solver. The argumentation is illustrated by treating a language for specifying goods flow problems in some detail. This language is based on timed colored Petri-nets.

1   Introduction

The present paper is based on two types of experiences. In the first place the experience that only in exceptional cases the step from the decision situation to a mathematical model is a simple and natural one. A case where the step is relatively natural is the problem of routing and loading of trucks for the distribution of baking flour from the mill to the bakeries. Although, even in such a case, quite relevant constraints do not fit in the mathematical language. However, in most decision making problems the step from problem to model is large and, usually, it is also one of the most essential steps in the development of decision support systems. In the second place, we have experienced that only for operational decision making does one encounter the situation that a fixed sequence of steps leads from problem to model. Only in operational decision making one sees that essentially the same problem recurs, only with different data. In tactical and strategic decision making, however, one usually gets different types of problems subsequently, although they may use more or less the same data and other knowledge about the functioning of the system.

Example: a distribution structure

In order to keep a distribution structure in good shape in a changing world, one needs frequent adaptations. That means that not only operational decisions have to be made, but also tactical and strategic questions arise frequently and because of the complexity of these questions it would be good to have some tools for supporting this tactical and strategical decision making. However, the requirements for these tools are to be developed for a loosely described set of questions. Only later the real questions and how they are formulated appear. So, a main characteristic of pre-fabricated tools should be their flexibility and adaptability with respect to the types of questions they can be used for. Let us list some of these questions regarding a distribution structure in order to get some idea about their diversity:

a. Do we still need a distribution structure with three levels in a united Europe with higher demands on speed and stronger price-competition?

b. Do we produce all our products in each factory or will the distribution structure still be able to satisfy the performance requirements if the less demanded products are only manufactured by one factory?

c. If the factories specialize, is it then still necessary that each central warehouse stores all products or is it sufficient to forward the products only to the closest central warehouse (possibly located on the same premises as the factory)?

d. Do we still need so many regional warehouses?

e. Are the regional warehouses located in the right areas?

f. From where does a warehouse get its products? Would it be better to keep fixed rules for this allocation or replace them by dynamic ones?

g. How is the transport organized (frequent combined shippings, less frequent direct shippings, or using an outside transport company)?

h. How do we fit a new line of products into the system?

i. How do we adapt the distribution structure after merging with a regional competitor who possesses his own distribution system?

etc.

This example clearly shows how the same data and the same mechanisms with regard to production, distribution, and demand form the basis for answering a great variety of questions. Most of these questions cannot be answered directly with the given data and mechanisms, but require some form of advanced modelling and model analysis. For different questions one will need different types of models and different modes of analysis. For the location of facilities or allocation of production, linear programming might be useful. However, for allocation of regional warehouses to central warehouses we need some form of integer programming. For determining order levels, inventory theory will be useful and for evaluating the time performance of (parts of) the system, a queueing model might be sensible. For detailed checking of proposed new structures, it will be necessary to use simulation and scheduling models to mimic the daily operations. In fact, different questions of the types listed above will require nonstandard models.

If one wants to make a decision support environment for the type of situation as described before, then the main arguments are *time* and *money*. *Time*, since without a decision support environment each new question would require a time-consuming analysis which easily takes more time than available. *Money*, because such complete decision analyses are expensive; they not only cost much time, but also require many man-hours. So the main reasons for investing in a decision support environment for tactical and strategic decision making are the speeding-up of analyses and the abatement of costs. Note that it is practically never the goal to make tools in such a way that the decision makers can do the analysis themselves. A reasonable aim is that the tools are such that analyses can be performed by staff members of the decision makers with only incidental help of computer programmers or analysts. Also for these seemingly modest aims it is not easy to design a decision support environment as will be clear from the list of components of such an environment:

a. A library of flexible algorithms for standard models.

b. Tools for making new algorithms for nonstandard models.

c. A way of storing the knowledge about the situation. This knowledge consists of: data, physical procedures, control processes, and external influences.

d. Tools for translating this knowledge into models of a selected type.

e. Tools for specifying alternative systems and scenarios, which may regard any of the aforementioned features: data, physical procedures, information processes, control processes, external influences.

Essential for obtaining the possibility of using the same knowledge for different types of models is the *decoupling of problem specification and model formulation* and not only decoupling of model formulation and analytic tools. Note that *problem specification* entails, on the one hand, the knowledge of the existing situation and possibly alternatives and on the other hand, scenarios for future behavior if desirable.

It would be more efficient if the aforementioned components were useful for more than one situation. In fact, the chances for a library of algorithms to be useful in a wide variety of situations are much better than for ways for storing knowledge about the situations. Such ways will usually be more specific. However, even for algorithmic libraries, a lot is still to be done in order to enhance the flexibility of programs and to obtain tools for making new algorithms for nonstandard models.

The critical components, however, are the components mentioned under the labels c and e, which have the task of the specification of the current situation and possibly some alternatives. As said before, one cannot hope that this task can be performed by general-purpose tools. However, one might hope that specification tools can be developed for sizable classes of situations. Indeed, several attempts have been made, particularly for classes of highly technical decision problems (for flexible manufacturing problems, see Silva and Valette, 1990 and Van der Aalst and Waltmans, 1990; for scheduling problems, see Carlier et al., 1984; Hatono et al., 1991, Tamura and Hatono, 1991; for cyclic job shops or assembly systems, see Harhalakis et al., 1991; for communication and computer systems, see Garg, 1985, Holiday and Venon, 1987, Magott, 1984, and Molloy, 1982).

The next four sections of this paper will be devoted to the introduction of an approach for specifying a broad class of goods flow situations. The exposition is not so much devoted to the technicalities, but rather emphasizes the possibilities and difficulties of such an approach. Section 2 states the problem and introduces the main tool, viz, Petri nets. In Section 3, time is added to increase the descriptive power. Several typical features for goods flow situations are described in this section. Section 4 is devoted to the use of hierarchical structures and parametrized modules. Section 5 introduces some ways of analyzing specifications of the type introduced in previous sections. Finally, Section 6 contains some comments and conclusions.

2   Specifying goods flow situations

In Sections 2-5, we will introduce an approach for specifying goods flow situations for use in decision support. The ideas for this approach were stimulated by a research project at the Technical University Eindhoven in cooperation with and sponsored by the Dutch Organization for Applied Research TNO. The author uses some of the technical developments in this project to illustrate his views on decision support environments. The author is greatly indebted to the other members of the project team for the discussions which helped in forming his view. In particular, his thanks go to Wil Van der Aalst and Kees van Hee. The presentation relies strongly on the specification tool ExSpect which has been developed at the Technical University Eindhoven. However, these sections should not be considered an exposition of this tool. For such an exposition the reader is referred Van Hee et al., (1989) and Van der Aalst, (1992). In the present exposition ExSpect is only a vehicle for explaining the author's views on decision support environments. The existence of ExSpect also serves as proof that these views are not completely unrealistic.

The presentation does not aim to convince the reader that this is the right approach. A lot of further development and testing will be required before the aproach is really in a mature state. After all, the conclusion might even be that this is not the right way to proceed. In fact, the main danger is that the approach is too ambitious. Even so, the great attractiveness of the approach is that it heads right to the center of the crucial aspect of decision support, namely, the specification of situations and alternatives. The main reason for presenting this research-in-progress here is that it shows that focussing on an approach for this crucial aspect gives a completely different view on the usual ways of treating decision support and also on the difficulties encountered with regard to the acceptance of such efforts.

The approach is ambitious in different ways. In the first this is place because of the large range of practical situations emphasized, namely, virtually all goods flow situations where the different stages of the process are separated by discrete events. This would mean that practically all batch-type chemical production situations would be included as long as the highest level of detail is the batch. Of course, it is technically possible to go into more detail by splitting the processing of a batch into phases which are separated symbolically by discrete events. However, this would not lead to a natural specification of the goods flow. The second aspect where the ambition of the approach becomes clear is the fact that it aims at giving specifications which are very close to the real-life process and allowing the user to translate this specification into different types of models. In the usual modelling languages the starting point is a fixed type of analysis (for instance, linear programming or simulation) and the task of the language is to avoid computer programming technicalities (in the simulation case) and to simplify the modelling task itself (in both cases). A third ambition is to develop a specification tool which is so unambiguous that it can be understood as an executable prototype with which experiments can be performed. A fourth ambition is that it can be used for a large range of decision problems including the selection of the basic distribution structure, on the one hand, and the control rules for a warehouse, on the other. A final ambition is to integrate the physical layer of the goods flow with the information flow and the process control.

A specification tool for goods flow problems has to be built on representations for the most elementary physical steps in the goods flow:

1. transformation of certain goods into others;

2. displacement of goods;

3. buffering of goods.

The elementary physical steps have a striking similarity with the elementary steps in information processing. If we combine this feature with the ambition to integrate *goods flow* and *information flow* into one specification, then it becomes obvious that approaches, which have shown some usefulness for specifying information flows at different levels, might provide a good starting point for our goal. Particularly, since *concurrency* and *parallelism* are essential aspects of goods flows, it seems sensible to determine if Petri nets would perform as a basic concept for a specification tool (compare for instance Magott, 1984, and Garg, 1985). In a recent paper Thomasma and Hilbrecht (1991) came to the conclusion that Petri nets provide the most useful approach for specifying material handling control algorithms in flexible manufacturing systems. Di Mascolo et al. (1991) show that Petri nets provide the right language to formulate all sorts of kanban systems in a unified way. So, also the goal of integrating the control process into the specification of physical flows and information flows seems to be best attainable by choosing Petri nets. For a recent overview of properties, analysis, and applications of Petri nets, the reader is referred to Murata, (1989). For the description of a specification tool for distributed information systems based on Petri nets, the reader is referred to Van Hee et al. (1989). The specification language described in the latter paper indeed leads to executable prototypes. Therefore, the project at the Technical University Eindhoven chose this language, which goes by the name of ExSpect, as basis for the specification tool for goods flow systems. An extensive report on the use of ExSpect for logistic systems together with some extensions for that purpose will be published as Van der Aalst (1992). A preliminary report is in Van der Aalst and Waltmans (1991). One of the extensions proposed is treated in Van der Aalst (1991).

In Section 3 the basic notations of Petri nets are introduced. This introduction is informal. We start with a graphical description, since one of the nice features is that Petri nets allow for a graphical representation:

The basic notations are

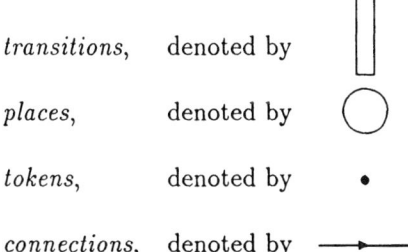

transitions,   denoted by

places,   denoted by

tokens,   denoted by

connections,   denoted by

A *connection* always connects one *place* with one *transition*. If a connection leads from place i to transition j, then place i is called an *input-place* for transition j. If a connection leads from transition j to place i, then place i is called an *output-place* for transition j (see Figure 1). A place may contain a number of tokens.

Each connection is supposed to have a *weight*, which is a positive integer. The weights can be inscribed alongside the connection (see Figure 2), however weights of size 1 are usually deleted. Any network constructed along these lines is called a Petri net. Petri nets have been introduced to describe dynamic phenomena. So it is important to have rules for the dynamics of Petri nets. In fact, the only dynamic aspect is in the tokens: the distribution of the tokens over the places can change. So the distribution of the tokens over places can be viewed as the state of the Petri net. The rules for a change of the state are the

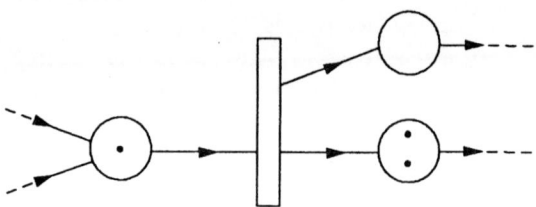

Figure 1: *Part of a Petri net with one token in the input-place of the depicted transition and two tokens in one of its two output-places.*

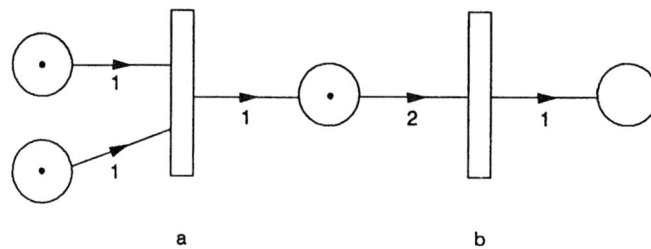

Figure 2: *A Petri net with weights indicated for the connections; transition a is enabled and transition b is not enabled.*

following. If each input place of a transition contains a number of tokens at least equal to the weight of its connection with that transition, then the transition is said to be *enabled* for *firing* (see Figure 2). When firing, a transition consumes from each input-place exactly the number of tokens equal to the weight of the connection. Furthermore, it produces for each output-place exactly the number of tokens required by the weight of the connection between the transition and the output-place (see Figure 3).

Indeed, the basic Petri net model as introduced so far reflects the elementary steps of the goods flow process: transitions reflect transformation and/or displacement of goods, whereas places reflect buffering possibilities. In this way the subsequent states of a Petri net may reflect the subsequent positions of a real goods flow system. However, time durations do not play a role. Therefore, we will introduce time more explicitly in the next section.

3 Specification by timed Petri nets

For making decisions about goods flow problems, the evaluation of time aspects is crucial in most cases. Time plays a role in essentially two ways: in the first place in the *duration* of transformations and displacement and in the second place in the *availability* of processors

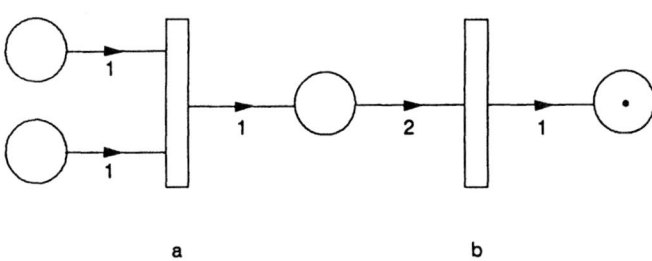

Figure 3: *The Petri net of Figure 2 after the firing of transition a and the subsequent firing of transition b.*

for transformation and/or displacement. The second way corresponds in a natural way to the usual way of introducing time in Petri nets (see Murata, 1989; Zuberek, 1980), namely, by specifying the time at which a transition might fire. However, in goods flow situations the duration of activities is the most relevant feature, therefore we will introduce time delays due to activities. In fact, also the availability feature can be modelled in this way, so we do not loose modelling power. In the simplest version, each transition gets an attribute *delay* which is a nonnegative real number. In order to process these delays in the right way, all tokens get a *time stamp*, indicating the moment they become available for consumption. It is supposed that a transition becomes enabled as soon as the right numbers of tokens are available in its input-places, however, it can only fire at the time indicated by the maximum of the time stamps of the tokens to be consumed. If there are more tokens than necessary, then the ones with the lowest times are consumed.

Note that a transition that is enabled is not necessarily fired: it is quite possible that another transition consumes some of its tokens, leading to disabling the first-mentioned transition. This way of treating time in Petri nets has been introduced by Van Hee et al. (1989). For an illustration of the time procedure see Figure 4, where tokens are represented by their time stamp rather than by a dot.

The timed Petri net of Figure 4 represents some operation which takes two time units and can only perform one execution at a time.

In Figure 5 the situation is depicted where the operation is a displacement and the transport vehicle needs some time to return. In Figure 6 the same transport operation is depicted taking into account the loading and unloading operations which require the presence of the vehicle, the load, and the forklift.

In a similar way a more complex production situation may be described. Figure 7 gives an example of an assembly situation with six types of components coming from the outside. Via three types of subassemblies the final product is reached, which is delivered to the outside world.

In the examples so far the control was of the push type. This means that as soon as jobs and tools are available, the operation is performed. The time stamps of incoming

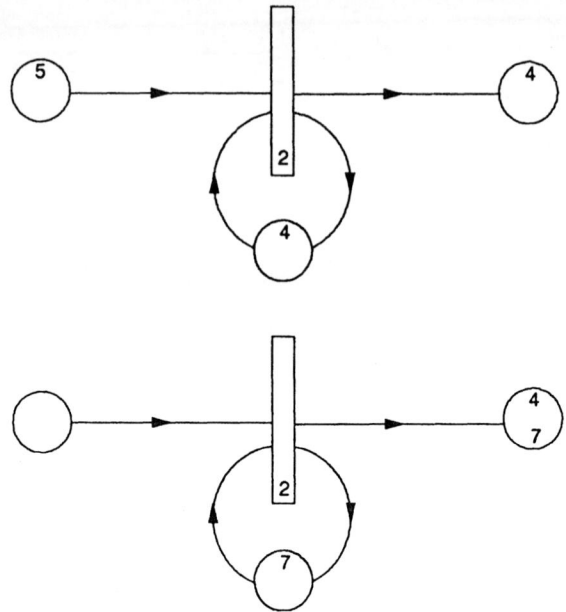

Figure 4: *The transition in the top timed Petri net is enabled to fire at time 5: the transitions will give a delay of 2 time units. The lower bottom Petri net depicts the situation after the firing.*

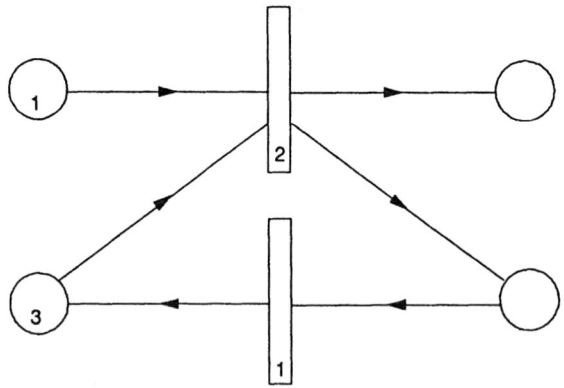

Figure 5: *An operation which requires some dead time after an execution before it can start the next execution, e.g., a transport operation requiring the return of the vehicle involved.*

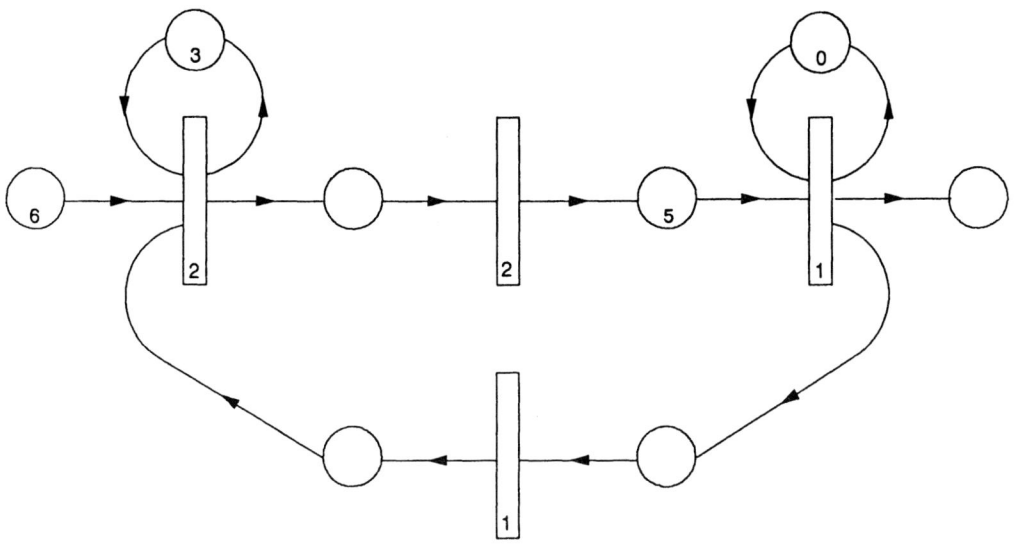

Figure 6: *A transport operation with loading and unloading. The unloading is enabled and may fire at time 5, which will make it possible to start the next loading at time 7.*

transport jobs (Figures 5 and 6) or of incoming components can be exploited to implement the type of control as used in MRP-environments: the time stamps of incoming tokens can then be interpreted as release dates. However, also within the described system there can be some sort of control other than just "push" or "produce if you can." In fact, the specification concept as described above is a very natural tool for specifying pull control in the form of kanbans. Figure 8 describes a situation of two subsequent production steps, where the second step needs three units from the first step together with one unit of an external component. The first step is triggered by the second step: each consumption of three units by the second step triggers an order release of three units for the first step. This order can only be executed if the components required are available. In this way the amount of intermediate stock is regulated. Actually, the number of tokens in the cycle determines the performance: with less tokens the second step has more risk of being unable to produce, with more tokens the amount of work in process will be higher (for an extensive treatment of specifying kanban systems by Petri nets, see Di Mascolo et al., 1991).

For other types of control it would be nice to have another feature in the Petri net concept, namely, the feature of having *values* attached to the tokens. In the Petri net literature this feature is usually referred to as *coloring* of the tokens (cf. Jensen, 1987; Murata, 1989). In fact, the introduction of coloring is not absolutely necessary, however, the price of not allowing it would consist in getting very large and relatively unnatural specifications in many situations. In this short overview, we will not treat coloring, but rather refer to Van Hee et al. (1989) and to Van der Aalst (1992) for an integrated treatment of the time and color concepts. In those references a model is introduced which makes it possible that

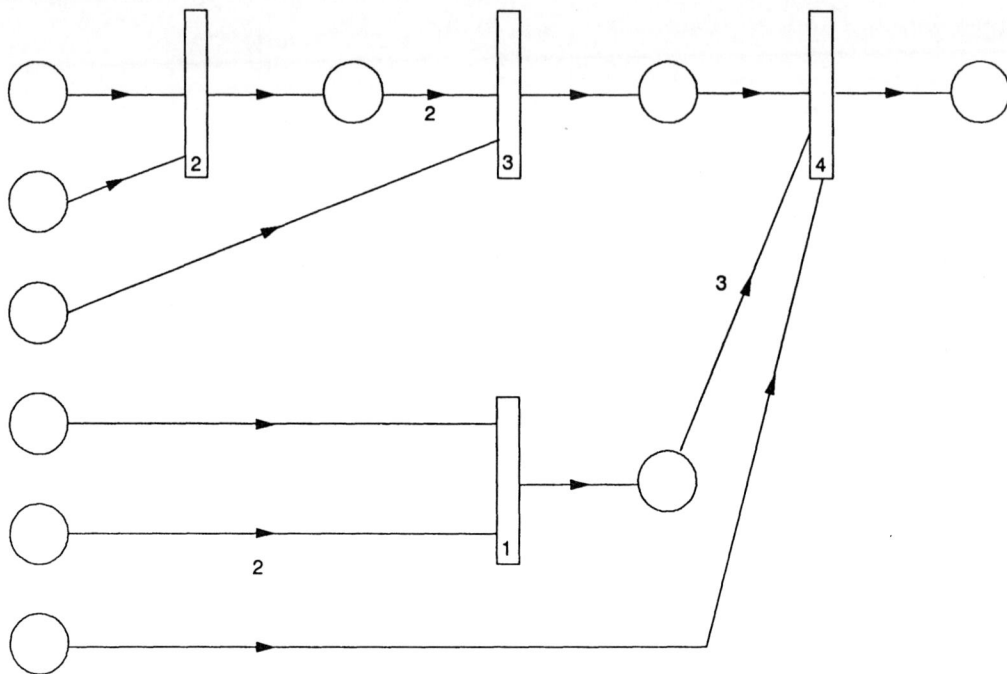

Figure 7: *An assembly situation with six incoming types of components and three types of subassemblies. In some stages there are more specimen's of one component or subassembly needed for the next step: this is represented by the weight of a connection; weights of size 1 are not depicted.*

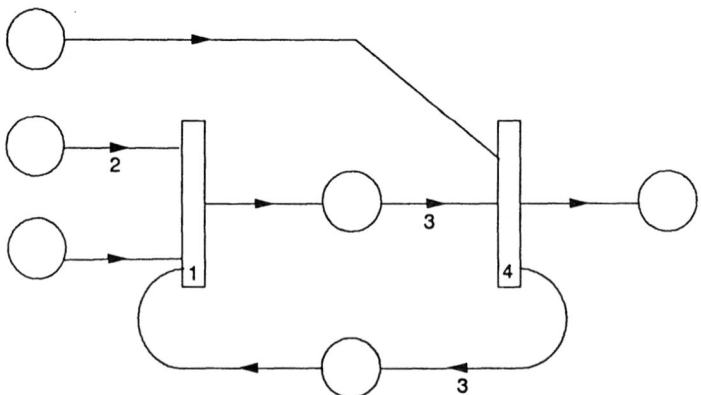

Figure 8: *An assembly system where the final assembly step releases orders (kanbans) for the preceding subassembly. The number of tokens in the cycle is an important design quantity.*

a transition produces tokens with values determined by a function of the values of the consumed tokens. Also the delay of a transition may be a function of the values of the consumed tokens. These features extend the expressive power of the Petri net concept considerably, particularly since the user may define the type of the value rather freely. In this way, it becomes indeed possible to make specifications which integrate the description of the material flow with the description of the control and of the information flow.

There is still one feature missing so far and that is the feature of uncertainty. Given the way that time has been introduced in Petri nets in this paper, the natural way to introduce uncertainty is to accept probability distributions for the time delay instead of the deterministic values as used so far. Indeed, uncertainty has been introduced in this way in ExSpect, see Van Hee et al. (1989). For other, strongly related, ways of handling uncertainty, see Murata, (1989) and Hatono et al. (1991). With this type of handling uncertainty one obtains Petri net models which are very well suited for simulation. However, as will be explained in Section 5, other ways of evaluation are only feasible for stochastic Petri nets under rather strict conditions. Therefore, it appears to be useful to introduce the possibility of only specifying an upper and a lower bound for each delay (see Van der Aalst 1991; 1992). Indeed, such a specification is not sufficient for a simulation, but it seems to be possible to develop other evaluation techniques for Petri nets with *interval time delays* (see Section 5).

4   Hierarchy and modularisation

The approach introduced in Section 3 provide's an expressive (partly graphical, partly functional) method for specifying goods flow processes including the information flows and the control processes. However, specifications for real systems have a tendency to become rather large and complicated. This is partly due to the preciseness with which the mechanisms can be specified. The main source, of course, is the inherent complexity of modern goods flow processes due to the high performance requirements. Therefore, thinking about goods flow processes always takes place in a more structured way. A specification tool for goods flow processes should support this structured way of thinking rather than hamper it. Support of a structured way of thinking requires the possibility of executing the specification process, or parts of it, in a top-down fashion. Tamura and Hatono (1991) describe a hierarchical approach for specifying flexible manufacturing systems by stochastic Petri net models based on a hierarchical structure emerging from the application area.

In the case of the assembly production in Figure 7, one can imagine that the two subassembly steps of the top level are performed in one department and the final assembly, together with the remaining subassembly step, is performed in another department. In that case the first two stages in a top-down specification process would result in the representations of Figures 9 and 10, where parts of Figure 7 are replaced by boxes or rectangles. In our case, where we started with Figure 7, Figure 10 may be seen as an aggregation of

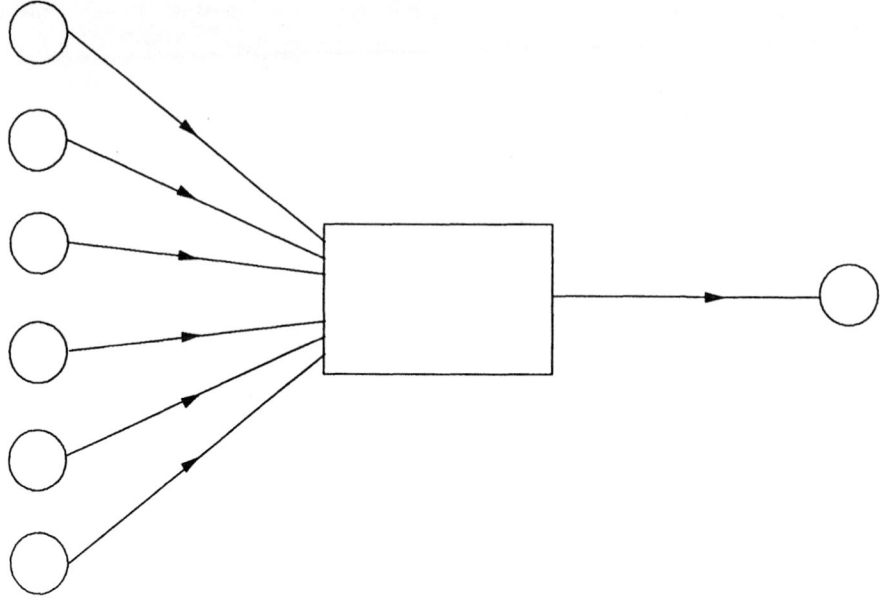

Figure 9: *The assembly situation of Figure 7 with the production itself represented by a "black" box.*

Figure 9 and Figure 9 as an aggregation of Figure 7. In a top-down specification process, Figures 9 and 10 may be seen as stages in this process with the boxes as provisionally unspecified parts.

The feature of specifying top-down in different stages is supported by the tool ExSpect and also the possibility of exhibiting a specification with diminished level of detail. The way of using these properties in specifying goods flow systems is particularly addressed in Van der Aalst (1992).

The possibility of distinguishing subsystems and treating them separately does help much in coping with the complexity and size of a specification. However, it does not diminish the amount of work which is required by a detailed specification. Therefore, it would be necessary to have standard modules available which can be plugged in. For a specific area of application, like goods flow systems, it seems possible to develop standard modules for several activities. When doing so, it is important to choose the right sort of parametrization. For keeping flexibility it is important to have primarily modules for the elementary activities with regard to handling goods, information, and control. For quick and easy use, it is better to have modules representing more complex activities. In fact, these approaches can be combined by providing in the first place modules for the elementary activities and using these modules to formulate modules for more complex activities. Now the user can choose whether he prefers the flexibility or the easiness of modelling. Eventually, the user can define his own modules for more complex activities.

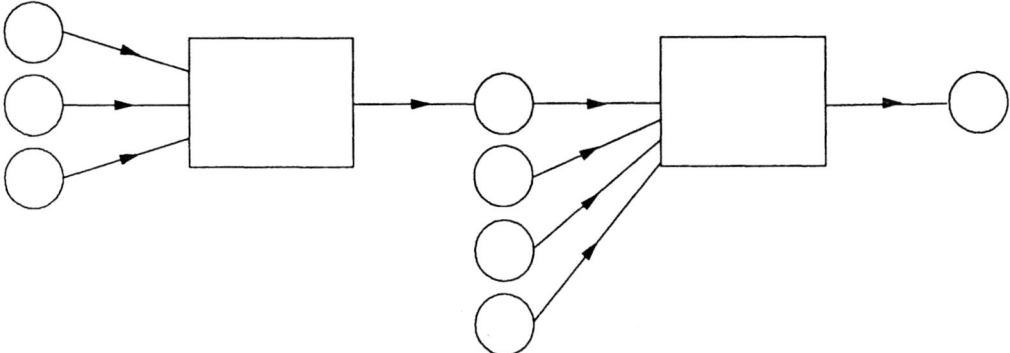

Figure 10: *The assembly situation in Figure 7 with boxes representing departments which perform two assembly steps each.*

For an elaborate treatment of this topic, the reader is referred to Van der Aalst (1992).

5   From specification to analysis

The material in Sections 2-4 show's that indeed it is possible to make a unified specification method for a large class of goods flow situations while integrating the flows of material and information with the control processes. However, the specification was not a goal in itself. We embarked on this tour, because we concluded in Section 1 that a specification method would be an indispensable tool in a decision support environment. What remains, however, is to show that such a specification method can be used for decision support. In down-to-earth terms this would mean that first it should be possible to formulate specifications for alternative scenarios and to compare their performances. Second, it should be possible to use the specification in a search process for alternatives, i.e., it should be possible to get suggestions for improvements. In Section 1 we concluded that it would be good to decouple specification and analysis. The specification method presented in Sections 2-4 is a result of such a decoupling. There are two main advantages to this method:

1. the specification method is closely related to the view of the user on the real world;

2. the specification method does not narrow the class of possible methods of analysis.

However, the price of the last advantage is that the step from specification to analysis becomes a real step, at least if one wants to use standard methods from the O.R.-literature.

In this section we deal with the possibilities of using the specifications as developed in Sections 2-4 for evaluating scenarios and possibly suggesting new ones. It should be emphasized that this treatment will be far from complete, primarily since much is still uncertain. So the goal of this section will be to indicate possibilities and difficulties rather than to present solutions.

Before starting with the exploration it seems sensible to be more precise on the results of the specification process. Using the tool ExSpect, the user specifies his processes in the form of the figures illustrated in this paper on an graphical screen. Via the windowing system the non-graphical information can be added in the form of attributes, tables, and functions. The user interface may be primarily graphical, but the resulting specification is represented in the ExSpect language. The user might specify directly in the language, but this is not advisable.

The representation in ExSpect specifies a complete model of reality. Indeed, in ExSpect it is possible to use this representation directly for a simulation. The tool is such that it allows for the addition of a measuring system and the execution of a simulation. Such a simulation may take real-time form, including stops with interventions by the user and restarts. Therefore, performing a simulation is the first and most natural way of evaluating the specified model. For an extensive treatment of this feature, see Van Hee et al. (1989) and Van der Aalst (1992). This is indeed a very nice feature even though the simulation is relatively time-consuming.

There are essentially two different approaches to analyzing the situations specified. The one described in the introduction suggests a library of standard algorithms and tools for making new ones. In that setting one would have to translate the model from the specification language into the form required for algorithmic treatment. In the other approach one tries to develop a library of algorithms which directly treat the specification and do not require a translation step. Until now, the first approach did not get much attention for Petri net type specifications. Therefore, it is not clear how feasible the approach would be. It is clear, however, that some types of translation entail more complications than others. It will definitely be a more straightforward task to translate a Petri net specification to a model for a standard simulation package, than to translate it to a model for a queueing network analyzer. However, translation to a model for a queueing network analyzer is definitely a more straightforward task than translation to a linear programming model. The closer the structure of the analytic model is to the structure of the specification, the more straightforward the translation and the more likely it becomes that translation can be done (semi-)automatically. It is hoped that these translation processes can be supported by a combination of rule-based procedures and user actions.

The second approach, namely, developing methods of analysis which use the specification as a an input, has received much more attention in the literature. Some methods are specifically designed for the evaluation of Petri net models and other methods are tailored for treatment of Petri net specifications. In Section 6 we sketch some of the efforts along both lines.

Within Petri net theory there is, apart from simulation, much attention being given to the computation of place invariants, transition invariants and reachability graphs (cf. Murata,

1989). For decision support with regard to goods flow problems, these notions have only a limited value. Moreover, for practical goods flow problems the computational effort required has a tendency to grow beyond reasonable bounds. Therefore, we will not treat these topics further.

With simulation it is possible to evaluate important aspects of the performance of a design. Simulation is appropriate for evaluating the treatment of a package of orders as well as for evaluating long-term behavior under the influence of a demand generator. In the first type of case, statistical aspects are treated by repeating the simulations, like in physical experiments. In the second type of case, one long experiment or simulation run may be used to evaluate the statistical aspects of the performance. For these topics see, Bratley et al. (1987). However, simulation has some disadvantages. First it is time-consuming and, second, it only evaluates one scenario, without giving much help in finding better scenarios. However, in the last few years results have been published under the heading *perturbation analysis*, which might be described as a method of estimating the gradient of some performance measure for a discrete event system with respect to a parameter θ, based on a simulation run for one value of θ only. For an introduction to the method and some of its technicalities, see Ho (1987), (1988) and Heidelberger et al. (1988). Perturbation analysis might provide a basis for developing techniques for helping in the search process for new and better scenarios or designs. This development is still in its infancy.

The other disadvantage of simulation is that it is rather time-consuming. Therefore it would be nice to have analytic methods for doing the same types of performance analysis. For timed Petri nets with deterministic times, there has been some research activity in determining cycle times (cf. Murata, 1989). Also the analysis of time properties via reachability graphs has received some attention (cf. Zuberek, 1980). However, for goods flow problems it is important to work with uncertain time delays. If all time delays are negative exponentially distributed, then the dynamic behavior of the Petri net is described by a Markov process and the analysis of the Markov process can be used for the performance evaluation. However, the condition on the distributions is rather severe and the Markov processes get very large state spaces which tend to make the analyses practically infeasible. For details of this approach and more precise conditions see Marsan (1990), and Molloy (1982). A new approach consists of representing the uncertainty of time delays by upper and lower bounds. This approach has been introduced by Van der Aalst (1991, 1992) in order to have models which are more realistic than purely deterministic timed Petri nets, but which remain open for analysis. Indeed, Van der Aalst proposes some methods of analysis which are quite executable and give useful bounds for the time performance. These ideas have been integrated in the tool ExSpect (see Van der Aalst, 1992). Another quite interesting development is the analysis of Petri net specifications with queueing network techniques. Di Mascolo et al. (1991) treat a restricted class of goods flow situations, namely, kanban systems, in this way. They provide an approximate analysis for the related queueing network. As a whole, the developments in this area seem promising.

6 Comments and conclusions

In the introduction we stressed the importance of having a specification method as a basis for a decision support environment. In the Sections 2-4 we described the specification method for goods flow problems. It appeared that it is quite well possible to develop such a specification method for a large class of goods flow situations and a large class of questions within each situation. Special features of the method included first, the integration of physical flows, information flows, and control processes in one specification and in the second place its completeness which made it possible to use the specification as a prototype with which one can execute experiments (i.e., simulations). Perhaps even more important are the expressive power and the close relation of specifications to reality. However, the price to be paid for the advantages consists of, on the one hand, the large size and complexity of a specification and, on the other hand, the difficulty in relating the specifications with algorithmic models of the usual type. The first difficulty is contested by introducing a hierarchical approach and parametrized standard modules. The second difficulty is obviously the most important one. Until now most effort in this respect has been devoted to attempts of avoiding the usual algorithmic approaches and developing analytical methods specifically tailored for this type of specifications. In doing so, the more usual algorithmic methods have served as a source of inspiration. However, particularly at this aspect, much research must still be done. One may expect that rule-based methods combined with user interaction will provide the means for translating specifications to typical algorithmic models.

References

Van der Aalst, W. M. P. (1991). Interval timed Petri nets and their analysis. *Computing Science Note* 91/09. Technical University Eindhoven, May 1991.

Van der Aalst, W. M. P. (1992). Timed coloured Petri nets and their application to logistic systems. Ph.D. thesis, Technical University Eindhoven (forthcoming).

Van der Aalst, W. M. P. and A. W. Waltmans (1990). Modelling flexible manufacturing systems with ExSpect. pp. 330-338. In: B. Schmidt (ed.), *Proceedings of the 1990 European Simulation Multiconference*. Simulation Councils.

Van der Aalst, W. M. P., and A. W. Waltmans (1991). Modelling logistic systems with ExSpect. pp. 269-288. In: H.G. Sol and K.M. van Hee (eds.), *Dynamic Modelling of Information Systems*. North-Holland, Amsterdam.

Bratley, P., B. L. Fox, and L. E. Schrage (1987). *A Guide to Simulation*. Springer-Verlag, New York (second edition).

Carlier, J., Ph. Chretienne, and C. Girault (1984). Modelling scheduling problems with timed Petri nets. pp.62-82. In: G. Rosenberg (ed.), *Advances in Petri nets 1984*. Springer-Verlag (LNCS 188), New York, NY.

Garg, K. (1985). An approach to performance specification of communication protocols using timed Petri nets. *IEEE Trans. Software Engin.* SE-11, 1216-1225.

Harhalakis, G., S. Laftit, and J.-M. Proth (1991). Event graphs for modeling and evaluating modern production systems. pp. 438-451 in: G. Fandel and G. Zäpfel (eds.), *Modern production concepts; theory and applications.* Springer-Verlag, Berlin.

Hatono, I., K. Yamagata, and H. Tamura (1991). Modeling and on-line scheduling of flexible manufacturing systems using stochastic Petri nets. *IEEE Trans. Software Engin.* SE-17, 126-132.

Van Hee, K. M., L. J. Somers, and M. Voorhoeve (1989). Executable specifications for distributed information systems. pp. 139-156. In: E.D. Falkenberg and P. Lindgreen (eds.), *Information system concepts an in-depth analysis.* North-Holland, Amsterdam.

Heidelberger, Ph., X.-R. Cao, M. A. Zazanis, and R. Suri (1988). Convergence properties of infinitesimal perturbation analysis estimates. *Management Science* 34, 1281-1302.

Ho, Y. C. (1987). Performance evaluation and perturbation analysis of discrete event dynamic systems. *IEEE Trans. Automatic Control* AC-32, 563-572.

Ho, Y. C. (1988). Perturbation analysis explained. *IEEE Trans. on Automatic Control* AC-33, 761-763.

Holiday, M. A., and M. K. Venon (1987). A generalized timed Petri net model for performance analysis. *IEEE Trans. Software Engin.* SE-13, 1297-1310.

Jensen, K. (1987). Coloured Petri nets. In: W. Brauer, W. Reisig, and G. Rosenberg (eds.), *Advances in Petri Nets 1986, Part I: Petri Nets, Central Models and Their Properties.* Springer (LNCS 254), New York.

Magott, J. (1984). Performance evaluation of concurrent systems using Petri nets. *Information Processing Letters* 18, 7-13.

Marsan, M. A. (1990). Stochastic Petri nets: an elementary introduction. pp. 1-29. In: G. Goos and J. Hartmanis (eds.) *Advances in Petri Nets 1989.* Springer-Verlag, New York (LNCS 424).

Di Mascolo, M., Y. Frein, Y. Dallery, and R. David (1991). A unified modeling of kanban systems using Petri nets. *Intern. J. Flexible Manufacturing Systems* 3, 275-307.

Molloy M. K. (1982). Performance analysis using stochastic Petri nets. *IEEE Trans. Comp.* C-31, 913-917.

Murata, T. (1989). Petri nets: properties, analysis and applications. *Proceedings of the IEEE* 77, 541-580.

Silva, M., and R. Valette (1990). Petri nets and flexible manufacturing. pp. 274-417. In: G. Goos and J. Hartmanis (eds.) *Advances in Petri Nets 1989*. Springer-Verlag, New York (LNCS 424).

Tamura, H., and I. Hatono (1991). Modeling and scheduling of flexible manufacturing systems using timed/stochastic Petri nets. pp. 96-101. In: *Proceedings of IFAC Workshop on Discrete Event System Theory and Applications in Manufacturing and Social Phenomena*. Shenyang, People's Republic of China, June 25-27, 1991.

Thomasma, T., and K. Hilbrecht (1991). Specification methods for material handling control algorithms in flexible manufacturing systems. *Intern. J. Flexible Manufacturing Systems* 3, 231-250.

Zuberek, W. M. (1980). Timed Petri Nets and Preliminary Performance Evaluation. In Proceedings of the 7th Annual Symposium on Computer Architecture, 8, (3) of the Quarterly Publication of the ACM Special Interest Group on Computer Architecture, 62-82.

Some New Results in Interactive Approach to Multicriteria Bargaining

Lech Kruś, Piotr Bronisz
Systems Research Institute
Polish Academy of Sciences, Poland

1   Introduction

A growing interest in methodology and applications of multiperson decision support systems is observed. In particular, approaches based on multicriteria optimization in group decision problems are subject of the papers by Korhonen, Moskowitz, Wallenius, Zionts (1986), Jarke, Jelassi, Shakun (1987), Kersten (1988), DeSanctis, Gallupe (1987), Korhonen and Wallenius (1989) and others. On the other hand there exists the developed theory of bargaining problem started by Nash (1950), continued by Raiffa (1953), Harsanyi, Selten (1972), Kalai, Smorodinsky (1975), Roth (1979), Thomson (1980), Imai (1983) and others. It seems to be reasonable to construct systems supporting negotiations combining the multicriteria optimization approach, and achievements of the theory and the practical experience in bargaining. However in this case new theoretical problems arise related to a generalization of the solution concepts and their axiomatization, a construction of interactive processes making easier the decision analysis of the bargaining situation and supporting the negotiation.

A multicriteria bargaining problem is a generalization of the classical bargaining problem introduced by Nash (1950). A group of individuals is engaged in a bargaining process. No group member is assumed to control any of the decision variables by himself, therefore each decision is the result of a negotiation. Each of the group member has his own criteria which value any decision. We assume that each member has his own utility function, but it is not known explicitly. The assumption follows from the fact that in many practical application it is very hard to construct such utility functions.

A multicriteria bargaining problem is described by a set of individuals involved in a bargaining process (players), a set of possible decisions, a set of functions mapping possible decisions to criteria spaces of the players, a set of feasible consequences of possible decisions in criteria space of the players called an agreement set, and a distinguished outcome in the agreement set called a status quo point. Any outcome in the agreement set can be the result of the bargaining process if the players reach unanimous agreement. If such agreement is not possible, the result is the status quo point adequate to the initial situation.

The paper presents our recent developments in an interactive approach to the multicriteria bargaining problem (see also Bronisz, Kruś, Wierzbicki (1989), Kruś (1991a),

Kruś (1991b), Kruś, Bronisz (1991)). Two interactive schemes are presented.

The first one is based on a generalization of the Raiffa-Kalai-Smorodinsky solution concept (see Raiffa 1953, Kalai-Smorodinsky 1975) to multicriteria case. Axiomatic characterization of the solution is presented as well as its properties.

The second interactive scheme was inspired by a certain bargaining process proposed by Raiffa (1982) called the Pareto improving a single negotiated text and by the limited confidence principle proposed by Fandel and Wierzbicki (1985). The convergence of the scheme is investigated. The scheme was implemented in the computer system MCBARG supporting multicriteria bargaining.

2 Formulation of the multicriteria bargaining problem

Let $N = \{1, 2, \ldots, n\}$ be the finite set of players, each player having m^i criteria. Let $m = \sum_{i=1}^{n} m^i$ and $M = \{1, 2, \ldots, m\}$. A **multicriteria bargaining (MCB) problem** is defined as a pair (S, d), where an agreement set S is a subset of m-dimensional Euclidean space, called R^m, and a disagreement point (status quo point) d belongs to S.

The MCB problem has the following interpretation: every point $x \in R^m$, $x = (x_1, x_2, \ldots, x_n)$, $x_i = (x_{i1}, x_{i2}, \ldots, x_{im^i})$, in the agreement set S represents payoffs for all the players that can be reached when they do cooperate with each other (x_{ij} denotes the payoff of the j-th criterion for the i-th player). If the players do not cooperate, the disagreement point is the result. The problem consists in supporting the players in reaching a nondominated solution, agreeable and close to their preferences.

Each criterion can be maximized or minimized. For simplicity of presentation we assume that all criteria are maximized.

Sometimes, when there is no reason to distinguish particular player and his particular criterion, we use for $x \in R^m$ the following notation $x = (x_1, x_2, \ldots, x_m)$. We employ a convention that for $x, y \in R^k$, $x \geq y$ implies $x_i \geq y_i$ for $i = 1, \ldots, k$, $x > y$ implies $x \geq y$, $x \neq y$, $x \gg y$ implies $x_i > y_i$ for $i = 1, \ldots, k$. We say that $x \in R^k$ is a **weak Pareto optimal** point in X if $x \in X$ and there is no $y \in X$ such that $y \gg x$; $x \in X$ is a **Pareto optimal** point in X if there is no $y \in X$ such that $y > x$. The **ideal (utopia)** point $I(S, d)$ is defined by $I_k(S, d) = \max\{x_k : x \in S, x \geq d\}$ for $k \in M$.

We consider MCB problems (S, d) satisfying some of the following conditions:

C1. S is compact, there is $x \in S$ such that $x \gg d$ and $I(S, d) \notin S$.

C2. S is comprehensive, i.e. for $x \in S$ if $d \leq y \leq x$ then $y \in S$.

C3. For any $x \in S$, let $Q(S, x) = \{k \in M : y \geq x, y_k > x_k \ for \ some \ y \in S\}$. Then for any $x \in S$, there exists $y \in S$ such that $y \geq x$, $y_k > x_k$ for each $k \in Q(S, x)$.

C4. S is convex.

Condition $C1$ states that the set S is closed, upper bounded, the problem is not degenerated and not trivial. Condition $C2$ says that criteria are disposable, i.e. that if the players can reach the outcome x then they can reach any outcome worse than x.

$Q(S, x)$ is the set of all coordinates in R^m, payoffs of whose members can be increased from x in S. Condition $C3$ states that the set of Pareto optimal points in S contains no "holes". Any convex set satisfies condition $C3$.

Let B denote the class of all MCB problems satisfying conditions $C1$, $C2$, B^* denote the class satisfying conditions $C1$, $C2$, $C3$ and B^{**} denote the class satisfying $C1$, $C2$, $C4$.

Looking for a solution in the MCB problem we have to consider jointly two decision problems: the first one – the solution should be related to the preferences of all the players, the second one – the solution should fulfill some basic fairness rules. The theory developed in the paper is thought as a background for construction of decision support systems aiding the players in both the decision problems. The first decision problem relates directly to multicriteria decision making done by each of the players. We propose application of aspiration–led approach, called also as a reference point approach (Wierzbicki 1982, Wierzbicki 1986). Considering the second decision problem, solution concepts of the MCB problem are proposed satisfying properties (called according to the theory of bargaining problem as axioms) that could be accepted by rational players.

The aspiration–led approach in multicriteria decision making includes a learning procedure in which the decision maker can analyzed nondominated outcomes with use of aspiration levels (reference points). It means, the decision maker specifies a reference point in his objective space, and the computer based system responses with nondominated outcome being close to the reference point. The decision maker explores various nondominated points by changing the reference points. The procedure is repeated until a satisfactory solution is found.

This approach is proposed here to be applied for analysis of the MCB problem by each player independently. Using the approach the player can explore set of nondominated points in his objective space (such a points we call as individually nondominated - the formal definition follows). We assume that the exploration should be finished with a selection of the preferred by the player nondominated outcome. Composition of the selected by all the players preferred nondominated outcomes we call as an utopia point relative to the players aspirations (RA utopia).

Definition 2.1 *For any MCB problem (S,d), a point* $x^i \in S$ *is* **individually nondominated** *by player* i, $i \in N$, *if* $x \geq d$, *there is no* $y \in S$ *such that* $y \geq d$, $y_i > x_i^i$. *A point* $u \in R^m$ *is an* **utopia point relative to the players aspirations (RA utopia point)** *if for each player* $i \in N$, *there is an individually nondominated point* $x^i \in S$ *such that* $u_i = x_i^i$.

The individually nondominated point is an outcome which could be achieved by a rational player i if he would have full control of the moves of the other players. Of course, we assume that no player agree on the payoff worse than following from the disagreement point. A RA utopia point significantly differs from the ideal (utopia) point. The ideal point reflects only some information about the bargaining problem — possible maximal values of criteria. The RA utopia point generated by the individually nondominated points selected by the player, $i \in N$, carries also information about the most preferable outcomes for all the players.

For any MCB problem (S, d), it is easy to notice that for any RA utopia point u, we have $u \geq d$. In the next sections we confine our consideration to a set $U(S, d)$ of all RA utopia points satisfying $u \gg d$, i.e. we assume that each criterion considered in the problem is essential.

3 Characterization of some solution concepts

We introduce the following definition.

Definition 3.1 *A solution for a multicriteria bargaining problem is a function* $f : B \times R^m \longrightarrow R^m$ *which associates to each problem* $(S, d) \in B$ *and each RA utopia point* $u \in U(S, d)$, *a point of* S, *denoted* $f(S, d)$.

A solution is a rule which assigns to each MCB problem — reflecting structure of the problem and each RA utopia point — reflecting preferences of the players, a feasible payoff.

3.1 Generalized Raiffa–Kalai–Smorodinsky solution concept

Raiffa (1953) proposed a solution concept which was axiomatically characterized by Kalai and Smorodinsky (1975). We present a generalization of the concept to MCB problems.
We impose on a solution the following four axioms:

A1. Weak Pareto optimality.
  A point $f(S, d, u)$ is weak Pareto optimal in S.

A2. Invariance under positive affine transformations of criteria.
  Let $T : R^m \longrightarrow R^m$ be an arbitrary affine transformation such that $T_k x = (a_k x_k + b_k)$, $a_k > 0$ for $k \in M$. Then $f(TS, Td, Tu) = Tf(S, d, u)$.

A3. Anonymity.
  For any permutation on M, π, let π^* denote the permutation on R^m. Then $\pi^* f(S, d, u) = f(\pi^* S, \pi^* d, \pi^* u)$.

A4. Restricted monotonicity.
  If $u \in U(S, d) \cap U(S', d)$ and $S \subseteq S'$ then $f(S, d, u) \leq f(S', d, u)$.

The first three axioms are usually imposed on solutions of axiomatic bargaining problem. Axiom $A1$ says that the players behave in a rational way. Axiom $A2$ demands that a solution does not depend on selected affine measure of any criterion. Axiom $A3$ says that a solution does not depend on the order of the players, nor on the order of the criteria. The last axiom $A4$ assures that all the players benefit (or at least not lose) from any enlargement of the agreement set, if the RA utopia point does not change.
We can prove the following result.

Theorem 3.1 *There is an unique function, $f : B^* \times R^m \longrightarrow R^m$, satisfying the axioms A1 – A4. For each $(S,d) \in B^*$ and each $u \in U(S,d)$ it is the function defined by*

$$f^R(S,d,u) = max_{\geq}\{x \in S : x = d + h(u-d) \ for \ some \ h \in R\}.$$

Proof. It is easy to verify that the function f^R satisfies the axioms A1 – A4. We prove that f^R is the unique solution. Let (S,d) be an arbitrary element of B^* and let $u \in U(S,d)$ be a RA utopia point. We show that $x^* = f^R(S,d,u)$ is the solution for (S,d) and u. Let T be the unique positive affine transformation mapping d to the origin $0 = (0,0,\ldots,0)$ and u to the point $1 = (1,1,\ldots,1)$. It is easy to notice that the point Tx^* has equal coordinates. We define now the bargaining problem $(S^S,0)$, where $S^S = \{x \in TS : for \ every \ permutation \ on \ M, \ \pi, \ the \ point \ y = \pi^*x \ is \ contained \ in \ TS\}$. It is easy to show that $(S^S,0) \in B^*$. Moreover we have $S^S \subseteq TS$, $Tx^* \in S^S$ and $1 \in U(S^S,0)$. By the axioms A1 and A3, $f(S^S,0,1) = Tx^*$. By the axioms A1, A3 and A4 it follows that $f(TS,0,1) = Tx^*$. By the axiom A2, $f(S,d,u) = x^*$. $\square$

Intuitively, the outcome $f^R(S,d,u)$ is the unique point of intersection of the line connecting u to d with the boundary of S. It is easy to notice that in the unilateral case, i.e. when $m^i = 1$ for $i \in N$, each bargaining problem (S,d) has an unique RA utopia point which coincides with the ideal point and the proposed solution coincides with the n–person Raiffa–Kalai–Smorodinsky solution. The following theorem pictures another connection between these two concepts.

For $(S,d) \in B$, let $x^i \in S$ be an individually nondominated point defined by player i and $u \in U(S,d)$ be the utopia point generated by $x^1, x^2, \ldots, x^n$. Because the problem is not trivial, the points $d, x^1, x^2, \ldots, x^n$ generate n–dimensional hyperplane H. It can be verified that each point $x \in H$ can be uniquely presented in the form $x = d + (a_1(u_1 - d_1), a_2(u_2 - d_2), \ldots, a_n(u_n - d_n))$, where $u_i \in R^{m^i}$ for $i \in N$. Let $S^H = S \cap H$ and T be the mapping from H to R^n defined by $T(d + (a_1(u_1 - d_1), a_2(u_2 - d_2), \ldots, a_n(u_n - d_n))) = (a_1, a_2, \ldots, a_n)$.

Theorem 3.2 *For each $(S,d) \in B$ and each $u \in U(S,d)$ if g^R denotes the n–person Raiffa–Kalai–Smorodinsky solution then $T(f^R(S,d,u)) = g^R(T(S^H), T(d))$.*

Proof. It is easy to notice that the n–person bargaining problem $(T(S^H), T(d))$ is normalized, i.e. the disagreement point $T(d)$ is equal to 0 and the ideal point $I(T(S^H), T(d)) = T(u) = 1$. We have
$g^R(T(S^H), T(d)) =$
$= max_{\geq}\{a \in T(S^H) : a = T(d) + h(I(T(S^H), T(d)) - T(d)) \ for \ some \ h \in R\} =$
$= max_{\geq}\{T(x) \in T(S^H) : x = d + h(u-d) \ for \ some \ h \in R\} = T(f^R(S,d,u)).$ $\square$

The theorem shows that the n–person Raiffa solution concept can be applied directly to the MCB problem (S,d) if we confine consideration to the outcomes in S^H, i.e. to intersection of the agreement set S with the hyperplane H.

3.2 Generalized Imai solution concept

The generalized Raiffa–Kalai–Smorodinsky solution can generally be only weak Pareto optimal. Imai (1983) has proposed in the case of the classical bargaining problem the lexicographic maxmin solution concept. This solution is Pareto optimal. Now we propose generalization of the Imai solution for a MCB problem in the class B^*.

For any bargaining problem $(S, d) \in B^*$ and any RA utopia point $u \in U(S, d)$, let $L : R^m \longrightarrow R^m$ be an affine transformation satisfying $L_k(x) = (x_k - d_k)/(u_k - d_k)$ for $k \in M$. The transformation normalize the problem, i.e. $L(d) = 0$ and $L(u) = 1$. Let $\succ^l$ be the lexicographic ordering on R^m, i.e. for any $x, y \in R^m$, $x \succ^l y$ if and only if there is $k \in M$ satisfying $x_k > y_k$ and $x_l = y_l$ for $l < k$. Let $P : R^m \longrightarrow R^m$ be a transformation such that for any $x \in R^m$, there is a permutation on M, π, with $P(x) = \pi^* x$ and $P_1(x) \leq P_2(x) \leq \ldots \leq P_m(x)$.

We introduce a stronger variant of the axiom $A1$.

$A1^*$. Pareto optimality. A point $f(S, d, u)$ is Pareto optimal in S.

For any bargaining problem $(S, d) \in B^*$ and any RA utopia point $u \in U(S, d)$, the proposed generalized Imai solution has the form

$$f^L(S, d, u) = \{x \in S : P(L(x)) \succ^l P(L(y)) \ for \ any \ y \in S\}.$$

Theorem 3.3 *The proposed generalized Imai solution is uniquely defined on the class* B^*. *It satisfies axioms* $A1$, $A1^*$, $A2$, $A3$.

Proof. Let (S, d) be an arbitrary element of B^* and let $u \in U(S, d)$. Because the set S is nonempty and compact, the functions P and L are continuous, so $f^L(S, d, u)$ exists. The uniqueness follows from condition $C3$. If $x \in f^L(S, d, u)$, $y \in f^L(S, d, u)$, $x \neq y$ then, there is $z \in f^L(S, d, u)$ with $P(L(z)) \succ^l P(L(x))$. Contradiction. It is easy to verify that the function f^L satisfies the axioms $A1^*$, $A2$ and $A3$. $\square$

To locate the generalized Imai solution, we propose the following procedure. For any bargaining problem $(S, d) \in B^*$, any RA utopia point $u \in U(S, d)$ and any $y \in S$, let $v(S, d, u, y) \in R^m$ be a vector satisfying $v_k(S, d, u, y) = u_k - d_k$ if $k \in Q(S, y)$, otherwise $v_k(S, d, u, y) = 0$.

Let $x(S, d, u, y) = \max_{\geq} \{x \in S : x = y + hv(S, d, u, y) \ for \ some \ h \in R\}$.

Let $\{x^t\}_{t=0}^{\infty}$ be a sequence with $x^0 = d$ and $x^t = x(S, d, u, x^{t-1})$ for $t = 1, 2, \ldots$.

Then, it can be shown, that there is a number t^* satisfying $x^{t^*} = x^{t^*+1}$, $t^* \leq m - 1$ and $x^{t^*} = f^L(S, d, u)$. It is easy to notice, that $x^1 = f^R(S, d, u)$.

Moreover, if x^1 is Pareto optimal, then the generalized Raiffa–Kalai–Smorodinsky solution point is equal to the generalized Imai solution point.

It is easy to notice that theorem 3.2 does not hold for the generalized Imai solution and the n–person Imai solution.

3.3 Generalization of other solution concepts

Any solution concept for classical bargaining problem can be generalized to MCB problem on the class B^{**} in a way following from the theorem 3.2. For any MCB problem $(S, d) \in B^{**}$ and any RA utopia point $u \in U(S, d)$, let T and S^H be as in theorem 3.2.

Definition 3.2 *A function* $f : B^{**} \times R^m \longrightarrow R^m$ *is a solution of multicriteria bargaining problem generalizing n–person classical solution* g *if for any multicriteria bargaining problem* $(S, d) \in B^{**}$ *and any RA utopia point* $u \in U(S, d)$,

$$T(f(S, d, u)) = g(T(S^H), T(d)).$$

It is easy to notice that if a classical solution has properties of weak Pareto optimality, invariance under positive affine transformations, anonymity then a generalized solution has them also. Let us notice that the generalized Imai solution in section 3.2 differs from proposed here generalization of the n–person Imai solution.

3.4 Continuity

The continuity of a solution is an important property from the point of view of applicability in interactive procedures of decision support. The model of bargaining problem is in general only an approximation of a real problem. We would like to have a property that the solution will not differ significantly in case of small change of the model. In an interactive procedure we would like also the solution to not change significantly under small changes of the players preferences. Therefore the following definition of the continuity is proposed.

Definition 3.3 *A solution of multicriteria bargaining problem* $f : B \times R^m \longrightarrow R^m$ *is continuous if for any sequence of multicriteria bargaining problems* $\{(S^t, d)\}_{t=1}^{\infty} \in B$ *and any sequence of RA utopia points* $\{u^t\}_{t=1}^{\infty} \in U(S^t, d)$ *such that, in the limit as t goes to infinity,* (S^t, d) *converges to* $(S, d) \in B$ *and* u^t *converges to* $u \in U(S, d)$, *we have* $f(S^t, d, u^t)$ *converges to* $f(S, d, u)$.

It is easy to prove that the generalized Raiffa–Kalai–Smorodinsky solution satisfies continuity property and the generalized Imai solution does not. Moreover, it is easy to notice that if an n–person classical solution has this property then a generalized solution proposed in section 3.3 has it also.

4 Iterative Solution Concept

The iterative solution concept presented and discussed in this point has been proposed under inspiration of the single negotiation text procedure described by Raiffa (1982), and of the principle of limited confidence Fandel (1979) and Fandel, Wierzbicki (1985).

The single negotiation text procedure has been originally proposed by Roger Fisher and is often employed in international negotiations. Raiffa (1982) has described the procedure on the example of Camp David negotiations. According to the procedure a negotiation process consists of a number of rounds. In each round a mediator prepares a package for the consideration of protagonists. Each package is meant to serve as a single negotiation text to be criticized by protagonists then modified and remodified. Typically the negotiation process starts from the first single negotiation text which is far form the expectations of the protagonists. The process is progressive for each of the protagonists.

The principle of limited confidence has been proposed as a result of observations of the players behavior in iterative gaming experiments. It has been observed that the players in particular rounds (iterations) of the experiments try to limit possible improvements of counterplayers outcomes. It is motivated by limited confidence of the players to the expected behavior of the counterplayers, to the model applied in the gaming experiment and finally to the outcomes obtained in the future.

We consider class B^* of the MCB problems, i.e. all problems (S, d) satisfying the conditions C1, C2, C3.

We assume, that the solution of the MCB problem (S,d) is looked for in some number of rounds (iterations) $t = 1, 2, \ldots, T$, in which outcomes $d^t \in S$ are determined. The final, admissible and accepted by the players, outcome d^T is the solution of the problem.

The following postulates that should be fulfilled by the process $\{d^t\}$, $t = 1, 2, \ldots, T$, are proposed:

P1. The process starts at the disagreement point, and all outcomes belong to the agreement set, i.e. $d^0 = d$, $d^t \in S$ for $t = 1, 2, \ldots, T$,

P2. The process is progressive, i.e. $d^t \geq d^{t-1}$ for $t = 1, 2, \ldots, T$,

P3. The final outcome is Pareto optimal, i.e. d^T ($= \lim_{t \to \infty} d^t$ if $T = \infty$) is a Pareto optimal point in S.

P4. *Principle of α-limited confidence.* Let $0 < \alpha_i^t \leq 1$ be a given confidence coefficient of the i-th player at round t. Then acceptable demands are limited by:

$$d^t - d^{t-1} \leq \alpha_{\min}^t [u(d^{t-1}) - d^{t-1}]$$

for $t = 1, \ldots, T$, where $\alpha_{\min}^t$ is a minimal confidence coefficient at the round t,
$\alpha_{\min}^t = \min \{\alpha_1^t, \ldots, \alpha_n^t, \alpha_{\max}^t\}$,
$\alpha_{\max}^t = \max \{a \in R : d^{t-1} + a[u(d^{t-1}) - d^{t-1}] \in S\}$,

$\alpha_{\max}^t$ defines a maximum value of the confidence coefficient at the round t, resulting from the requirement, that the outcome d^t should belong to the set S.
$u(d^{t-1})$ is the RA utopia point which reflects the preferences of the players in the subset $\{ x \in S : x \geq d^{t-1} \}$ of the set S .

P5. *Principle of rationality.* Each player is assumed to behave in rational way, trying to maximize his outcomes in particular rounds according to his preferences expressed with use of RA utopia points. It is assumed that at each round t, each player i explores a set of his individually nondominated points in the set $S^t = \{x \in S : x \geq d^{t-1}\}$, and defines his preferred point $x^{it}, i \in N$. Let $u(d^{t-1})$ denotes RA utopia point of the set S^t defined on the base of the preferred by the players individually nondominated points in the round t, i.e.
$u(d^{t-1}) = (u_1(d^{t-1}), u_2(d^{t-1}), \ldots, u_n(d^{t-1})), \quad u_i(d^{t-1}) = x_i^{it}, i \in N$.

Formally, the players rationality is formulated as follows:
For any d^t, at each round t, there is no such an outcome $x \in S$, $x > d^t$ that fulfills the condition $x - d^{t-1} \leq \alpha^t_{min} * [u(d^{t-1}) - d^{t-1}]$.

Theorem 4.1 *For any multicriteria bargaining problem $(S, d) \in B^*$ and for any confidence coefficients α^t_i such that $0 < \varepsilon \leq \alpha^t_i \leq 1$, $t = 1, 2, \ldots, T$ there is a unique process d^t, $t = 0, 1, \ldots, T$, $T \leq \infty$, satisfying the postulates P1, P2, P3, P4, P5. The process is defined as follows:*

$$(*) \qquad \begin{aligned} d^0 &= d, \\ d^t &= d^{t-1} + \alpha^t_{min} * [u(d^{t-1}) - d^{t-1}] \qquad for \quad t = 1, 2, \ldots, T \end{aligned}$$

where

T is a minimal number t for which $d^t = d^{t-1}$ or $T = \infty$.

Proof. Let us consider the sequence d^t, $t = 0, 1, \ldots$. From P3, P4, P5 it follows, that sequence is defined uniquely. The sequence is monotonically increasing and limited, so it is convergent. Let $d_{lim} = \lim_{t \to \infty} d^t$. From the construction (*) it follows, that $d_{lim} \in S$. Let d_{lim} be not Pareto optimal in S. Then for any round t, the following relations hold:
$\| d^t - d^{t-1} \| = \| \alpha^t_{min} * [u(d^{t-1}) - d^{t-1}] \| \geq \varepsilon \| u(d_{lim}) - d_{lim} \| = \gamma > 0$. It follows from the above, that the sequence $\{d^t\}^{\infty}_{t=0}$ is not convergent, what is contradictory to the assumption. So, we have shown, that d_{lim} is Pareto optimal in the set S. $\square$

5 Interactive mediation procedures

5.1 Interactive procedure based on the generalized Raiffa-Kalai-Smorodinsky solution

The generalized Raiffa-Kalai-Smorodinsky solution concept can be a base for construction of an interactive procedure supporting negotiation process over the bargaining problem. The procedure proposed herein is assumed to be performed in some number of iterations (rounds) t, $t = 1, 2, \ldots$. At each iteration two stages can be marked out.

In the stage one each player $i \in N$ independently explores nondominated outcomes in the agreement set S, with an application of the aspiration–led approach. That is, the player i assumes his reference point $r^t_i \in R^{m^i}$, and evaluates reference points for the counterplayers $r^t_j \in R^{m^j}$ $j \neq i$. The system calculates respective to the player reference point individually nondominated outcome and then anticipated compromise outcome based on the generalized Raiffa-Kalai-Smorodinsky solution concept. The exploration process is repeated for some number of different reference points of the player. In this way, the player obtains a characterization of his individually nondominated outcomes set and of the nondominated outcomes in the agreement set S for the evaluated references of the counterplayers. The exploration is continued as long as the player can decide to select the reference point and the compromise outcome being close to his preferences. The stage one ends when all the players have selected their preferred reference points and anticipated outcomes.

In the second stage a result of the round is derived. System calculates first RA utopia point u^t according to the selected preferred reference points of the players. After that, the generalized Raiffa-Kalai-Smorodinsky outcome $f^R(S, d, u^t)$ is derived. The outcome is proposed to the players as a cooperative solution. This solution should be considered as a mediation proposal to be discussed by the players. If the players agree, the procedure is finished, otherwise it goes to the next iteration $t = t + 1$.

Comments:

In particular round, the anticipated outcomes of the particular player are calculated for predicted by the player reference points of the counterplayers. Therefore, the result of the round is in general different than the anticipated outcomes obtained during the exploration stage. It seems be reasonable to predict the counterplayers reference point on the base of their preferred selections from the previous round. In such a case the counterplayers preferences can be better evaluated in consecutive iterations.

The cooperative outcomes proposed as a mediation ones in particular iteration of the procedure guarantee "fairness" in sense of axioms characterizing the generalized Raiffa-Kalai-Smorodinsky solution concept.

Le us denote by λ_i^t a direction of the outcome improvement in the objective space assumed by the player i in a round t with use of the reference point. The improvement direction is defined by $\lambda_i^t = r_i^t - d_i$ The cooperative outcomes denoted by $\hat{x}_i^t$ are compatible with the improvement directions of all the players. That is, for each $i \in N$ there is a number $\beta > 0$ such that $\hat{x}_i^t - d_i = \beta \lambda_i^t$. The property prevents direction manipulations by the players. The generalized Raiffa-Kalai-Smorodinsky solution is continuous on the class B^*. It is also continuous as regards the reference points. These properties should guarantee convergence of the procedure assuming rational behavior of the players.

The interactive procedure has been applied and tested in a negotiation support system prepared in the case of bargaining problem on joint development project (Kruś, Bronisz 1990).

Other solution concepts proposed in subsection 3.3 can be also implemented in a form of the interactive procedure in the similar way.

5.2 Interactive procedure based on the iterative solution concept

The iterative solution concept has been also used for construction of an interactive mediation procedure. The procedure consists of a number of rounds t, $t = 1, 2, \ldots$. In each round there are two phases of decision support. The first phase deals with unilateral, interactive analysis of the MCB problem with stress on learning, organized through system response to the player specified reference points for objective outcomes and confidence coefficients. The second phase includes calculation of the multilateral, cooperative solution outcomes according to the players preferences.

The first phase is performed by each of the players independently. For the particular player i the system begins this phase presenting information about the range of possible outcomes and reasonable reference points. The system generates also some initial values serving the player introductory information. It is called the neutral outcome, a solution obtained by the system under the assumption that the improvement direction is defined according to the ideal point.

The player learns about possible outcomes assuming different reference points $r_i^t > d_i^t$ for his objectives and confidence coefficients α_i^t. The player is asked also to assume reference points for counterplayers r_j^t, $j \neq i$. The reference points for the counterplayers can be also assumed on the base of the previous round. For each reference point $r_i^t > d_i^t$ and r_j^t, $j \neq i$ and confidence coefficient α_i^t given by the player at the round t, the system calculates:

RA utopia point:

$$u_i(S, d^{t-1}, r_i^t) = \max_{\geq} \left\{ x_i \in R^{m_i} : x \in S, \ x \geq d^{t-1}, \ x_i = d_i^{t-1} + a[r_i^t - d_i^{t-1}] \text{ for some } a \in R \right\},$$

one-shot solution:

$$x^t = \max_{\geq} \left\{ x \in S : x = d^{t-1} + a * [u(S, d^{t-1}, r^t) - d^{t-1}] \quad \text{for some} \quad a \in R \right\},$$

maximal confidence coefficient:

$$\alpha_{\max}^t = \max_{\geq} \left\{ a \in R : d^{t-1} + a * [u(S, d^{t-1}, r^t) - d^{t-1}] \in S \quad \text{for some} \quad a \in R \right\},$$

anticipated solution:

$$y^t = d^{t-1} + \alpha_{\min}^t * [u(S, d^{t-1}, r^t) - d^{t-1}]$$

where $r^t \in R^m$, $r^t = (r_1^t, r_2^t, \ldots, r_n^t)$, r_j^t, $j \neq i$ are given as have been assumed by the player i, or $r_j^t = r_j^{t-1}$ for $j \neq i$,
$u(S, d^{t-1}, r^t) \in R^n$ is the RA utopia point relative to the reference point r^t.

This learning process is continued as long as the player can specify his preferred outcome and reference point. The first phase is finished when all the players have selected their preferred reference points. Then the system goes to the second phase.

In the second phase, on the base of the preferred points r_i^t selected by all the players, the RA utopia point u^t and the mediation proposal d^t of the negotiation round is derived. The result is calculated according to the iterative solution concept, following the limited confidence principle (the minimal confidence coefficient is used for all players), trying to improve outcomes for all the players in the directions specified by their reference points. Thus, the system acts as a neutral mediator proposing a single-text provisional agreement improving the initial situation and forming a basis for the next round of negotiations.

The results are presented to the players, and the players can begin the next round assuming the obtained result as a new status quo point. The process terminates when the Pareto optimal solution in the agreement set is reached.

The presented above procedure has been implemented in MCBARG system supporting multicriteria bargaining (Kruś, Bronisz, Lopuch 1990). A security of the players information is one of the important issues that should be assured by a decision support system.

In the MCBARG system, information of particular players is protected by a system of passwords. However, it is possible that the players can jointly agree on a limited access to information of each other. Three levels of such an access have been assumed.

In the first level any player has no access to the counterplayers information at all. in this case in the first phase of the interactive procedure the player can not simulate reference points of the counterplayers, and the reference points are calculated by the system on the base of the counterplayers preferred selection at the previous round (in the first round the reference points are assumed on the base of the ideal point).

In the second level a player can simulate (assume) different reference point of the counterplayers, and analyze what is their influence on his results.

In the third level a player can simulate different reference points of the counterplayers and analyze their influence on his results but also can simulate the impact of his reference points on the results of the counterplayers.

6 Final remarks

The paper presents new theoretical results within multicriteria bargaining problem. The bargaining problem is considered in the case when each player has different set of objectives and the utility functions of the players are not given explicitly. The theory is being developed as a background for construction of interactive procedures supporting players in analysis of the problem and supporting also a mediation procedure. In the procedures we would like underline the importance of learning processes of the players, and the aspiration led approach is a suitable in this case. The outcomes calculated by the system as the results of particular rounds of the both procedures should be treated as mediation proposals to be discussed by the players, modified and remodified. The ideas of the mediation support in the case of the procedure based on iterative solution concept are close to the single negotiation text procedure frequently used in international negotiations.

References

Axelrod, R. (1985). The Evolution of Cooperation. Basic Books, New York.

Bronisz, P., L. Kruś, A.P. Wierzbicki (1989). Towards Interactive Solutions in Bargaining Problem. In A. Lewandowski, A.P. Wierzbicki (eds.): Aspiration Based Decision Support Systems. *Lectures Notes in Economics and Mathematical Systems*, Vol. 331, Springer Verlag, Berlin.

Dreyfus, S. (1985). Beyond Rationality. In M. Grauer, M. Thompson, A.P. Wierzbicki (eds.): Plural Rationality and Interactive Decision Processes. Proceedings Sopron 1984, Springer-Verlag, Heidelberg.

Fandel, G. (1979). Optimale Entscheidungen in Organizationen. Springer-Verlag, Heidelberg.

Fandel, G., A.P. Wierzbicki (1985). A Procedural Selection of Equilibria for Supergames. (private unpublished communication).

Harsanyi, J.C., R. Selten (1972). A Generalized Nash Solution for Two–Person Bargaining Games with Incomplete Information. Management Sciences, Vol. 18, pp. 80–106.

Imai, H. (1983). Individual Monotonicity and Lexicographical Maxmin Solution. *Econometrica*, Vol. 51, pp. 389–401.

Jarke, M., M.T. Jelassi, M.F. Shakun (1987). Mediator: Towards a negotiation support system. *European Journal of Operation Research* 31, pp. 314–334, North-Holland.

Kalai, E., M. Smorodinsky (1975). Other Solutions to Nash's Bargaining Problem. *Econometrica*, Vol. 43, pp. 513–518.

Kersten, G.E. (1988). A Procedure for Negotiating Efficient and Non–Efficient Compromises. *Decision Support Systems* 4, pp. 167–177, North-Holland.

Korhonen, P., H. Moskowitz, J. Wallenius, S. Zionts (1986). An Interactive Approach to Multiple Criteria Optimization with Multiple Decision–Makers. *Naval Research Logistics Quarterly*, Vol. 33, pp. 589–602, John Wiley & Sons.

Korhonen, P., J. Wallenius (1989). Supporting Individuals in Group Decision–Making. Helsinki School of Economics, Finland, (forthcoming).

Kruś, L. (1991a). Some Models and Procedures for Decision Support in Bargaining. In P. Korhonen, A. Lewandowski, J. Wallenius (Eds), Multiple Criteria Decision Support. *Lectures Notes in Economics and Mathematical Systems*, Springer Verlag, Berlin.

Kruś, L. (1991b). Methods of Multicriteria Decision Support in Bargaining. Monography. Systems Research Institute, Polish Academy of Sciences. Warsaw, Poland. (to be appeared in Polish)

Kruś, L., P. Bronisz (1991) . Multicriteria Mediation Support. in: Proc. of IIASA Scoping Conference on "Systems Analysis Techniques for International Negotiations", October 9-10, 1991, Laxenburg, Austria.

Kruś, L., P. Bronisz (1990). Decision Support in Negotiations on Joint Development Program. Proc. of Polish - Chinese Conference, Systems Research Institute, Polish Academy of Sciences, Warsaw.

Kruś, L., P. Bronisz, B. Łopuch (1990). MCBARG - Enhanced. A System Supporting Multicriteria Bargaining. CP–90–006, IIASA, Laxenburg, Austria.

Nash, J.F. (1950). The Bargaining Problem. *Econometrica*, Vol. 18, pp. 155–162.

Raiffa, H. (1953). Arbitration Schemes for Generalized Two-Person Games. *Annals of Mathematics Studies*, No. 28, pp. 361–387, Princeton.

Raiffa, H. (1982). The Art and Science of Negotiations. Harvard Univ. Press, Cambridge.

Roth, A.E. (1979). Axiomatic Models of Bargaining. *Lecture Notes in Economics and Mathematical Systems*, Vol. 170, Springer-Verlag, Berlin.

DeSanctis, G., R.B. Gallupe (1987). A Foundation for the Study of Group Decision Support Systems. *Management Science*, Vol. 33, No. 5, pp. 589–609.

Thomson, W. (1980). Two Characterization of the Raiffa Solution. *Economic Letters*, Vol. 6, pp. 225–231.

Wierzbicki, A.P. (1982). A Mathematical Basis for Satisficing Decision Making. *Mathematical Modelling*, Vol. 3, pp. 391–405.

Wierzbicki, A.P. (1983). Negotiation and Mediation in Conflicts I: The Role of Mathematical Approaches and Methods. WP–83–106, IIASA, Laxenburg. Also in H. Chestnat et al. (eds.): Supplemental Ways to Increase International Stability. Pergamon Press, Oxford, 1983.

Wierzbicki, A.P. (1985). Negotiation and Mediation in Conflicts II: Plural Rationality and Interactive Decision Processes. In M. Grauer, M. Thompson, A.P. Wierzbicki (eds.): Plural Rationality and Interactive Decision Processes. Proceedings Sopron 1984, Springer-Verlag, Heidelberg.

Wierzbicki, A.P. (1986). On the Completeness and Constructiveness of Parametric Caracterization to Vector Optimization Problems. *OR-Spectrum*, Vol. 8, pp. 73-87.

A Configuration of Intelligent Decision Support Systems for Strategic Use: Concepts and Demonstrations for Group Decision Making

Fumiko Seo and Ichiro Nishizaki

Kyoto Institute of Economic Research, Kyoto University

Sakyo-ku, Kyoto, 606, Japan

1 Introduction

This paper presents a new concept of Intelligent Decision Support Systems for Strategic Use (IDS^3) and its demonstration for constructing group utility functions based on possibility distributions.

Decision environments which decision makers face are increasingly complex in the present world. Decision makers are confronted with the diversification and variation of human preferences under ambiguity. Decision Support Systems (DSS) have been developed to aid in making better decisions as quick responses to the complex and changeable decision environments. In recent development of DSS, decision analysis has been discussed, intentionally or unintentionally, as a device for searching a satisfactory alternative or a latent prospect for it under uncertainty. (Simon, 1969; Mittra, 1986; Mittra, 1988; Holzman, 1989). On the other hand, from the theoretical point of view, criticism to decision analysis has been raised by many authors such as March, (1978) and Simon, (1983), etc.. However, decision makers are still eager to be rational for avoiding to fall into the decision trap when they are embarrassed with compound decision situations. Eventually the criticism will lead to a combined construction of decision analysis and DSS. While the prescriptive approach in decision analysis would be a complementary substitute for the normative approach as a "willful choice model" and bridge between that and the descriptive approach as Bell, Raiffa and Tversky, (1988) suggested, the prescriptive approach in decision analysis should be combined properly with DSS in order to be effective in practical use.

This paper concerns constructing an intelligent DSS for strategic use (IDS^3) as a device for interactively improving decision processes under ambiguous and changeable conditions, which is based on disciplinary research results such as decision analysis. A particular intention is devoted to cope with group decision environments.

In Section 2, a history of DSS is retrospected and the characteristics of Intelligent Decision Support Systems (IDSS) are discussed in comparison with the preceding DSS. In Section 3, a configuration of an IDSS composed of several Shells is presented. Section 4 is devoted to a demonstration of the implementation of an IDSS for group decision making and shows a device for constructing group utility functions based on the possibility distributions in fuzzy set theory. In Section 5, the concept of IDSSs for strategic use (IDS^3) is discussed along with concluding remarks.

2 History of DSS and Characteristics of IDSS

1. Development of CBIS

The main concepts of computer-based information systems (CBIS) can be distinguished in seven types which correspond to the stages of development with some overlap. (1) Operation systems (OS), (2) electronic data processing (EDP), (3) scientific and business calculations with proper program languages such as Fortran, (4) database management systems (DBMS), (5) master-transaction processing (MTP), (6) management information systems (MIS), and (7) decision support systems (DSS). EDP has been developed during the 1950s. By the mid-1960s, (3)-(5) have advanced dramatically with the assistance of the progress in hardware systems such as the IBM 360 Series. MIS has pervaded as "managerial information systems to help managers with decision making" (Parker, 1989) through the late 1960s and the early 1970s, but intentionally in combination with MTP which entirely concerns empirical tasks. The failure of MIS to cope with the contingencies in the 1970s such as the "oil shock", however, has caused people to disbelieve MIS. The reason might be the inefficiency of MIS in aiding decisions in changeable environments.

DSSs were developed in the 1970s as user-friendly, or easy-to-use computing and communication devices for aiding in decision making. An intention also is devoted to construct database information systems embodying "adaptive approaches" for spatial and land-use planning support in developing countries (e.g. Van der Meulen, 1988). In the 1980s, personal computers have spread remarkably along with the new era of OS such as MS-DOS. This is a beginning of the down-sizing phenomena for personal use in computer systems. In this period, concern for uncertainty and incomplete or semantic knowledge databases as a part of DSS has increased (Sowa, 1984; Brodie, 1984; Negoita, 1985; Kanal and Lemmer, 1986; Mittra, 1988). Strategic information systems (SIS) have been proposed during the late 1980s. While SIS intends to use the information systems more strategically, its main concern is in the transaction processing with efficient computer network and thus the managerial viewpoint of decision making is going to lose. We intend to reconstruct DSS on more advanced perspectives for decision processes in practice.

2. Characteristics of IDSS

The main characteristic of traditional DSS is in the integration of input-output capabilities with analytical tools, via the control module, as suggested by Mittra, (1986), which mainly use optimization and simulation techniques such as mathematical programming, statistical analysis and forecasting, and other quantitative data manipulation tools, based on the optimization principle with substantial rationality.

In the recent complex real world, however, the properties of databases which decision makers should manage and problems for decision makers to solve have changed. They are multiobjective, which embody (i) more nonquantifiable, (ii) more nonstatistical and (iii) more societal properties, and thus they include (a) more linguistic, (b) more noncrisp (fuzzy), and (c) more human factors, which should be treated on the satisficing principle with the procedural rationality. For coping with these problems, the new era of DSS would include (1) interpretative (logical inference-oriented), (2) intelligent (versatile and "soft"), (3) interactive (process-oriented) and (4) preferential (human decision-oriented) properties; it means that the DSSs should be based on multiple disciplinary research fields corresponding to the multiple objectives analysis. We call the DSSs embodying these characteristics Knowledge-Based Intelligent Decision Support Systems, or in short Intelligent Decision Support Systems (IDSS).

3 A Configuration of IDSS

A configuration of an IDSS is depicted in Figure 1. The main framework of an IDSS is composed of three phases. Phase I is the Data Structure Shell, which concerns relational data-base engineering. Phase II is the Expert System Shell concerning knowledge-based engineering and Phase III is the Human Decision Shell concerning managerial decision making. The Expert System Shell is not simply based on empirical, or ad hoc knowledge of experts, but mainly related to disciplinary knowledge-bases; it is a main characteristic of the IDSSs over the traditional DSSs. Each Shell is composed of several Cells, which are arranged into three stages according to the development processes of decision making: (1) primitive, (2) elaborated and (3) illuminated. The ultimate purpose of managerial decision making in the last phase is conflict solving in hazardous management. The main framework of an IDSS is adaptable and versatile for any disciplinary approach as long as it is constructed as a minor (special) Shell in accordance with the conceptual framework of the Shell systems of the IDSSs.

In the beginning, the problem owner has a perception about the real world and, to analyze it, requests experts to assist him with their knowledge and experiences. This primal and intuitive recognition by the problem owner is elaborated in the utilization of IDSS. Empirical databases in the real world are scrutinized in the Data Structure Shell which is constructed as an evolution process from the data collection in the primitive stage (Stage 1), via the relational database structuring (vertical and horizontal) in the elaborated stage (Stage 2), to the database management in the illuminated stage (Stage 3). The role of experts is operationally developed in the Expert System Shell evoluting from the knowledge acquisition in Stage 1, via the dialog inference in Stage 2, to the knowledge representation in Stage 3. The primal recognition of the problem owner is also elaborated in the Human Decision Shell, which is composed of evolution processes from the problem setting in Stage 1, via the inference and reasoning in Stage 2, to the conflict solving in Stage 3, and finally the effectiveness of the decision results to hazardous management is examined from the point of view of the original problem recognition.

Interactions between the Data Structure Shell and the Expert System Shell construct an *information/solver interface*. Output from this part of IDSS is interacted with the Human Decision Shell; this interaction constructs *a man/machine or decision interface* which means that the judgement of decision makers plays a more essential role in the final phase of IDSS than in the preceding phases.

In practice, a special Shell for an analytical method is constructed as a minor Shell

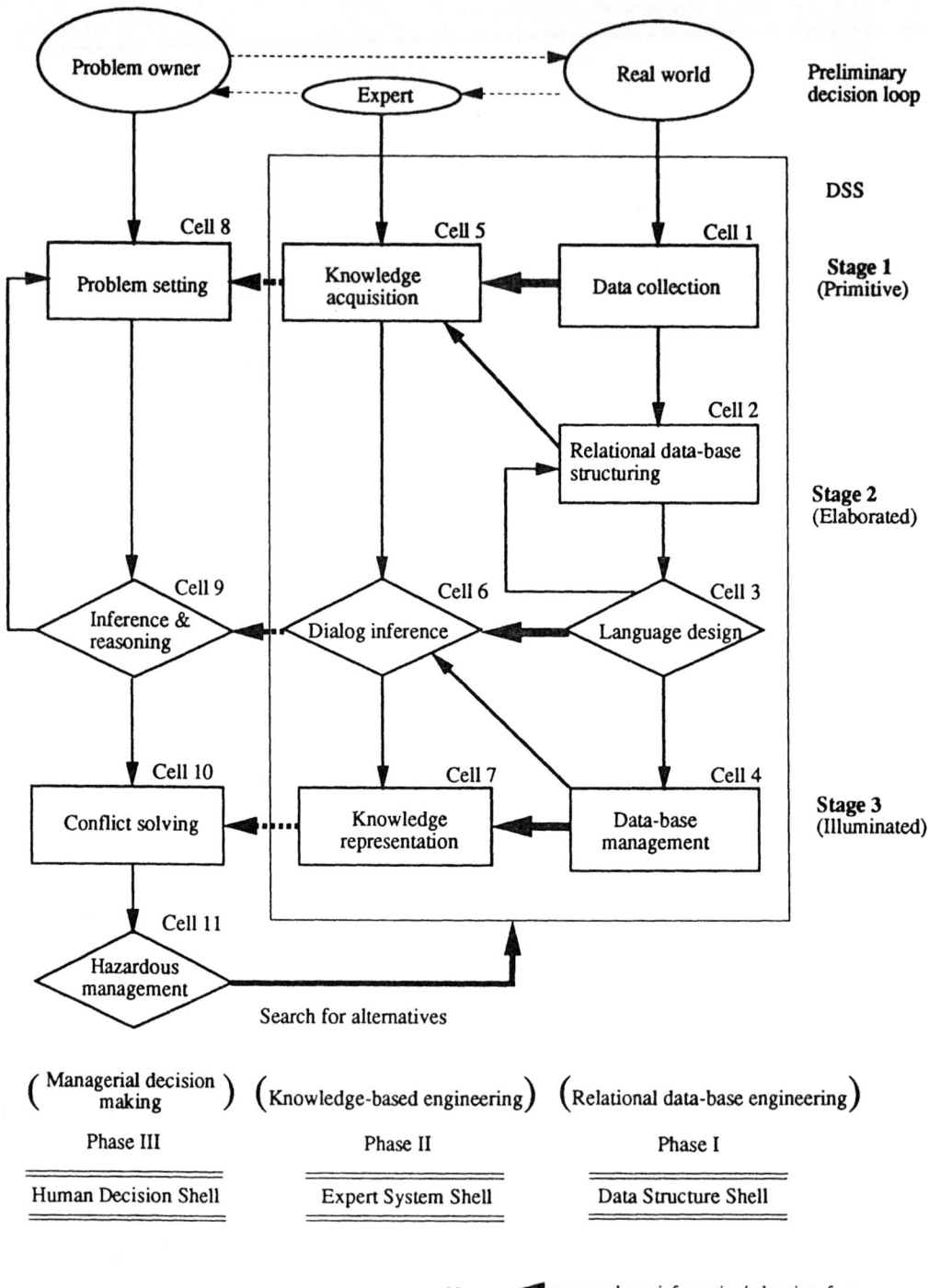

Figure 1. **Main frame of the intelligent DSS**

and combined with the main framework (major Shells) via its Cells (or Steps for an algorithm). A "decomposition" of a method as Cells of the special (minor) Shell and their combination with the main framework leads to the construction of a particular IDSS for practical use, which is based on a specific disciplinary field or a compound as a multidisciplinary approach, such as decision analysis and fuzzy set theory.

4 Demonstration for Group Decision Making

Decision analysis, developed by Raiffa, (1968), Schlaifer, (1969) and many others, has concerned the construction of utility functions. Some people are persisting in the following question: "who is the decision maker?", or "whose utility function is it ?" While the "knowledgeable" person is presumed as the decision maker, a more reasonable reply would be to assume an "as-if" rule which means that, as a result of intensive brain-storming by group discussions, a compromise for constructing a utility function will be realized and the utility function is presented as if a single decision maker has assessed it. This interpretation seems not to be unreasonable as long as the procedural rationality holds. The group decision problems, however, are still one part of decision analysis because, due to the complexity in group decision environments, an articulation of variety in the utility evaluation will be required for the conflict resolution among interest. For this purpose, a possibilistic approach in fuzzy set theory (Zadeh, 1965, 1978) is used for representing a variation of utility assessment along with traditional decision analysis (Seo, 1990). The IDSS is applied to the construction of group utility functions.

The Data Structure Shell is constructed in order to proceed from raw data (Stage 1), via spread sheet calculation (Stage 2), to graphical representation (Stage 3), with some elementary statistical processing. Stage 2 and Stage 3 are shown in Figure 2.

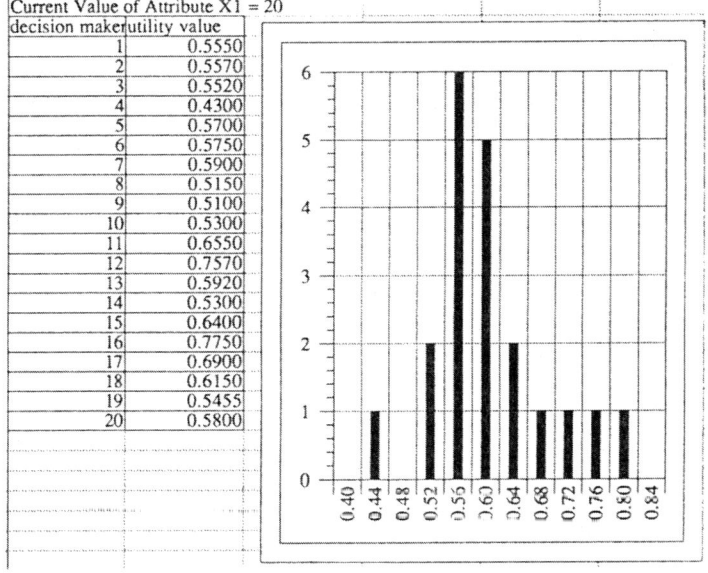

Current Value of Attribute X1 = 20	
decision maker	utility value
1	0.5550
2	0.5570
3	0.5520
4	0.4300
5	0.5700
6	0.5750
7	0.5900
8	0.5150
9	0.5100
10	0.5300
11	0.6550
12	0.7570
13	0.5920
14	0.5300
15	0.6400
16	0.7750
17	0.6900
18	0.6150
19	0.5455
20	0.5800

Figure 2. Spread sheet and graphic representation

The construction of group utility functions is performed in four cases for different group decision situations in the Expert System Shell:

Case 1. Point assessment. Each decision maker(DM) assesses his/her utility value for an attribute as one point which is set in continuous numbers.

Case 2. Interval assessment. Each DM assesses his/her utility value in an interval between discrete numbers which are arranged in an adjacent order.

Case 3. Wide interval assessment. Each DM assesses his/her utility value in an arbitrary wide interval on the real line.

Case 4. Fuzzy number assessment. Each DM assesses his/her utility value as a fuzzy number.

The database of an attribute x for which utility values are assessed has beenconstructed in the Data Structure Shell (Phase I) with necessary processing such as statistical analysis. Then the Expert System Shell (Phase II) is operated.

For Case 1 to Case 3, personal utility values $u_i(x)$ are assessed in Stage 1, in which any diversifications are treated as a possibility distribution and represented by possibility distribution functions. The possibility distribution Π_u associated with the utility assessment u is represented by the possibility distribution function π_u which is defined to be numerically equal to the membership function $\mu_G(u)$, and is shown as

$$\Pi_u = \mu_G(u_1)/u_1 + \mu_G(u_2)/u_2 + \cdots + \mu_G(u_n)/u_n. \tag{1}$$

A utility diversification in an n-person group decision making is shown with Eq.(1) as a possibility distribution. For assisting in making a compromise between the diversified assessments, the group utility assessment is represented as an "aggregated" membership function through the dialog inference in Stage 2 with a linear or nonlinear regression analysis when it is necessary.

In Type 4, utility values are assessed as the fuzzy numbers in Stage 1. In Stage 2, the conjunction (min) operation is performed for deriving group decision making as a membership function. These processes are demonstrated in Figure 3 to Figure 6.

In Stage 3, a fuzzy group utility function (FGUF) is derived from the membership functions. For aiding the construction of FGUF based on personal utility values, a nonlinear utility function, e.g.,

$$u = a + be^{-cx}, \tag{2}$$

is used and a utility value $u(x)$ is calculated and revised for different attribute levels x. Various utility values for a preliminary assessment point, for a reference point such as a mean value of the fuzzy number representation of the utility values and for revised points are calculated and presented numerically to each assessor interactively. Finally, a FGUF is constructed on the membership functions which are derived from the preceding utility assessment. Figure 7 to Figure 9 depict an example.

The analysis of this result is the work in the Human Decision Shell, depending on the problem which decision makers face.

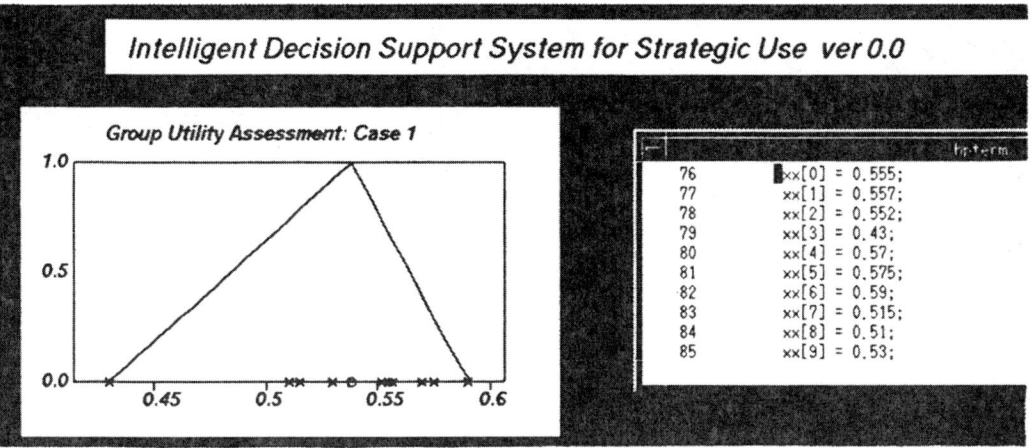

Figure 3. Group utility assessment: Case 1

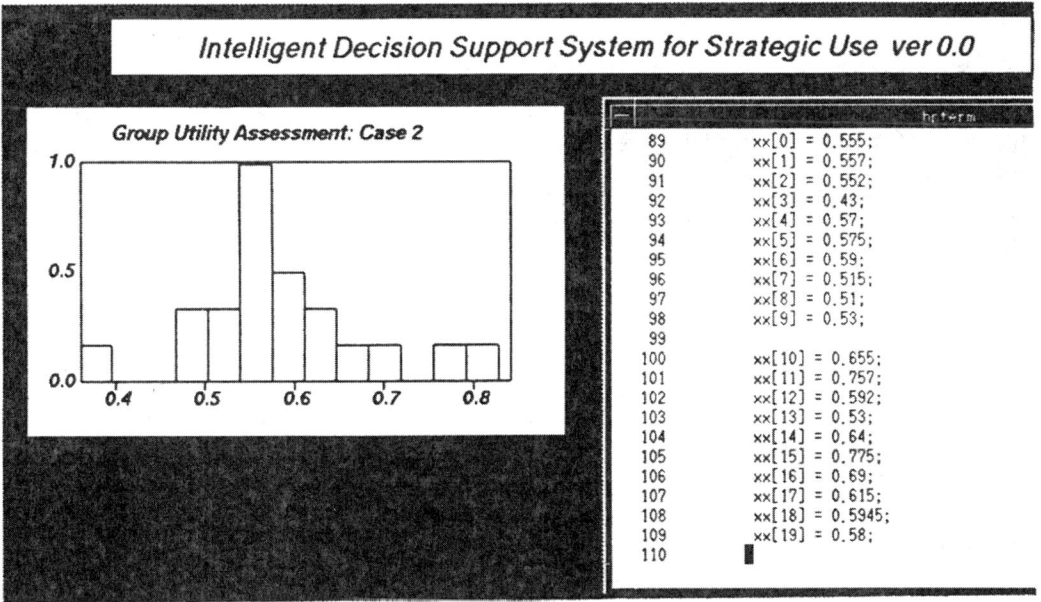

Figure 4. Group utility assessment: Case 2

Intelligent Decision Support System for Strategic Use ver 0.0
Group Utility Assessment: Case 3
1.0
0.5
0.0
0.4
0.5
0.6
0.7
0.8
hnterm
76 xxl[0] = 0.555;
77 xxl[1] = 0.557;
78 xxl[2] = 0.552;
79 xxl[3] = 0.43;
80 xxl[4] = 0.57;
81 xxl[5] = 0.575;
82 xxl[6] = 0.59;
83 xxl[7] = 0.515;
84 xxl[8] = 0.51;
85 xxl[9] = 0.53;
86
87 xxr[0] = 0.655;
88 xxr[1] = 0.757;
89 xxr[2] = 0.592;
90 xxr[3] = 0.63;
91 xxr[4] = 0.64;
92 xxr[5] = 0.775;
93 xxr[6] = 0.69;
94 xxr[7] = 0.615;
95 xxr[8] = 0.5945;
96 xxr[9] = 0.58;
97

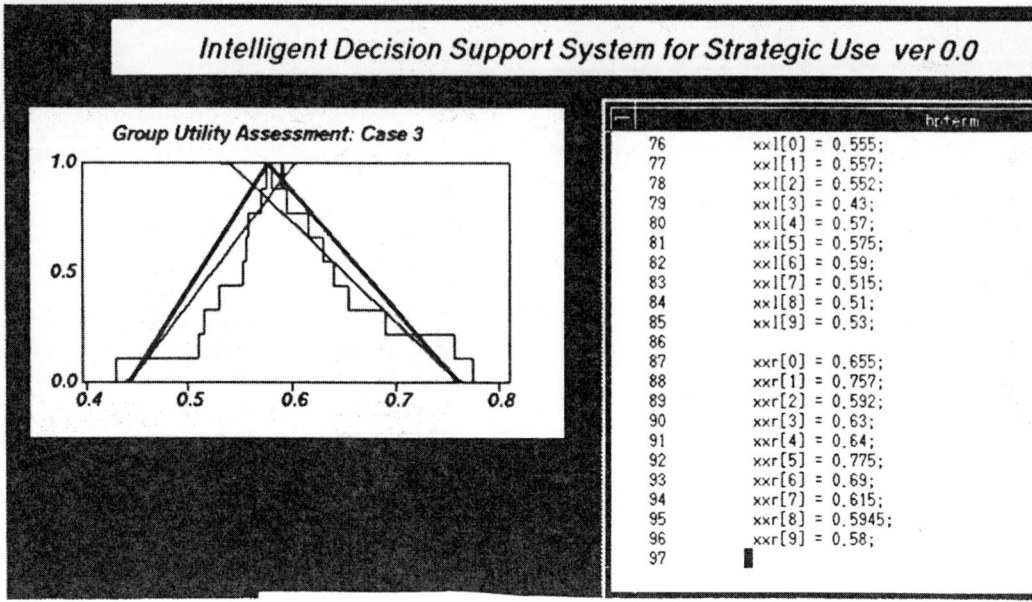

Figure 5 Group utility assessment: Case 3

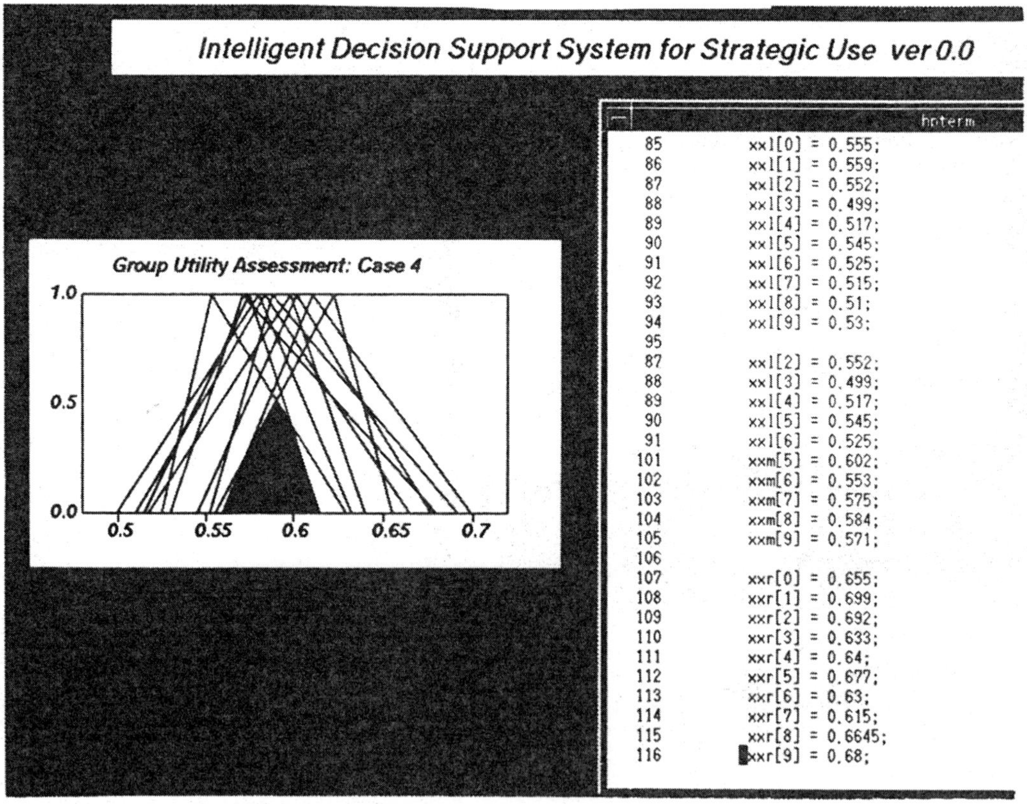

Intelligent Decision Support System for Strategic Use ver 0.0
Group Utility Assessment: Case 4
1.0
0.5
0.0
0.5
0.55
0.6
0.65
0.7
hnterm
85 xxl[0] = 0.555;
86 xxl[1] = 0.559;
87 xxl[2] = 0.552;
88 xxl[3] = 0.499;
89 xxl[4] = 0.517;
90 xxl[5] = 0.545;
91 xxl[6] = 0.525;
92 xxl[7] = 0.515;
93 xxl[8] = 0.51;
94 xxl[9] = 0.53;
95
87 xxl[2] = 0.552;
88 xxl[3] = 0.499;
89 xxl[4] = 0.517;
90 xxl[5] = 0.545;
91 xxl[6] = 0.525;
101 xxm[5] = 0.602;
102 xxm[6] = 0.553;
103 xxm[7] = 0.575;
104 xxm[8] = 0.584;
105 xxm[9] = 0.571;
106
107 xxr[0] = 0.655;
108 xxr[1] = 0.699;
109 xxr[2] = 0.692;
110 xxr[3] = 0.633;
111 xxr[4] = 0.64;
112 xxr[5] = 0.677;
113 xxr[6] = 0.63;
114 xxr[7] = 0.615;
115 xxr[8] = 0.6645;
116 xxr[9] = 0.68;

Figure 6. Group utility assessment: Case 4

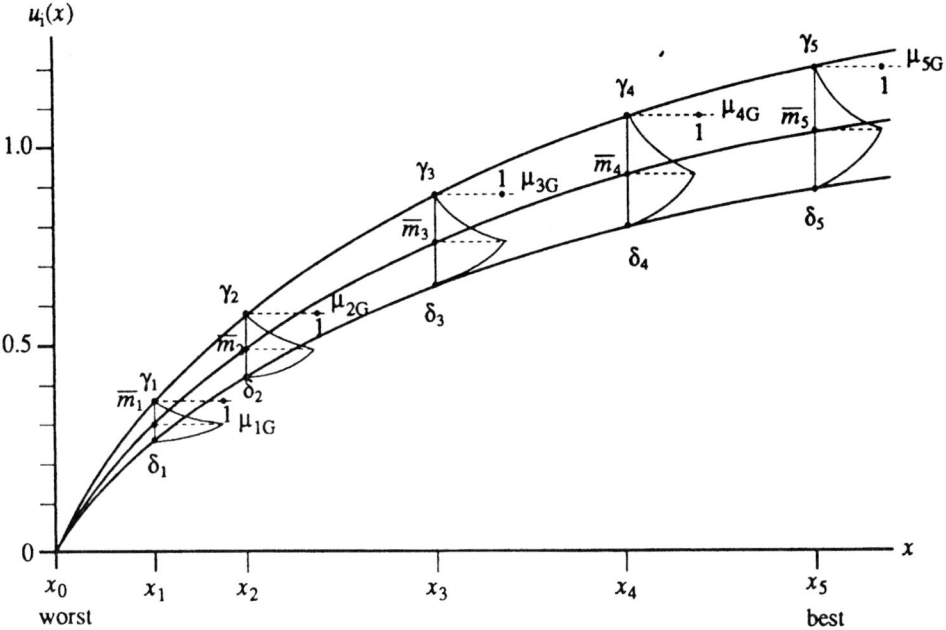

Figure 7. Construction of the fuzzy group utility function (FGUF)

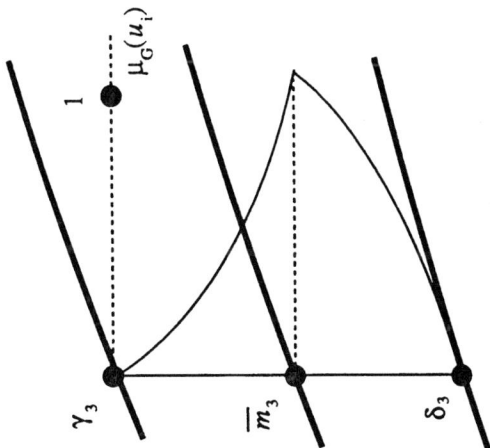

Figure 8. A membership function of a utility value: The L-R type fuzzy number representation $\tilde{u}_i \triangleq (\overline{m}, \gamma, \delta)_i$

5   Construction of IDS³: Concluding Remarks

The term "intelligent" of IDSS is mainly concerned with the following functions, distinct from the traditional DSS.

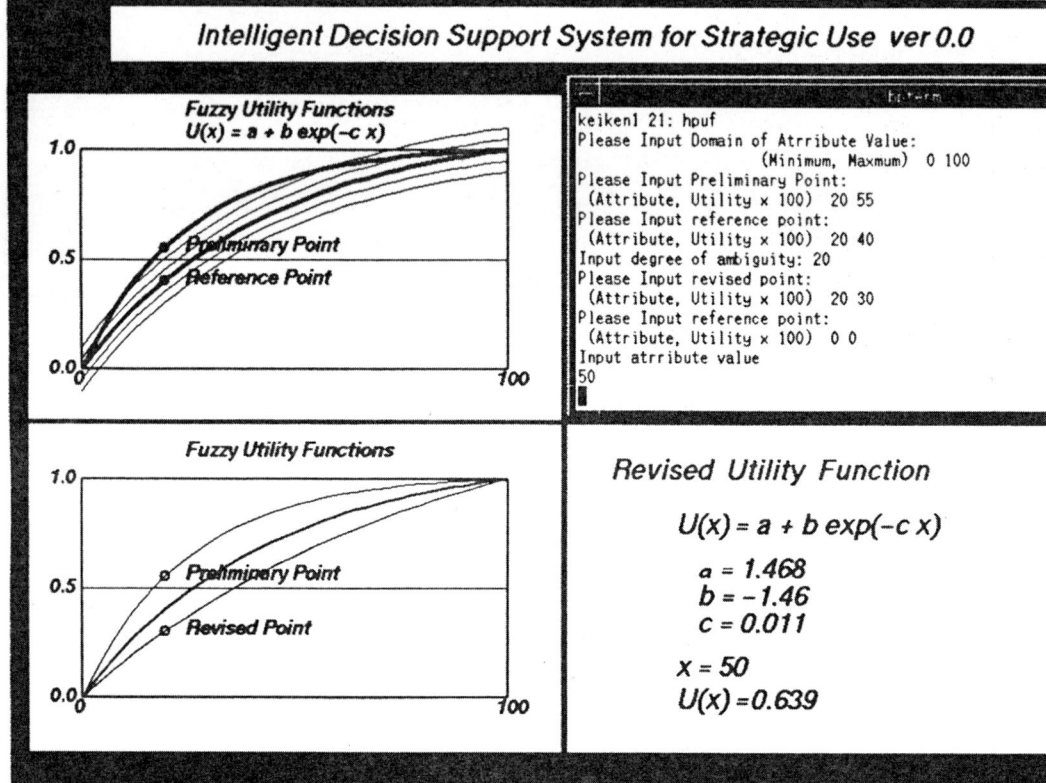

Figure 9. Representation of utility function

(1) Illuminating power, which is promoted by (i) multiobjective decision making, (ii)interactive processing, (iii)interpretative capabilities and (iv) conceptual clarity.

(2) Preference analysis based on perceptive recognition for treating the judgmental phase of decision processes.

(3) Utilization of axiomatic powers developed in multiple disciplinary fields.

(4) Manipulation of uncertainty or ambiguity including decision diversification.

The IDSS is extended as IDS³ for strategic use, which is a major purpose of the construction of IDSS.

The main characteristic of the strategic use is the adaptability for contingencies which has the following functions.

(1) Efficiency for "what-if" analyses.

The IDSS should be made as easy to adapt to any change in the database and decision environments and it should be possible to use the IDSS efficiently for decision makers who are ready to change. The monitoring and the changeability of database in the Data

Structure Shell, the revisability of identifications and analyses of problems in operational terms in the Expert System Shell, and the rapid responses of decisions to the changed output from the information/solver interfaces in the Human Decision Shell should construct a sequential and bidirection interactive cycle. In this process, an efficient configuration of the Data Structure Shell is important. The spread sheet packages should be able to recalculate promptly the entire worksheet and to send the database to graphical representation packages via some statistical processing as pretreatment for easy understanding. In the Expert System Shell, advanced packages for quantitative modeling, simulation and optimization should be efficiently combined.

(2) High-grade informative properties with networking.

This request corresponds to the recent development of computer systems, which embodies a large capacity for RAM and storages (disks, tapes) and, at the same time, the "down-sizing" is performed in connection with network systems for distributed processing, in which multitask and multiuser systems are included. A computer network can be constructed with a mainframe, workstations, and personal computers, which are connected with each other and also with other domestic and overseas computer systems, via internetwork (e.g. Ethernet) with a protocol such as TCP/IP. A LAN (local area network) can be constructed for an organization according to a specific type of problems. Figure 10 depicts an example of the composition of a network of IDS[3], which is constructed as a LAN in Kyoto Institute of Economic Research, Kyoto University. Recent development of OS as an open system, such as UNIX, greatly contributes to the construction of the network because of its good transplantability in different types of machines. The server/client systems are developed on the UNIX systems, which are composed of the database server, the X server (for visual display management) and application program units as the clients, for instance.

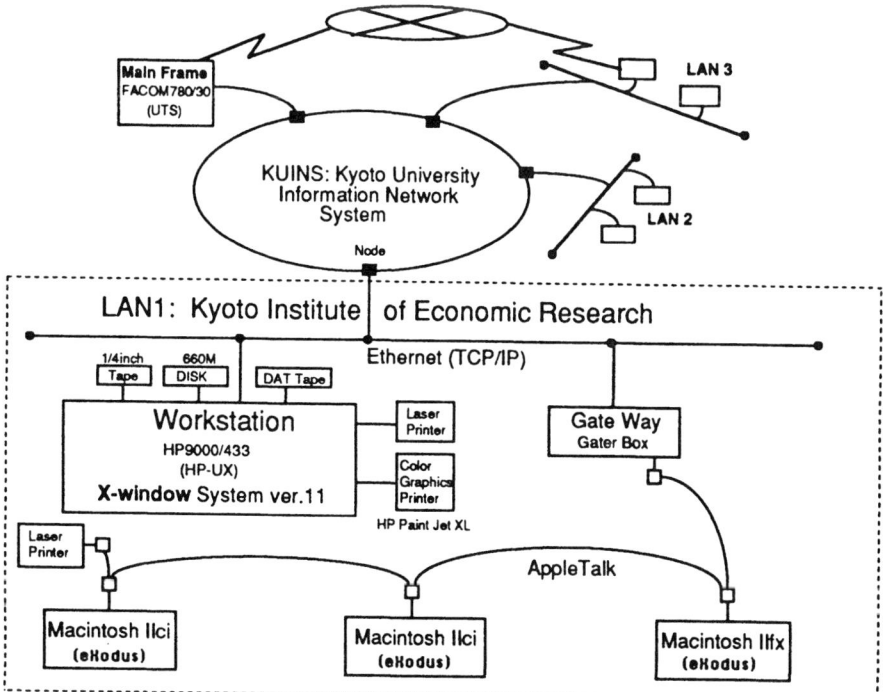

Figure 10. Computer network in KIER

(3) User-friendly properties.

The merit of "down-sizing" is its "user-friendly" property. For adapting to the contingencies promptly, the understanding of problems newly occurred and the acquisition of the necessary knowledge to cope with them should be easy. For assisting the DM with IDSS, accessibility to the devices is crucial because the DM and experts for the problems are not necessarily experts on computer systems. The qualification as an eligible expert for assisting a DM is placed on knowledge and discernment in the specific fields of the problems which the DM faces. The X-window system recently developed is a promising device for this purpose, which is a multiwindow display software in a network and on which the GUI (graphical user interfaces) such as OSF/Motif can be operated. The workstation systems with advanced capacities are recommendable to work it in connection with personal computers using proper softwares such as the Exodus for end users.

Acknowledgement. The authors are indebted to the Kikawada Foundation for supporting this study with a research grant for 1991-1992.

References

Bell, D.E., Raiffa, H., and Tversky, A. (1988). Descriptive, normative, and Prescriptive interactions in decision making. In Bell, Raiffa and Tversky (eds.), *Decision Making: Descriptive, Normative, and Prescriptive Interactions*, Cambridge University Press, Cambridge. 1988.

Brodie, M.L., Mylopoulos, J., and Schmidt, J.W. (1984). On Conceptual Modelling. Springer-Verlag, Berlin.

Kanal, L.N. and Lemmer, J.F. (1986). Uncertainty in Artificial Intelligence. North Holland, Amsterdam.

March, J. (1978). Bounded rationality, ambiguity and engineering of choice. *Bell Journal of Economics*, 9(2), 587-608.

Mitra, G. (1988). Models for decision making: an overview of problems, tools and major issues. In G. Mitra (ed.), *Mathematical Models for Decision Support*, Springer-Verlag, Berlin.

Mittra, S.S. (1986). Decision Support Systems, Tools and Techniques. John Wiley & Sons, New York.

Negoita, C.V. (1985). Expert System and Fuzzy Systems. The Benjamin/Commings Inc. Menls Park, California.

Raiffa, H. (1968) Decision Analysis. Addison Wesley, Reading, Mass.

Schlaifer, R. (1969). Analysis of Decision under Uncertainty. McGraw-Hill, New York.

Seo, F. (1990). On construction of the fuzzy multiattribute risk function for group decision making. In Kacprzyk and Fedrizzi (eds.), Multiperson Decision Making, Models Using Fuzzy Sets and Possibility Theory, Kluwer, Dordrecht.

Simon, H. (1969, 1981). The Sciences of the Artificial. The MIT press, Cambridge.

Simon, H. (1983). Reason in Human Affairs. Stanford University Press. Stanford, California.

Sowa, J.F. (1984). Conceptual Structure: Information Processing in Mind and Machine Addison-Wesley, Mass.

Zadeh, L.A. (1965). Fuzzy sets. *Information and Control*, 8, 338-353.

Zadeh, L.A. (1978). Fuzzy sets as a base for a theory of possibility. *Fuzzy Set and Systems*, 1, 3-28.

Parker, C.S. (1989). Management Information Systems, Strategy and Action. McGraw-Hill, New York.

Holtzman, S. (1989). Intelligent Decision Systems. Addison-Wesley, Reading, Mass.

Van der Meulen, G. (1988). Computer aided urban planning, management and decision making in developing countries. *Regional Science Studies*, Japanese Regional Science Association, Tokyo.

Interactive Multicriteria Decision Support: Combining Rule-based and Numerical Approaches*

Kurt Fedra, Chunjun Zhao, and Lothar Winkelbauer
Advanced Computer Applications
International Institute for Applied Systems Analysis
A-2361 Laxenburg, Austria

Abstract

The main objective of the project is to develop and test, in a practical case study, new methods for multicriteria decision support that seek to combine approaches and methods from both traditional OR and new and emerging methods of AI and expert systems in particular. As a consequence of this experimental approach, the actual computer implementation of the concepts developed is a necessary condition for their practical testing in the field.

Currently existing DSSs are only providing support for an analysis that requires a numerical representation of objectives, and thus more or less directly measurable criteria dimensions. To allow the decision maker to base the analysis and selection from a set of alternatives on numeric **and** symbolic criteria the hybrid multicriteria optimization tool HYDAS (HYbrid Discrete Analysis System) has been adapted and extended. HYDAS combines numerical, symbolic, graphical, and statistical methods to support the decision maker when exploring the solution space and enables the user to arrive at a well-informed decision.

The case study for testing the methodological research is based on an environmental decision making problem, i.e., investment in air pollution control at the city level, but also at the national level, considering the distribution of investment over a large number of cities in China. The different scenarios which can be created by the decision maker using the air pollution control model form his set of alternatives in the multi-stage decision-making process.

The basic simulation model for alternative strategies for development and the decision support tool HYDAS have been integrated within one framework system, called XDSS (extended Decision Support System). XDSS also incorporates a geographic information system module, a number of dedicated editors that allow the easy manipulation and modification of data sets, as well as auxiliary software for

*The research project described in this paper is funded by the Austrian *Fonds zur Förderung der wissenschaftlichen Forschung (FWF)* under Project No. P7415-PHY.

file handling, model control, automatic example generation using Monte Carlo techniques, etc.

All components of XDSS use a common graphics user interface which is completely menu driven and makes extensive use of visualization techniques to provide for easy control and coordination of the modules of XDSS. The system is programmed in C, using the X Windows libraries for graphics display.

1 Introduction

Interactive decision support for complex, multi-dimensional (multi-objective, multicriteria) problems requires the representation of the decision problem (which is ultimately a trade-off among conflicting objectives with non-commensurate criteria), that makes the trade-offs and compromises obvious and well understood.

On the one hand, the necessary formalism to treat a complex problem within one logically consistent framework should allow a *natural* representation, based upon natural units and problem-adequate classification of criteria values. On the other hand, the manipulation of the usually complex and multi-dimensional decision space (for continuous problems) or set of alternatives (for discrete problems) should be easy and convenient, so as not to distract the user from his ultimate task of comparative evaluation with the technical aspects of the tool.

A natural representation which helps to understand and gain insight can be achieved with a largely symbolic, graphical, and geometrical representation; the interactive control over the problem description (the data describing the alternatives), e.g., interactive selection of display parameters, projections, scaling, and rotation of the data similar to a CAD system, can make the analysis of a rather abstract hyperspace (multi-dimensional data and decision space) come very close to the manipulation of a physical object. This clearly supports the intuitive understanding of the structural properties of the problem, its *Gestalt*.

At the same time, a natural representation require's the description of alternatives in quantitative as well as qualitative terms. Criteria values are thus described numerically on cardinal scales, symbolically with numerical ranges on ordinal scales, or symbolically using logical labels (TRUE or FALSE), indicating the presence or absence of specific properties.

To manipulate and process such a hybrid representation format in turn requires a hybrid approach. The combination of numerical methods of multicriteria analysis (for an overview see Zhao, Winkelbauer, and Fedra, 1985) and AI technologies (Barr and Feigenbaum, 1981, 1982; Charniak, Riesback, and McDermott, 1980; Cohen and Feigenbaum, 1982; Savory, 1985) such as knowledge bases of logical/symbolic representations of decision rules or rule-based heuristic procedures can provide the necessary flexibility and at the same time the *natural* representation of a complex decision process.

The integration of AI methods, expert systems technology in particular, allows advantage to be taken of new methods and approaches. Rules provide a convenient and easily understandable format and mechanism to represent and process decision rules. They allow a formulation that is very close to natural language, easy to understand, and thus easy to formulate by the decision maker. A construct such as IF *an alternative has a certain set of properties*, THEN *classify it*, is easy enough to understand and manipulate. It adds the possibility of using logical rather than only numerical descriptions and operations. The logical descriptions and operations are very often more familiar, and thus acceptable, to a decision maker in comparison to classical optimization paradigms.

Also, basic inference mechanisms such as forward and backward chaining provide powerful methods to manipulate symbols describing alternatives, and also provide the possibility for a natural language explanation of logical conclusions and results. In conjunction with numerical data manipulation methods such as statistical analyses and traditional optimization methods, the use of rule-based logical and symbolic extensions adds both realism and ease of use to a hybrid approach.

Finally, by restricting the application of a decision support tool to a well-defined domain, it is possible to incorporate some general information about the domain in a knowledge base as well. Providing meta rules to structure the decision support tool itself could allow provision of a convenient vocabulary of symbols, plausibility checking mechanisms, and expert guidance for the efficient use of the tool in a given situation.

Project objectives and approach

In an experimental rapid prototyping approach, the project described in this report seeks to integrate AI and OR methods, by developing rule-based extensions of multicriteria DS tools (Majchrzak, 1984, 1985; Wierzbicki, 1979, 1980; for a discussion of several alternative methods see Zhao, Winkelbauer, and Fedra, 1985). To allow for an experimental evaluation of these methods, main features of the approach developed here must be implemented on the computer in an operational interactive decision support system. An important component of this implementation is an interactive user interface to enable the easy manipulation of symbolic data in a graphical representation.

The philosophy underlying this hybrid approach emphasizes the role of the human decision maker in a tightly coupled man–machine system, the role of information versus data, the role of intuitive understanding and insight versus formal solution, and the role of learning versus rigid formalism.

Real-world decisions are always made under various degrees of uncertainty and with a considerable, if not dominant, intuitive component that also includes unspecified (and probably unspecifiable) criteria, subconscious decision rules, hidden objectives, or sloppiness, ignorance, and stubbornness.

Any strictly formalized approach with explicit rules only and a predefined procedure that is based on a strict mathematical concept of *optimality* ignores these components. It is based on a highly idealized, grossly simplified, and somewhat naive model of decision processes and decision makers. Acceptance of formal decision support methods, in practice, for all but a few very simple and very well-formalized problems is therefore very low.

The approach used here is interactive, heuristic, adaptive, and experimental. Foregoing the undeniable beauty of mathematical rigor for practical applicability, it tries to understand the decision-making process as an adaptive learning exercise rather than as a formal search operation.

For the given problem, i.e., urban air pollution control, represented as a set of alternatives described in terms of a set of criteria, the user goes through iterative cycles of analysis and formulation of specifications. The computer presents the current problem, allowing the user several options of display, manipulation, and analysis of the data to explore the problem. The user then simulates the previous decisions, such as defining subsets of criteria, constraints on criteria values, composite elimination or evaluation rules, and ranking and scoring procedures. Each simulated decision results in either a reduction of the number of candidate alternatives to choose from or a restructuring of the problem and a subsequent new round of analysis.

Since the user will go back and forth, tightening and relaxing his requirements or applying different elimination rules, the process amounts to an *interactive definition of the requirements* (or, in a colloquial sense, optimality) the final solution will have to fulfill. This definition of optimality, however, does not have to be given *a priori*, but evolves with growing insight into the problem structure. The user learns about what is possible, the implied trade-offs, or the effects of his preferences and decision rules on the result. Rather than being the result of a set of rigid and literally blind external requirements, his choice can be based on insight gained into the problem structure as much as his external, *a priori* requirements.

As an extension to the traditional approach, symbolic criteria are added to the set of numeric criteria which can then be evaluated using a specific rule base for each problem area. The evaluation of the rules as such can either restrict the set of alternatives to be evaluated by the numeric optimization part or help to select the optimal solution from the set of pareto–optimal alternatives which have been calculated using the numerical OR methods.

A general data representation format

A basic prerequisite for the inclusion of qualitative, symbolic methods together with the basic numerical representation and treatment is the development of a general data format for the representation of alternatives that provides enough flexibility to capture the nature of a problem, and at the same time lends itself to efficient computerized manipulation.

A general description language for alternatives, where each of the criteria values of a given alternative can be represented by either a numerical value, a range, a symbol, or a combination of numerical range and symbol, is presented below:

```
DESCRIPTOR
total_investment
T O
U Million Yuan
V 0 / 1000 /
Q What is the total level of investment for
Q pollution control technology  and measures
Q for this city in the current year
ENDDESCRIPTOR

DESCRIPTOR
traffic_increase
T D
V low / medium / high /
Q how would you rate the expected increase
Q in motor vehicle traffic for this city
ENDDESCRIPTOR

DESCRIPTOR
filter_efficiency
T P
U percent
V very_small[0,25] / small[25,50] / medium[50,75] /
  high[75,95]   / very_high[95,100] /
Q how would you classify the average efficiency of
Q this filter technology for dust reduction
ENDDESCRIPTOR
```

The generic descriptor definition includes the name of the descriptor as it is used in the system's interface (underscores are filtered out for display; they are, however, required to simplify the file input procedures). The following records (lines) are all preceded by a one- or two-character symbol, namely:

- T: type of the descriptor; one of D (decision variable), B (base year data), O (model output), or P (model parameter);

- U: the unit of measurement for the descriptor;

- V: the range or list of values the descriptor can take; this is either a range of numbers for a numerical descriptor or a list of symbols for a symbolic descriptor; in hybrid cases, the symbol name is followed by two numbers indicating the numerical range associated with the symbol;

- R: a list of rules that can be used to construct a derived dimension or affect the coupling of symbolic and numeric descriptions;

- TB: a list of decision tables used to derive a descriptor value;

- Q: the question to ask in the interactive editor, providing information for the user to set the value properly.

In the case of a hybrid descriptor, the user can specify either its numerical or symbolic value, depending on the context. The numerical ranges corresponding to a given symbolic value are defined in the generic descriptor definition (see above). If a symbol needs to be derived from a number, the corresponding interval/symbol pair is determined. If the number needs to be derived from the symbol, the mean value of the range corresponding to the symbol is used. Internally, for the computation of the Pareto set and the reference point analysis, symbols are translated into the median value of their range and ranks, respectively. Obviously, an extension to the use of fuzzy sets would improve the system at least in conceptual and theoretical terms.

The rules can be used for two specific purposes:

1. assigning a value to a derived descriptor, depending on the values of primary descriptors directly calculated by the model;

2. affecting the coupling of symbols and numerical ranges in the case of hybrid descriptors.

A derived descriptor can be constructed with a simple IF ... THEN rule format that allows for assignment of new values, testing of equality and relative magnitudes with operators such as $<<, <, \leq, >, \geq, >>$, and simple arithmetic operations such as addition, subtraction, or multiplication.

A derived descriptor could thus represent the sum, average, or median value from a group of related descriptors; within a city, this could, for example, be the per capita emission total (i.e., the sum of industrial-, residential-, and traffic-generated emissions divided by the number of residents. At the national level, this could be a weighted average from a number of representative cities for a given descriptor. In either case, the value of the new descriptor is constructed from the available information about other descriptor values. If any of the necessary pieces of information is missing, the inference engine will ask the user to supply this information directly.

The second use of rules at this level is a context-specific modification of the default association between numerical ranges and symbols set in the descriptor definitions. Here special conditions, such as the value of other descriptors or the city, and reference year, can be used to change the default range. This adds a considerable amount of flexibility to an otherwise restrictive vocabulary, and allows for a context-dependent interpretation. For example, it is natural to assume that large as a symbolic label describing some quantity has a different meaning in Beijing or Shanghai as compared with one of the smaller coastal cities.

In terms of the discrete decision support system, each alternative is then described by a list of descriptor values. Choosing any of the descriptors as a dimension in the criteria space also requires the selection of the intended direction of optimization, i.e., minimization or maximization of the respective values.

2 The Hybrid Decision Support System: XDSS

Rather than developing concepts in the abstract and looking for subsequent possible application—usually to find that the real world is somewhat more complex than the method developed allows for, so that the problem has to be reshaped until it fits the solution method—we start with a given problem, i.e., air pollution control in the major cities of the PRC.

At both the national level and the level of the individual city, funds for investment have to be allocated for pollution control as part of an overall integrated development policy. For a discussion of this *integrated development* concept in a regional Chinese context see Fedra, Li, Wang, and Zhao, 1987.

The main mechanisms related to the problem are industrial development, the development of traffic, and the domestic use of energy, all leading to increased levels of pollutant emissions. On the other hand, this economic development also provides the necessary means to tackle urgent environmental problems such as the introduction of pollution control technologies or refurbishing old, inefficient technological systems that generate disproportionally large amounts of pollution.

Depending on the decision maker's point of view, he may attempt to satisfy environmental standards while maximizing economic growth; alternatively, and somewhat naively in the case of a developing country, he may want to minimize environmental pollution by aiming at just the necessary degree of economic growth. In practice, the decision maker will strive for a balance between economic and environmental objectives, subject to a number of technological, and economic, and physiographic constraints and conditions. The ultimate decisions will certainly depend on the relative importance of economic versus environmental criteria, which is basically a political, socio-cultural, and maybe ethical, rather than scientific, problem.

The case study for testing the methodological research is based on an environmental decision-making problem, i.e., investment in air pollution control at the city level and at a national level, considering the distribution of investment over a large number of cities in China. To assess the level of air pollution in a city an urban air pollution control model has been jointly developed by the Academy of Environmental Science of the PRC and the School of Economics and Management of the Tsinghua University, Beijing, PRC within the framework of a project sponsored by the NNSFC (Wang and Mao, 1990).

To support the decision maker in selecting the *optimal* scenario or alternative from the set of alternative scenarios according to his preferences, objectives, and constraints, a

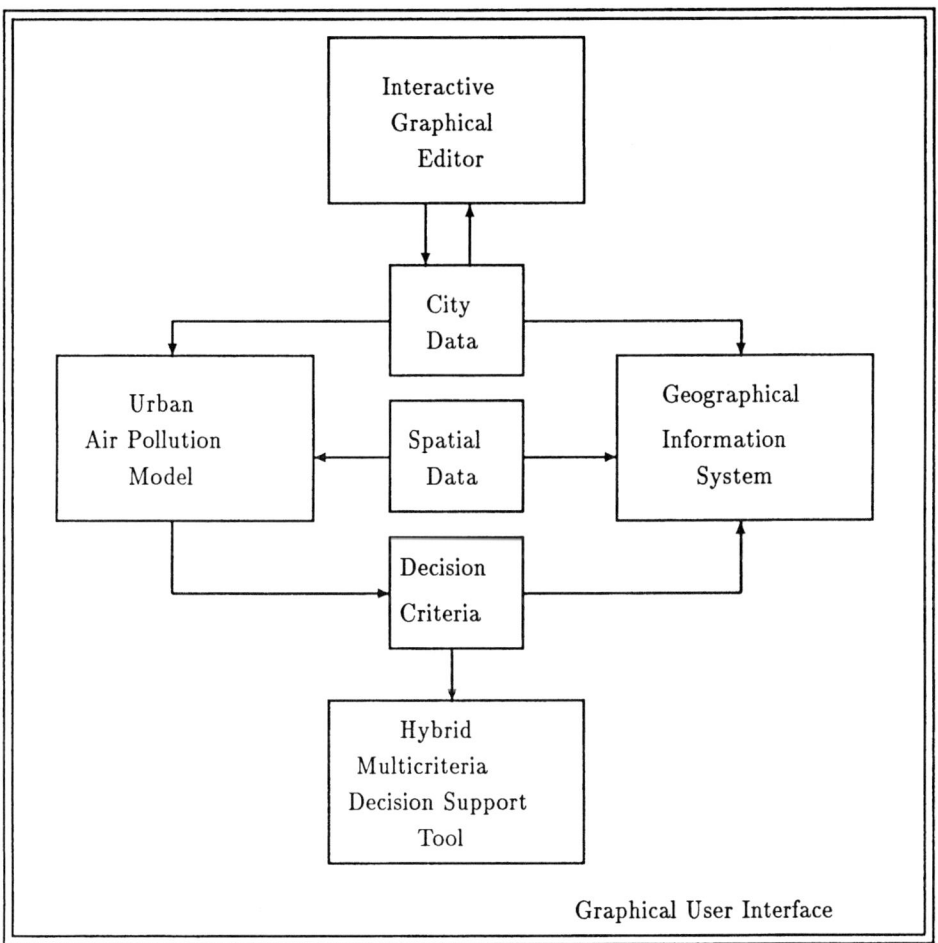

Figure 1: XDSS components and structure

prototype tool for hybrid multicriteria alternative selection (HYDAS) (Winkelbauer and Markstrom, 1990) has been adapted and extended in the project.

Both modules, the simulation model for alternative strategies and the decision support tool HYDAS, are being integrated within one framework system, called XDSS (extended Decision Support System) (Figure 1). XDSS also incorporates a geographic information system module, which allows presentation and analysis of spatial data used for the air pollution control model and the model results used in HYDAS. In addition, the system provides a number of dedicated editors that allow the easy manipulation and modification of data sets, as well as auxiliary software for file handling, model control, automatic example generation using Monte Carlo techniques, etc.

All components of XDSS use a common graphics user interface which is completely menu driven and makes extensive use of visualization techniques to provide for easy control and coordination of the modules of XDSS.

The prototype implementation of XDSS integrates the basic emission estimation model, a first version of HYDAS integrating qualitative, ordinal descriptors, and the geographical information system for the display and analysis of data and model results together with a number of auxiliary software components.

3   HYDAS: A Hybrid DSS tool

To support the decision maker in the task of selection of one *optimal* alternative according to his preferences, the hybrid multicriteria optimization tool HYDAS (HYbrid Discrete Analysis System) has been adapted and extended (Winkelbauer and Markstrom, 1990). It allows the decision maker to specify his preferences in absolute (constraints) as well as relative (reference points) terms, and provides an easy-to-use interface which makes extensive use of graphics, allowing the decision maker to concentrate on the decision itself, rather than the manipulation of his information base.

HYDAS combines numerical, symbolic, graphical, and statistical methods to support the decision maker when exploring the solution space and enables the user to arrive at a well-informed decision. The numerical analysis components are based on the numerical multicriteria alternative selection tool DISCRETE (Majchrzak, 1984, 1985; Zhao, Winkelbauer, and Fedra, 1985).

HYDAS consists of four basic components (Figure 2). These are numerical, symbolic, graphical, and statistical. Each component is developed currently to differing degrees. A discussion of each follows.

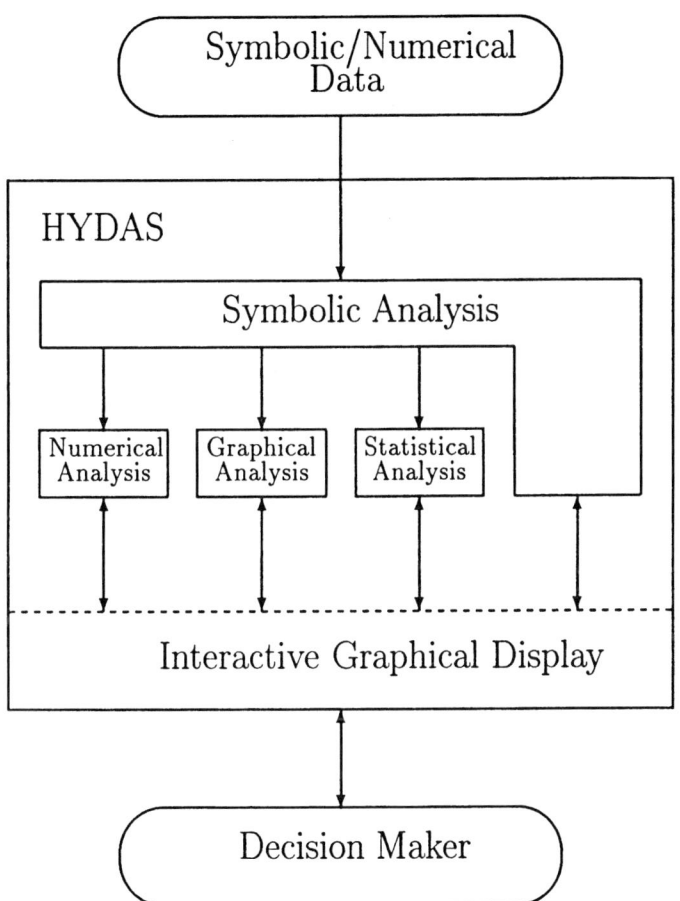

Figure 2: Components of HYDAS

Symbolic analysis

The first component of the analysis is symbolic. Symbols such as *small*, *moderate*, or *major* can all be transformed into ordinals if a fixed vocabulary exists. The information that is gained by ordinal data is that of position or rank. The position of an alternative is known only in relation to those next to it. It is fixed and defined by its neighbors.

Symbolic descriptors have the advantage that the user may have a difficult time describing preference in terms of specific numerical levels. The decision maker might not be sure about the exact numerical definition of *low* or want others to know what *low* might be. In these cases it might be preferable to describe an aspiration level or set a reference point with a statement such as *costs should be low*. In addition, the constraints of the system could be set this way as well. In these situations, the necessary vocabulary of rules could be developed through the use of rule editors.

Obvious methods for this type of data are the ranking techniques. Ranking techniques also have the advantage in that they can be used on both ordinal and cardinal types of data. Ranking methods are straightforward and easy to interpret.

A simple ranking of the alternatives by different criteria is an important first step for screening the input. By determining which alternatives perform well and which alternatives perform poorly, the user can gain insight into how the other components of the analysis will behave. In addition, at this point, significant outliers can be identified, and if desired, removed.

Numerical analysis

A standard numerical analysis with HYDAS would consist of the following three procedures: (1) feasible subset and Pareto determination, (2) efficient point determination, and (3) reference-point sensitivity. (For a formal discussion of these methods see, e.g., Grauer, Lewandowski, and Schrattenholzer; 1982, Lewandowski and Wierzbicki, 1987.)

All of the techniques used in HYDAS for solving the multicriteria problem involve determination of the Pareto set. The Pareto set is a subset of the feasible solutions such that none of its members are dominated by any of the other members of the feasible set. An alternative is considered dominated if there is at least one other alternative which is preferable in every one of the criteria. Of course feasible solutions which are dominated by other feasible solutions may exist. These are eliminated before any interactive problem solving begins. It is assumed that the decision maker would never want to use an alternative that is not contained in the Pareto set, although, in the graphical display options of the system, these alternatives are still shown. If the case arises that a dominated alternative is preferable, then the problem should be reformulated by adding or deleting the responsible decision criteria.

A discrete reference-point approach is used to solve this multicriteria problem where the decision set is not continuous (see above). In this case, the decision space is not defined by a smooth function, but by disconnected discrete points.

For the two-dimensional case, it is easy to visualize a scatter of the alternatives as points in the decision space. One method of finding the optimal alternative would be to find the point that is closest to utopia which serves as the default reference point. Utopia is defined by the optimal value for each criteria over all of the alternatives and does not necessarily correspond to any existing alternative. Most of the time, however, utopia is relatively far away from the cloud of alternative points.

Rearranging the decision space by transforming the natural value representation into one normalized between nadir (worst possible) and utopia (best possible) value for a dimension, allows us to view the distance in this metric as a measure of achievement, which makes the different natural measures commensurable. Thus, the selection of a point in this graph represents an implicit trade-off between the criteria, which is made more explicit when the distance is calculated not from the utopia point, but from some other reference point which can be interactively set by the user.

As the problem expands to more and more criteria, the distance is calculated as a simple n-dimensional Euclidean distance as the square root of the sum of the squares in each dimension. In principle, there is no limit to the number of criteria that could be analyzed. In practice, however, only a few criteria can be dealt with at a time in a meaningful way. In fact, in most circumstances, a problem can be effectively dealt with involving less than seven or eight criteria.

Also, there can be a significant correlation between the criteria. Correlated criteria should be eliminated in favor of others which can be equally effective.

This analysis is very dependent on the set of criteria that is selected. The inclusion of several highly correlated criteria, for example, can easily bias the recommendation of the model. However, the implementation of the system allows criteria to be toggled "active" or "inactive" very easily. This allows the user to examine the sensitivity of the solution to his choice of the decision criteria rather effectively.

Graphical display and analysis

With the use of high-resolution graphics, dynamic plots are available to the decision maker (Zhao, Winkelbauer, and Fedra, 1985). These offer the advantage of the familiarity of X-Y graphs. Almost every decision maker is comfortable with the Cartesian coordinate systems and the concept of a particular point having different values given by reading the appropriate axis. The dynamic graph retains this degree of familiarity yet adds another dimension of information. The user can move within the graphs and obtain more information about any point, i.e., criteria values for those criteria not represented by the axes in the current plots. The only problem is that as the problems get bigger, particularly with more dimensions, it is difficult to assimilate so many graphs.

A valuable tool to help the analyst examine the multicriteria problem is the two-dimensional scattergram (see Figure 3). Although each criterion adds another dimension to the solution space, the multidimensional problem can be broken down into a series of two-dimensional plots, which directly represent trade-offs between the two criteria involved.

Statistical analysis and data transformations

Traditional statistics are also very important to the decision-making process. Simple statistics such as mean, range, and variance, as well as correlation and some ranking statistics, can provide the user with an idea of the underlying properties of the problem. These are well known, widely understood, and well accepted. This makes them a good choice for decision support tools.

There are several reasons for providing the user with information about correlation of the criteria. These are: (1) bias of the results by placing heavy weights on criteria that are measures of the same thing, (2) reducing the complexity of the problem by allowing the user to remove redundant (correlated) criteria, and (3) providing the decision maker with information pertaining to the nature of the data.

There are many reasons for transforming data; the use of (parametric) statistical tests is based on the assumption of normality of the parent distribution.

A second-order Taylor power transformation (Hoaglin, Mosteller, and Tukey, 1983) is used in HYDAS to help with some of the non-symmetry that can be problematic in statistical analysis.

Alternatively, several non-parametric methods such as rank correlation, non-parametric linear regression, and contingency tables, can be used (Randles and Wolfe, 1979; Conover, 1980). These easy-to-use statistical methods allow for approximate data analysis which can give the user a *feeling* about the data with much less computational effort.

Current implementation

The version developed so far allows for numerical and qualitative ordinal data representation, limited statistical analysis, and the graphical display of the alternatives projected from the decision space onto two-dimensional scattergrams.

The interface provides the decision maker with a listing of the decision criteria, together with possible constraint and reference-point specifications. The decision maker can select and order the criteria he wants or needs to consider, toggling them as active or inactive and reordering their sequence in the display listing. He can also modify, for each criterion, the specifications of constraints and the reference value interactively using a graphics display shown in a pop-up window and then run the solver to obtain the *efficient* alternative which is—again graphically as well as numerically—displayed in the criteria listing.

The graphical display option provides for a scattergram of any two user defined criteria,

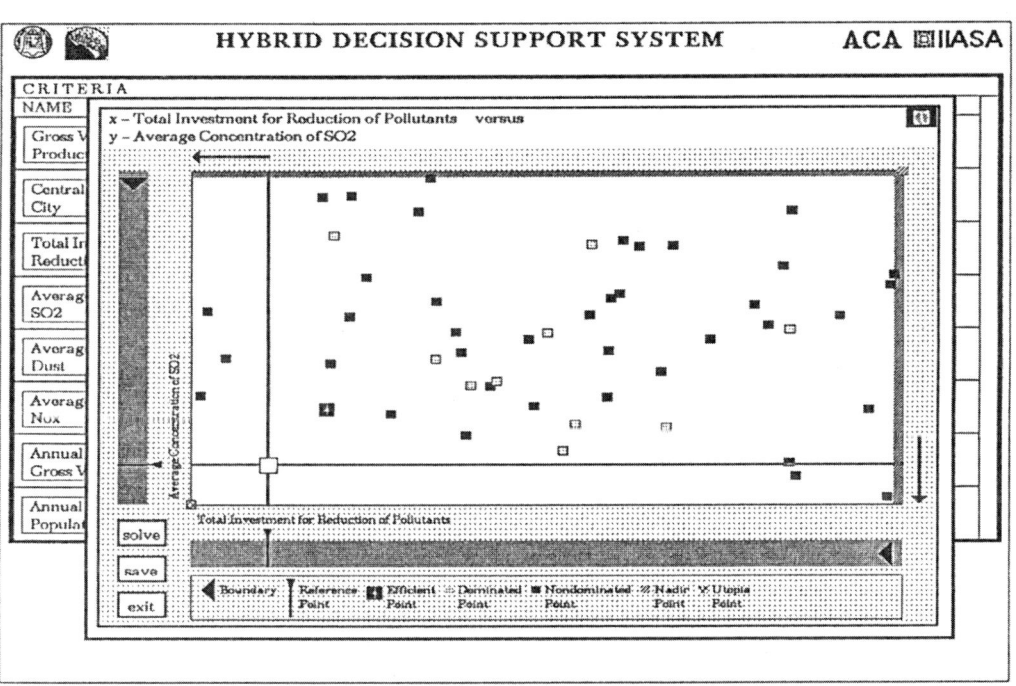

Figure 3: Graphics Interface for Modification of Constraints and Reference Points

and again allows both constraints and reference values to be set in this context (Figure 3).

4 Toward a fully integrated Hybrid decision support tool

The research presented in the previous sections is certainly only a first step toward a fully integrated hybrid approach to decision support. The underlying idea is to provide a more natural representation of the decision problem, the alternatives, and the decision-making process. However, we believe that the first results are encouraging and that the strategy and direction of research is promising.

There are several ways that the hybrid approach developed so far can be extended and improved. In the current version, numerical and symbolic descriptions are coupled through simple intervals with sharp boundaries. Fuzzy set theory (Schmucker, 1984; Zimmerman, Zadeh, and Gaines, 1984) is a very attractive and effective alternative for coupling numerical and symbolic techniques. An interesting option now under development is fuzzy reference point analysis. If fuzzy reference points are specified, the result is a fuzzy set of solution alternatives. However, several user interface issues need to be addressed before the integration of this option can take place. The most pressing of these is how to graphically display and define these membership functions.

The most attractive extension, however, is in the integration of rule-based methods as part of the reference point approach.

The basic reference point approach consists of two stages. First, we determine a feasible sub-set from the set of all alternatives available by formulating constraints or, more general, rules of exclusion. One such exclusion principle is the principle of *dominance* resulting in the Pareto set for further consideration.

The second step is to select one optimal or preferred alternative from this subset of alternatives. This can be achieved by normalizing the decision space, and measuring the distance from an ideal or reference point of each remaining alternative: the alternative nearest to the reference point is then selected as the efficient (optimal) solution.

The symbolic and logical extensions to this basically numerical approach again take two forms, corresponding to the two basic steps in the approach. The first step is the formulation of exclusion rules, which are straightforward and easily understandable in terms of classical production rules.

The second step is more complex, and, in fact, more general since it can also be construed to include the first. Both the filtering procedure (including the dominance principle) and the distance calculations in the reference point approach are sensitive to the set of dimensions considered. Adding or deleting criteria (dimensions) will obviously affect the result of both procedures.

Rather than simply adding or deleting directly measurable criteria, we can *reconstruct the decision space by introducing derived dimensions.* The values for these criteria represent some logical combinations of other dimension values, and can be used in addition to or instead of any or all of the directly measurable criteria.

For both procedures, exclusion rules or logical filters as well as the reconstruction of decision criteria, we can use a simple rule syntax that can operate on numerical as well as symbolic criteria values.

References

Balestra, G., and Tsoukias, A. (1990). Multicriteria Analysis Represented by Artificial Intelligence Techniques. *Journal of the Operational Research Society*, Vol. 41, No. 5. pp. 419-430.

Barr, A., and Feigenbaum, E.A. (1981). *The Handbook of Artificial Intelligence*. Volume I. Pitman. London. pp. 409.

Barr, A., and Feigenbaum, E.A. (1982). *The Handbook of Artificial Intelligence*. Volume II. Pitman. London. pp. 428.

Charniak, E., Riesback, C.K., and McDermott, D.V. (1980). *Artificial Intelligence Programming*. Lawrence Erlbaum Associates Publishers. Hillsdale, New Jersey. pp. 323.

Cohen, P.R., and Feigenbaum, E.A. (1982). The Handbook of Artificial Intelligence. Volume III. Pitman. London. pp. 639.

Conover, W.J. (1980). *Practical Nonparametric Statistics.* Second Edition. John Wiley & Sons. New York. pp. 493.

Fedra, K., Li, Z., Wang, Z., and Zhao, C. (1987). Expert Systems for Integrated Development: A Case Study of Shanxi Province, The People's Republic of China. SR-87-1. International Institute for Applied Systems Analysis, A-2361 Laxenburg, Austria. pp. 76.

Grauer, M., Lewandowski, A. and Schrattenholzer, L. (1982). Use of the Reference Level Approach for the Generation of Efficient Energy Supply Strategies. WP-82-19. International Institute for Applied Systems Analysis, A-2361 Laxenburg, Austria.

Hoaglin, D.C., Mosteller, F., and Tukey, J.W. (1983). *Understanding Robust and Exploratory Data Analysis.* John Wiley & Sons. New York. pp. 447.

Levine, P., Pomerol, M.-J., and Saneh, R. (1990). Rules Integrate Data in a Multicriteria Decision Support System. *IEEE Transactions on Systems, Man and Cybernetics*, Volume 20, Number 3, pp. 678-686.

Lewandowski, A., and Wierzbicki, A. (1987). Theory, Software and Testing Examples for Decision Support Systems. WP-87-26. International Institute for Applied Systems Analysis, A-2361 Laxenburg, Austria. pp. 247.

Li, Han-Lin (1987). Solving Discrete Multicriteria Decision Problems Based on Logic-based Decision Support Systems. *Journal of Decision Support Systems*, Volume 3, No. 2, pp. 101-119.

Majchrzak, J. (1984). Package DISCRET for Multicriteria Optimization and Decision Making Problems with Discrete Alternatives. IIASA Conference on Plural Rationality and Interactive Decision Processes. 16-26 August. Sopron, Hungary.

Majchrzak, J. (1985). DISCRET–A Package for Multicriteria Optimization and Decision Problems with Discrete Alternatives. In M. Grauer, M. Thompson, and A.P. Wierzbicki (eds.), *Plural Rationality and Interactive Decision Processes*. Lecture Notes in Economics and Mathematical Systems. Springer-Verlag. Berlin. pp. 319-324.

Randles, R.H., and Wolfe, D.A. (1979). *Introduction to the Theory of Nonparametric Statistics*. John Wiley & Sons. New York.

Savory, S. (ed.) (1985). *Künstliche Intelligenz und Expertensysteme*. (Artificial Intelligence and Expert Systems). Oldenbourg, Vienna. pp. 248. In German.

Schmucker, K.J. (1984). *Fuzzy Sets, Natural Language Computations, and Risk Analysis*. Computer Science Press. Rockville, Maryland. pp. 192.

Wang, Jingan, and Mao, Bo (1990). Final Report of the National Environmental Quality Decision Support System. Project No. (75) 60-03-06-03.

Wierzbicki, A.P. (1979). A Methodological Guide to Multiobjective Optimization. WP-79-122. International Institute for Applied Systems Analysis, A-2361 Laxenburg, Austria.

Wierzbicki, A.P. (1980). Multiobjective Trajectory Optimization and Model Semiregularization. WP-80-181. International Institute for Applied Systems Analysis, A-2361, Laxenburg, Austria.

Winkelbauer, L., and Markstrom, S. (1990). Integration of AI with Quantitative Methods for Decision Support. *Expert Systems with Applications: An International Journal*, Volume 1, No. 4. pp. 345-358.

Zhao, C., Winkelbauer, L., and Fedra, K. (1985). Advanced Decision-oriented Software for the Management of Hazardous Substances. Part VI: The Interactive Decision-Support Module. CP-85-50. International Institute for Applied Systems Analysis, A-2361 Laxenburg, Austria. 39p.

Zimmerman, H.J., Zadeh, L.A., and Gaines., B.R. (eds.) (1984). *Fuzzy Sets and Decision Analysis*. North Holland, Amsterdam.

Application of min/max Graphs in Decision Making

Edward Nawarecki
Institute for Computer Science
Academy of Mining and Metallurgy
Kraków, Poland

Grzegorz Dobrowolski
Joint System Research Department
of the Institute for Control and Systems Engineering,
Academy of Mining and Metallurgy, Cracow,
and the Industrial Chemistry Research Institute, Warsaw.
Kraków, Poland

Abstract

The paper deals with *decision situations* for which the *optimization model* can be formulated. The optimization model is computerized in the form of *decision support system* that assures easy simulation experiments and supports all necessary auxiliary actions.

For the problem with uncertain data and a wide out-of-model sphere the optimality looses its sense when a solution is tried to be realized. It may be even a limiting factor when attractive optimal states of the model draw the decision maker's attention to unrealistic states of reality. A natural tactics of the decision maker is to consider states of the model that are pre-defined (variants) on the basis of some states of reality and ought to be close (in some sense) to the state which is optimal with regard to the criterion applied.

In the paper the concept of *min/max graphs* is proposed to devise management of the variants. Basic definitions and theorems important for the application are reproduced. A central position is occupied by theorems concerning approximation of the graphs. A graph represents a set of solutions (variants) of the problem obtained in the procedural way. Such representation has the following valuable features:

- creates a common platform for a number of variants, especially for comparison;

- opens way for graphical representation of the set of variants;

- can influence the decision maker to carry out the analysis in more precise and systematic way;

- utilizing the estimation features of the min/max graph, allows for avoidance of detailed investigation of some variants.

The min/max graphs can be used in the DSS in the model management and dialog management subsystem. No sooner than the variant generation procedure is established the only link between the problem and the min/max graph is a set of values of the objective function. Then the concept can be encapsulated in the separate module of the DSS that is subsequently described.

The strategy of the variants generation is a crucial point of the concept. It decides about the semantics of the min/max graph and finally about the usefulness of the concept for the particular decision problem.

Some exercises are presented for the problem of programming development of the chemical industry. Results have been obtained with a prototype module that is concisely presented. A short discussion about usefulness of the approach closes the paper.

1 Introduction

The paper deals with decision situations that may be characterized as follows:

- Input data are uncertain and it is hard or even impossible to evaluate deviation introduced by them.

- The optimization model is used as a basic mathematical tool.

- A bunch of phenomena staying beyond the optimization model is numerous.

- The influence of unmodeled elements is taken into consideration by simulation scenarios produced by a human.

- The optimization model is computerized in the form of a Decision Support System **DSS** that assures easy simulation experiments and supports all necessary auxiliary actions.

The decision situations can be classified according to (Lewandowski and Wierzbicki, 1988) as the centralized single-actor (decision maker) situations. They involve: a user and a DSS. The user personifies a whole staff with the decision maker itself, it means: experts, analysts. The decision maker has the authority and experience to reach the decision; the experts and the analysts are responsible for the analysis of the decision situation.

A role of the optimization model in the DSS, characterized above, is different than the one in technical problems where the optimal solution is almost directly applicable. The solution enriched with results of the postoptimal analysis or other validation technics ends the analysis as soon as consensus of adequacy of the model is gained.

For the problem with uncertain data and wide out-of-model sphere optimality looses its sense when the solution is tried to be realized. Application of the optimization algorithm meets another goal than searching for the best alternative. Generally, it plays a role of a data processing algorithm featured by drastic reduction of input versus output data.

Optimality in DSS of the type may be even a limiting factor when attractive optimal states of the model draw the decision maker's attention to unrealistic states of reality. A natural tactics of the decision maker is to consider states of the model that are pre-defined (variants) on the basis of some states of reality and ought to be close (in some sense) to the state which is optimal with regard to the criterion applied.

Let us assume that the way in which the variants arise has no direct links to the supported part of the decision process (uncomputerized human creative factor). A schema of dealing with optimality in the DSS discussed can be proposed:

1. Find the optimal solution.

2. List the distinctive variables of the model (definition of the variants by the model variables).

3. Evaluate a consecutive variant — the optimal state of model under assumed behavior of the distinctive variables. The state is usually far from the optimal solution.

4. Verify and validate the variant also by comparison with the optimal solution.

It is assumed that the shape of the model is stable with respect to its structure and parameters what creates a platform for the final choice from among few variants.

The schema is a *what-if* tactics flexibly controlled by the optimal solution.

In the paper that is an extended and improved version of (Dobrowolski and Nawarecki, 1990) the concept of min/max graphs (Kluska-Nawarecka and Nawarecki, 1982) (Kluska-Nawarecka and Nawarecki, 1987)is proposed to devise management of the variants that can arise in the process described above.

Basic definitions and theorems important for the application are reproduced that create a base for the step-by-step representation of incomplete knowledge and a procedure for completing the information vital for the decision making process. A central position is occupied by theorems about estimation of the graphs that allow to avoid investigation of most variants. A simple procedure is sketched out.

2   A model of decision process

Let us build the model on a basis of optimization problem with a real-valued objective function:

$$F(x) = F(x_1, \ldots, x_N) \tag{1}$$

defined on an admissible domain of realizations of an argument vector:

$$x \in X = \prod_{i=1}^{N} X_i$$

where: X_i a set of realizations of a component $x_i, i = 1, \ldots, N$.

The admissible sets X_i can be sets of real numbers (values of physical parameters) or freely defined elements representing a class of phenomena, or even strategies (chains of decisions and incidents). For the sake of simplicity, it is assumed that X_i is a set of numbers representing realizations $x_i \in X_i$. The decision process searches for such $x^* \in X$, that the minimum (or maximum) of the objective function is gained:

$$F(x^*) = \min F(x_1, \ldots, x_N) \tag{2}$$

To model the informational uncertainty, it is assumed that only some components of $x = [x_1, \ldots, x_N]$ are controlled by a decision maker (their realizations can be freely chosen). The remaining components stay independent from decisions to be taken. Their realization can be observed by the decision system and only their domains X_i are predictable.

Moreover, it is assumed that composition of the set X in the sense of chosen X_i as well as a dichotomy into these under control, and those subordinated to the partner's activity or independent part of the reality can vary in the consecutive stages of the decision process. The above assumptions reflect features of a wide class of real decision problems, such as economic decision making, technical and medical diagnosis.

In the paper some representation of knowledge in the decision problem as above is proposed together with an inference engine that allows effective decision procedures. The dichotomy can be formulated as follows:

$$U \times V = X \tag{3}$$

where: U a set of parameters under the decision maker's control;
V a set of parameters unknown and independent from the decision system.

After reindexing the sets U, V is done:

$$[u_1, \ldots, u_n] \in U \quad ; \quad [v_1, \ldots, v_m] \in V$$

and $n + m = N$ holds. The objective function changes its denotation: $F(x) \equiv F(u, v)$.

For given U and V the objective function $F(u, v)$ gains its saddle point (Kluska-Nawarecka and Nawarecki, 1982) what can be written as:

$$\min_{u \in U} \max_{v \in V} F(u, v) = \max_{v \in V} \min_{u \in U} F(u, v) = F(u^*, v^*) \tag{4}$$

where: u^*, v^* min/max strategies of, respectively, the decision maker and an opponent.

In practice, the assumption of a saddle point existence does not create a limiting factor because there is a margin of flexibility in X construction and, in consequence, in uv-dichotomy. A triple $\langle F, U, V \rangle$ is called a decision situation.

The loss (decreasing of the objective function F) that comes from the opponent's activity or independence of realization of parameters belonging to the set V can be expressed as a difference between solutions of the problems (2) and (4).

$$\Lambda(U, V) = \min_{u \in U} \max_{v \in V} F(u, v) - \min_{x \in X} F(x) = F(u^*, v^*) - F(x^*) \tag{5}$$

Obviously, the biggest loss is observed when the set of controlled parameters is empty $U = \emptyset$.

For a given objective function F the value of $\Lambda(U, V)$ depends only on the uv-dichotomy and can characterize the decision situation $\langle F, U, V \rangle$.

The loss, defined as above, in real-life decision problems for the defined decision situation $\langle F, U, V \rangle$ is lesser than theoretical $\Lambda(U, V)$ given by (5). In average $v \neq v^*$ holds and

$$F(x^*) < F(u^*, v) < F(u^*, v^*) = F(x^*) + \Lambda(U, V) \tag{6}$$

Having two decision situations $\langle F, U, V \rangle$ and $\langle F, U', V' \rangle$ so that:

$$U' = U \times X_k \quad ; \quad V = V' \times X_k \tag{7}$$

(it means that the parameter X_k is excluded from V and joins U), the following inequality holds:

$$\Lambda(U', V') \leq \Lambda(U, V) \tag{8}$$

and the estimation can be done:

$$\lambda(X_k, U, V) = \min_{u \in U} \max_{v \in V} F(u, v) - \min_{u \in U'} \max_{v \in V'} F(u, v) \tag{9}$$

Defined that way $\lambda(X_k, U, V)$ is interpreted as information value of the set X_k. It means that if the parameter represented by X_k was measured and controlled the value of the objective function in that case could be approximated using $\lambda(X_k, U, V)$. It is worth while to underline that $\lambda(X_k, U, V)$ depends not only on X_k but the decision situation (the uv-dichotomy) as well.

3 A graph of min/max solutions

3.1 Definitions

A graph uniquely described by the objective function $F(x)$ and the admissible set of the parameter realizations $x \in X$ is called a graph of min/max solutions or shortly a min/max graph and denoted by $\Gamma(F, X)$. For each solution to the problem (4) corresponding to a given decision situation (the uv-dichotomy of X) a node $A(U, V)$ of the min/max graph exists.

A number of parameters grouped in the sets U versus V introduces partition of the min/max graph $\Gamma(F, X)$ into levels, as follows:

- A starting node A^0 of the min/max graph $\Gamma(F, X)$ corresponds to the solution of problem (2), i.e., to the dichotomy $U = X; V = \emptyset$. The node A^0 is the only one on the level 0.

- A final node A^N corresponds to the case of maximization with respect to all parameters; uv-dichotomy is $U = \emptyset; V = X$. It is the single node on the level N.

- The remaining nodes of the min/max graph $\Gamma(F, X)$ correspond to the solutions of (4) for all uv-dichotomy. A node A^m belonging to the level m represents a solution of (4) when $U = \prod_{i=1}^{m} X_i$ and $V = \prod_{i=m+1}^{N} X_i$.

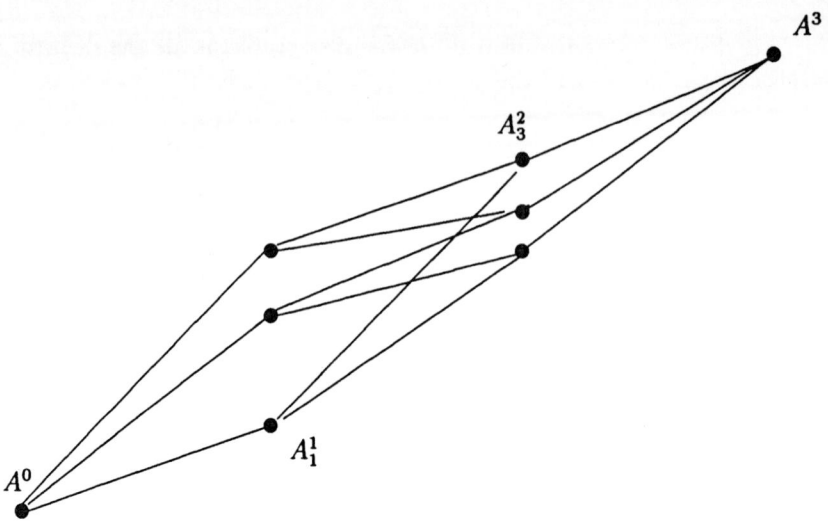

Figure 1: A min/max graph for $N = 3$

A min/max graph has $N + 1$ levels. Levels 0 and N have a single node. A number of nodes on remaining levels is equal to a number of possible uv-dichotomies; on the *1st* level set V contains a single parameter each time and a number of nodes is equal to N, on the *2nd* level V contains a pair of parameters, etc. Generally, on the level m there are $\Omega_m = \begin{bmatrix} N \\ m \end{bmatrix}$ nodes. Total number of nodes of the min/max graph is given by the formula:

$$\Omega = \sum_{m=0}^{N} \frac{N!}{m!(N-m)!} \tag{10}$$

To point at the particular node (except A^0 and A^N) the level index is insufficient. It is necessary to enumerate all the parameters in the U or V set and, in fact, m or $N - m$ indexes are needed. Below, whenever it does not cause misunderstanding, the simplified indexing will be used: a superscript will point at the level of the node and a subscript will differentiate the nodes of the same level. In the case that differentiation does not matter the subscript will be neglected. A min/max graph $\Gamma(F, X)$ is shown on figure 1 for the case of $N = 3$.

Two nodes A^m, A^n of the min/max graph $\Gamma(F, X)$ are joined by an arc $a\,(A^m, A^n)$ if and only if they belong to the neighboring levels (i.e., $n = m + 1$ or $n = m - 1$) and, simultaneously, the uv-dichotomies corresponding to the nodes differ only in a parameter X_i that is excluded from the set U and bound to V when one steps down the levels. It may be written as an implication:

$$a\,(A^m, A^n) \in \Gamma(F, X) \longrightarrow U^m = U^n \times X_i \vee V^n = V^m \times X_i \tag{11}$$

Such two nodes are also called the neighboring and $\sim$ denotes the neighborhood relation. Each node has N neighbors.

From among paths of the min/max graph a special subclass is important for further considerations. They are called s-paths. The s-path consists of arcs joining a sequence of neighboring nodes that belong to consecutive levels of the min/max graph. Two nodes A^m, A^n are joined by the s-path if and only if there exists the sequence of arcs of the shape:

$$A^m \sim\sim A^n \longleftrightarrow \{a(A^i, A^{i+1}), i = m, m+1, \ldots, n-2, n-1\} \tag{12}$$

where: $\sim\sim$ a symbol denoting the s-path relation.

A set of nodes belonging to the same level $\{A_i^m, A_j^m, \ldots, A_r^m\}$ is ordered if the relation is stated:

$$F(A_i^m) \leq F(A_j^m) \leq \ldots \leq F(A_r^m) \tag{13}$$

If the ordered set $\{A^m\}$ contains all nodes of the level the first and the last elements with respect to the relation are called the lower A_0^m and the upper A_M^m node of the graph $\Gamma(F, X)$ on the m-th level.
The value of:

$$\sigma^m = F(A_M^m) - F(A_0^m) \tag{14}$$

will be called a diameter of the min/max graph.

The set grouping all upper nodes of the graph for the consecutive levels $\{A_M^m\}$, $m = 0, \ldots, N$ described an upper envelope of the graph $\Gamma(F, X)$. In the analogous way a lower envelope is defined $\{A_0^m\}$, $m = 0, \ldots, N$.

It is comfortable, especially when one constructs a graphical form of the min/max graph $\Gamma(F, X)$, to introduce a normalization according to the following assumptions:

- the value F^* obtained as $\min_{x \in X} F(x)$ is taken as the reference level;

- values $\underline{F}(x)$ for nodes are calculated according to:

$$\underline{F}(U_i^m, V_j^m) = \underline{F}(A_i^m) = \frac{\max_{v \in V_j^m} \min_{u \in U_i^m} F(u, v) - \min_{x \in X} F(x)}{\max_{x \in X} F(x) - \min_{x \in X} F(x)} \tag{15}$$

- following (15) $\underline{F}$ is equal to 0 for the starting node and to 1 for the final one.

Because the normalization means change of values associated with the nodes (a scale of the graph) and does not affect features of the min/max graph the normalization symbol will be neglected.

Sometimes the notion of a subgraph of the min/max graph is useful. A subgraph $\Gamma_0^N(F, X)$ is uniquely defined by pointed at its starting $0 = A_i^n$ and final $N = A_j^m$ nodes. It consists of all nodes $\{A_k^s\}$ such that $n < k < m$, and $A_i^n \sim\sim A_k^s$ and all arcs joining them.

3.2   Relations

For two neighboring nodes $A^m \sim A^{m+1}$ of the min/max graph $\Gamma(F, X)$ the following inequality holds:

$$F(A^m) \leq F(A^{m+1}) \tag{16}$$

The feature directly follows definition of the min/max graph as a graph of solutions to the problem (2). Applying the inequality to consecutive pairs of the neighboring nodes the analogous inequality can be obtained for two nodes joining with the s-path $A^m \sim\sim A^n$:

$$F(A^m) \leq F(A^n) \quad ; \quad m < n \tag{17}$$

The inequality (17) is an extension of (16) using definition of the s-path and the transitive relation $\sim$.

For a triple of nodes belonging to different levels no matter the joining path exists or not the following, weaker estimation is always true:

$$F(A_0^m) \leq F(A^s) \leq F(A_M^n) \quad ; \quad m < s < n \tag{18}$$

where: A_0^m the lower node of level m,
A_N^n the upper node of level n.

The estimation is obtained by combining (17) with definition of the lower and upper nodes of the level.

The estimation (18) can be improved in the way:

$$F(A_i^m) \leq F(A^s) \leq F(A_j^n) \quad ; \quad m < s < n \tag{19}$$

where: A_i^m the node of level m with $F(x)$ lesser than for a node of the same level joined with A^s with the s-path,
A_j^n the node of level n with $F(x)$ greater than for a node of the same level joined with A^s with the s-path.

3.3 An approximated min/max graph

Let us introduce a notion of an approximated min/max graph for which some of values $F(A^n)$ are unknown.

The approximated min/max graph is built of two types of nodes:

Approximated for which interval approximation is provided instead of the value of F. The interval will be described as $[F_I, F_S](A^n)$ (infimum and supremum value) and, for the normalized graph, is a subset of $[0, 1]$.

Fixed for which the characteristics is as in basic definition of the min/max graph. Whenever it will be comfortable a fixed node can be regarded as an approximated one for which $F_I = F_S = F$.

A value of F for a fixed node is obtained externally as a result of the expertise or calculation using a model while values F_I, F_S for an approximated node result from values of the fixed nodes based on dependencies that are obligatory in the min/max graph. Inequalities (16), (17), (18), (19) provide the basis for approximation.

The approximated node $[F_I, F_S](A^n)$ can be calculated according to (17) using the following formulas for infimum:

$$F_I(A^n) = \max_{m,i} F(A_i^m) \quad ; \quad m < n \quad ; \quad A_i^m \sim\sim A^n \tag{20}$$

and supremum values:

$$F_S(A^n) = \min_{m,i} F(A_i^m) \quad ; \quad n < m \quad ; \quad A^n \sim\sim A_i^m \qquad (21)$$

while i indexes all the fixed nodes of the approximated min/max graph.

The lower and upper envelope can be determined for the approximated min/max graph straight from the definition, using modified (14):

$$\sigma^m = F_S(A_M^m) - F_I(A_0^m) \qquad (22)$$

after calculation of all the approximated nodes or applying (19) for the levels for which all the fixed nodes are known.

4 A min/max graph in the DSS

An impression can arise based on theoretical preliminaries that all the specified variables of the model are taken into account, are uv-dichotomized, and the min/max graph acquires enormous number of nodes (the decision maker is bound to deal with plenty of variants) and finally the question arises: Is the concept really efficient? Fortunately, in the technical problems the dimensionality is not so large and moreover many variables stay beyond the influence of the decision process. The evaluation methods and algorithms are often at hand that allow for excluding some of the variables from the analysis by their approximation with a given accuracy. For the decision problems with uncertain data and a wide out-of-model sphere the limiting factor in the subject is ability of the decision maker to deal with a number of variants in the comparison regimè.

In practice, there is no need to generate and analyze the *full* min/max graph. Instead a subgraph (built with respect to the chosen subset of the variables) is pondered and, of course, theoretical results hold for that case. The trick well corresponds to the idea of the optimization problem with distinctive variables (Gass and Dror, 1983) or with the assumption of existence of the decision maker's preference in the variable space (Dobrowolski, 1990).

As it will be shown, generating all nodes even of the partial graph is not necessary. The decision analysis can be successfully completed with the approximated min/max graph, some nodes of which are estimated using the rules presented in the previous section.

4.1 Manipulation of the graph

Let us assume that the approximated min/max graph sufficiently represents the decision making process. Its fixed nodes are variants already modeled while approximated nodes reflect variants that can be simulated on request.

The process unfolds from the beginning state, when the graph contains only the approximated nodes until the moment when the decision is reached because the graph is sufficiently approximated (or fully determined) on the base of obtained fixed nodes. That is a step-by-step process as the consecutive receiving a new fixed node improves approximation of the graph enriching the knowledge base of the problem. Three stages can be distinguished here:

1. Performing normalization of the graph by fixing starting and final nodes.

2. Building the first approximation when all nodes of 1-th or $N-1$-th levels are fixed. It depends on the application whether both levels or one of them are investigated. One must be aware that omiting some of those nodes results in nodes of trivial approximation i.e. $F_I = 0$ or $F_S = 1$.

3. Approaching the decision when fixing of nodes depends on the decision maker's intellect. Some tactics can be suggested:

   - improvement of approximation of interesting nodes (semantics of the decision process),

   - improvement of approximation of nodes of significant uncertainty (relatively great $[F_I, F_S]$),

   - narrowing the decision field by precising the envelopes,

   - clustering the nodes in order to point at the distinctive parameters and switch the analysis to subgraphs of the relatively great diameter.

During the decision process, the decision maker disposes two sources of information that influence each other all the time. These are: original information about the problem and the approximated min/max graph at its current state. Some inference rules independent from the semantics of the problem can be deducted from the graph features:

- pointing at a subset of nodes (variants) with the assumed (acceptable) value of F,

- pointing at a subset of neighbors (variants) of the given node that produces increasing (decreasing) F,

- evaluation of the information value of the given parameter X_i,

- for the given node (variant) finding out a uv-dichotomy that maximizes (minimizes) F.

4.2   Semantics of the decision process

The strategy of the variants generation during the third stage of the decision process is a crucial point of the concept. It corresponds to the semantic of the min/max graph and finally decides about its usefulness for the particular decision problem. From the general point of view, the following strategies are possible:

Sensitivity analysis The analysis can be carried out both with respect to variables and parameters of the model.

The first subcase directly arises from the formulation of the min/max graph concept. A variable is under control when its value is fixed otherwise it is free. For the linear optimization model a slight different schema can be interesting: a variable is fixed on its infimum versus supremum in an admissible set.

In the second subcase it can be assumed that influence of deviation of parameters in an interval is investigated. An appropriate schema is: a parameter is fixed on the lower bound of the interval next to the upper one.

The greatest value of (9) calculated for all the uv-dichotomies can serve here as a sensitivity measure. Graphical representation of the measure is the greatest jump for a variable or a parameter between neighboring levels of the min/max graph.

Improving the model The procedure is oriented towards minimization of uncertainty introduced by elements of the model (admissible intervals for the variables, assumed deviations of the parameters). Under assumption that enhancement means additional cost, the results from the sensitivity analysis can indicate what elements of the model ought to be refined.

In its simplest form (controllable variable is fixed) the procedure corresponds to looking over the min/max graph to find a subgraph for which the sensitivity measure (appropriate value of (9)) is the smallest or acceptable.

Unlimited what–if analysis Because there are no limitations imposed on the variants generation strategies completely free schema can be used as long as the decision maker can manage it.

4.3 Remarks on architecture

The concept of the min/max graph is introduced as a consequence of the optimization problem. The graph represents a set of solutions (variants) of the problem obtained in the procedural way. Such representation has the following valuable features:

- creates a common platform for a number of variants, especially for comparison;

- opens way for graphical representation of the set of variants;

- can influence the decision maker to carry out the analysis in more precise and systematic way;

- utilizing the estimation features of the min/max graph, allows for avoidance of detailed investigation of some variants.

The above suggests that the concept can be used in the DSS in the model management and dialog management subsystem (Sprague, 1980).

No sooner than the variant generation procedure is established the only link between the problem and the min/max graph is a set of values of the objective function. Then the concept can be encapsulated in the separate module of the DSS. It is called the UV module. Because no assumptions were made with respect to the shape of the objective function, the UV module may be built independently from the DSS means.

In the interactive mode communication between the UV module and the remaining part of the DSS is as follows: the module asks for the solution of the optimization problem corresponding to the node approximated so far, as an answer it obtains the value of the

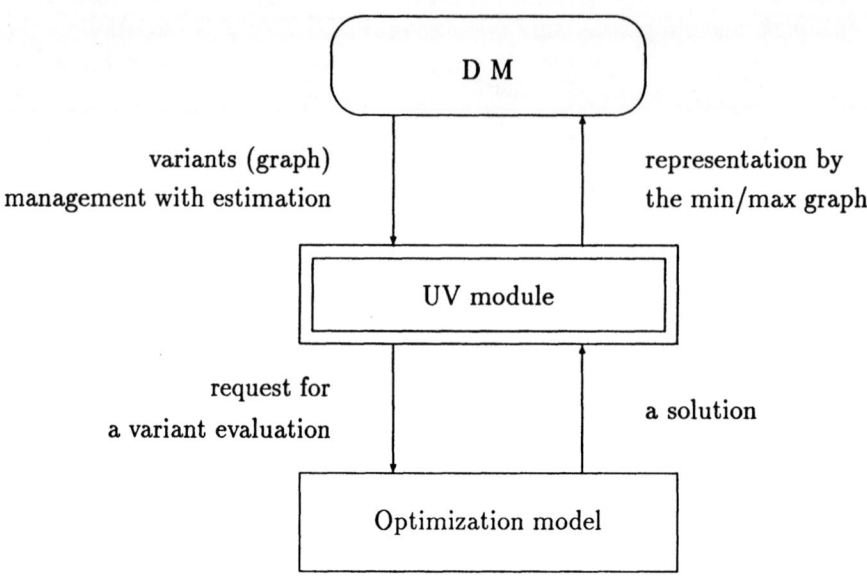

Figure 2: How the UV module can be incorporated into DSS

objective function and the new approximation of the min/max graph may be calculated. A function of the UV module is sketched out on figure 2. Each time the approximation is analyzed by the decision maker and the process may be interrupted when the decision field (the current state of the min/max graph) becomes clear to him.

5 Exercises

The idea presented in the paper and, built on its base, the UV module was tested in cooperation with the MIDA — Multiobjective Interactive Decision Aid (Dobrowolski and Ryś, 1989) that is a computer system supporting activities in the programming development of the chemical industry. A core of the system constitutes an optimization model (Dobrowolski and Żebrowski, 1989) that is used for generation of variants. The model reflects basic functions of the industry in its processing as well as distribution layer and is a means for studies upon the future structure of a chosen area of the industry with dense interrelations.

Particularly, production of polyvinyl plastics from coke, lime, and chlorine is modeled. Ten processing units interchanging twenty seven intermediates (carbide, acetylene, monomers,...) and utilities (steam, electricity, water,...) can produce acetic acid, polyvinyl acetate, polyvinyl chloride emulsion, and polyvinyl chloride. Some by-products (magnesium fertilizer, carbide oven gas,...) are also obtained. The area is rather small but it groups all important features of the chemical industry and well represents the class of decision problems.

In the case the analysis is controlled by a single criterion that is the yearly profit while investment and volume of production is monitored.

Two exercises will be reported here:

1. Sensitivity analysis with respect to prices of final products.

2. Variation of production program under investment constraints.

The point of departure for the first case is that market analysis with respect to prices often provides their values together with intervals of their most probable fluctuations. Assuming that prices of products are under consideration, the following interpretation of the min/max graph is useful. The central values of prices are the base of all activities while their lower values characterized possible negative impact of the environment. The node A^N is the basic result. Nodes of $N - 1$ can be interpreted as hypotheses that each time only a single price worsens. The rest of nodes represent mixed hypotheses and finally A^0 is the worst variant. The UV module can help to create a full diagram of uncertainty imposed by chosen prices. In the third stage of analysis with the min/max graph (see previous section) each hypothesis can be analyzed as the approximated node without necessity of its calculation with the model.

The second case is an example of what-if analysis carried out using utilities of the UV module. Let us assume that a limited investment is at disposal. Then searching for the best production program is an important problem. In the case a choice among four products is possible. The starting node reflects situation that nothing is produced. Nodes of the 1-th level represent production programs with a single product. The final node represents the situation when full bunch of products is allowed. The min/max graph concept makes it easy to choose product mix that is enough profitable, slightly worser than the optimum, but interesting from another (out of model) point of view. In the case the bi-objective analysis can also be suitable because the investment limitation has some elasticity in practice.

6 Summary

The concept of the min/max graph and its utilization in the DSS is practically verified now. Some results in the field of the programming development of the chemical industry were reported in the paper. The UV module can be incorporated into the DSS and can successfully help the decision maker to deal with a number of variants generated prior the final choice is made. The function of clarifying managing of the variants strengthened by visualization of the process is the greatest advantage.

A very important fact is that the UV module can be used also as a stand-alone computer tool and can cooperate with a DSS in presence of a mediate decision maker and import/export data options.

For the above application the concept badly needs extensions for the multiobjective case that is more natural for decision problems in the field.

In (Nawarecki et al., 1991) even straighter application of the concept can be found. The analysis of a decentralized multiprocessor structure was successfully supported in the absence of analytical description of the objective function. Uncertainty came to light in impresiced effect of some arguments on the decision process and incomplete information about optimal solutions.

References

Dobrowolski, G. (1990) Interactive Multiobjective Optimization Method in the Space of Model Variables. In P. Korhonen, A. Lewandowski and J. Wallenius (eds.): Multiple Criteria Decision Support, *Lect. Notes in Econ. and Math. Syst.* **356**, pp. 10–16, Springer-Verlag, New York–Heidelberg–Berlin, ISBN 3-540-53895-X.

Dobrowolski, G. and Nawarecki, E. (1990) Application of min/max estimates for decision making under informational uncertainty. In A. Ruszczyński, T. Rogowski and A. Wierzbicki (eds.): Contribution to Methodology and Techniques of Decision Analysis (First Stage), *IIASA Collaborative Paper* **CP-90-008**, pp. 121–129, Laxenburg, Austria.

Dobrowolski, G. and Żebrowski, M. (1989) Basic Model of an Industrial Structure. In A. P. Wierzbicki and A. Lewandowski (eds.): Aspiration Based Decision Support Systems, *Lect. Notes in Econ. and Math. Syst.* **331**, pp. 287–293, Springer-Verlag, New York–Heidelberg–Berlin, ISBN 3-540-51213-6.

Dobrowolski, G. and Ryś, T. (1989) Architecture and Functionality of MIDA. In A. P. Wierzbicki and A. Lewandowski (eds.): Aspiration Based Decision Support Systems, *Lect. Notes in Econ. and Math. Syst.* **331**, pp. 339–370, Springer-Verlag, New York–Heidelberg–Berlin, ISBN 3-540-51213-6.

Gass, S. I. and Dror, M. (1983) An interactive approach to multiple objective linear programming involving key decision variables. *Large Scale Systems* **5**, pp. 95–103.

Kluska-Nawarecka, S. and Nawarecki, E. (1982) Problème d'evaluation et de choix d'information dans les systèmes de commande. In Symposium Components and Instruments for Distributed Control Systems, Paris.

Kluska-Nawarecka, S. and Nawarecki, E. (1987) Decision Making in a System with Insufficient Input Data. In ACIAM'87, Paris.

Lewandowski, A. and Wierzbicki, A. (1988) Aspiration Based Decision Analysis and Support. Theoretical and Methodological Backgrounds. *IIASA Working Paper* **WP-88-03**, International Institute for Applied System Analysis, Laxenburg, Austria.

Nawarecki, E., Cetnarowicz, K., Marcjan, R. and Zygmunt, M. (1991) A min-max graph application to the base of knowledge organization of the expert system. In Proceedings of the IASTED International Symposium, Artificial Inteligence and Neural Networks, 1–3.07.1991, Zurich.

Sprague, R. H. (1980) A framework for the development of decision support systems. *MIS Quarterly* **4**(4), pp. 1–26.

Graphical Interaction in Multi-Criteria Decision Support
Some Implementation Issues

Rudolf Vetschera, Heinz Walterscheid
Faculty of Economics and Statistics
University of Konstanz, FRG

1 Introduction

Graphical User Interfaces (GUI's) are rapidly becoming a standard feature in modern software design. Modern graphical interfaces provide immediate access to all functions available in the software system involved in an intuitive, easy to use form. In this paper, we report on the current development of a graphical user interface, which can be used as a front end to multi-criteria decision methods. We first give (in section 2) an introduction to the theoretical concepts underlying that interface. Section 3 discusses several technical aspects of implementation and section 4 describes the current status of a software prototype. Section 5 provides an outlook to possible future developments.

2 A Projection-Based Representation for MCDM Problems

Several graphical techniques have been used to represent data values in MCDM problems, including standard displays like bar charts (e.g. in DIDAS: Kreglewski et al., 1988 or VIG: Korhonen, 1987), holistic images like star diagrams (Kasanen et al., 1989), Chernoff faces (Chernoff, 1973) or "Harmonious Houses" (Korhonen, 1991). Another stream of research is based on projection techniques taken from statistical principal component analysis (e.g. Lehert, de Wasch, 1983; Mareschal, Brans, 1988; Lewandowski, Granat, 1991). These techniques make it possible to represent data on a large number of alternatives graphically on screen in an easily interpretable format.

Formally, projection techniques can be summarized as follows (a more comprehensive description is given in Vetschera, 1991): We consider a decision problem concerning N alternatives, which are evaluated according to K criteria. The performance of these alternatives is represented by a suitably scaled $N \times K$ data matrix $\mathbf{X}$. Matrix $\mathbf{X}$ is transformed into a 2-dimensional graphical representation by postmultiplying it with a $K \times 2$ projection matrix. We denote the projection matrix obtained from principle component analysis by $\mathbf{P}$. Criteria can also be represented in this display as projections of unit vectors, i.e.

by **IP**. Data points of alternatives therefore are shown as linear combinations of vectors representing criteria, where the weights used in the linear combination correspond to the data values.

While this projection preserves as much information as possible on criteria values, it does not incorporate any information on the overall evaluation of alternatives. To overcome this problem, we define a second projection matrix **Q** in such a way that projection with matrix **Q** will cause alternatives to be lined up along one axis in increasing order of preference. Matrix **Q** can either be derived directly for linear utility functions or as an approximation to more complex, nonlinear forms of evaluation. A flexible representation, preserving the advantages of both types of projection, is achieved via a linear combination of the form

$$\mathbf{T} = \lambda \mathbf{P} + (1 - \lambda)\mathbf{Q} \tag{1}$$

The user can dynamically vary parameter λ of (1) to obtain different graphical representations of the problem.

For a finite data range $\underline{x_k}$ to $\overline{x_k}$ in each criterion k, it is possible to determine an indifference region in the projection space, in which possible equivalent alternatives to a given alternative might be located. The boundaries of that region are obtained by solving the following parametric linear program:

$$\text{maximize/minimize } z_1 + \theta z_2$$
$$\text{s.t.}$$

$$
\begin{aligned}
z_1 &- \sum_{k=1}^{K} t_{k,1} y_k &= 0 \\
z_2 &- \sum_{k=1}^{K} t_{k,2} y_k &= 0 \\
&\sum_{k=1}^{K} w_k y_k &= \sum_{k=1}^{K} w_k x_{n,k} \\
&\underline{x_k} \le y_k \le \overline{x_k}
\end{aligned}
\tag{2}
$$

In model (2), z_1 and z_2 represent the screen coordinates of possible equivalent points having criteria values y_k. The constants $t_{k,1}$ and $t_{k,2}$ are the respective elements of projection matrix **T**. The weights w_k are used to represent a simple, linear utility function for the evaluation of alternatives and at the same time form one column of matrix **Q**, which leads to a graphical representation according to preference values. For more complex preference structures, the w_k's must be calculated as linear approximations to the actual evaluation function. θ is varied from $-\infty$ to $+\infty$ by parametric analysis to obtain one half of the boundary of the indifference region. The entire boundary is obtained by performing both minimization and maximization of the parametric objective function.

It can also be shown that, if the horizontal axis is used to represent preference, points "to the right" of the indifference region will always be strictly better and points "to the left" strictly worse than the alternative for which the indifference region is calculated. An example of such an indifference region is shown in section 4.

3   Implementation Aspects

To provide an intuitive and user-friendly interface, fast response of the system to any action of the user is highly desirable. For the graphical representation method developed above, this means that the calculations involved in the projection operation itself, as well as the determination of indifference regions, must be performed in real time whenever the user changes the control parameters of the system.

Fortunately, changes in the user's view, achieved through modification of parameter λ, do not necessitate a recalculation of the projection matrices $\mathbf{P}$ and $\mathbf{Q}$, but only of their linear combination and the projected data values. These calculations can be performed in real time, so the display can be constantly updated while the user dynamically changes the value of λ.

The determination of indifference regions, however, requires the solution of linear programming problems and thus poses a more difficult problem for implementation. Computational experience, gained during the experimental implementation of this approach, has shown that using standard parametric programming algorithms, instead of restarting the solution process from scratch for every corner point of the indifference region, reduces computing time by about a factor of five for a test problem of 20 alternatives and 6 criteria. It is possible to generate an indifference region in about one second for this size of problem on a 25 MHz 386-based PC. While this is too slow to provide real-time updating during interactive changes of parameter λ, it is possible to redraw indifference regions reasonably fast after the user has fixed the new parameter value.

Furthermore, it might be possible to redraw indifference regions more quickly within the sensitivity range for parameter λ. To further investigate this possibility, we note that model (2) may be rewritten using the definition of projection matrix $\mathbf{T}$ as:

maximize / minimize
$$\sum_{k=1}^{K} [\lambda p_{k,1} + (1 - \lambda)q_{k,1} + \theta [\lambda p_{k,2} + (1 - \lambda)q_{k,2}]] \, y_k$$
s.t. $\qquad\qquad\qquad\qquad\qquad\qquad\qquad\qquad\qquad\qquad\quad$ (3)
$$\sum_{k=1}^{K} w_k y_k = \sum_{k=1}^{K} w_k x_{n,k}$$
$$\underline{x_k} \le y_k \le \overline{x_k}$$

The objective coefficients of this model depend simultaneously on both θ and λ. Therefore it is possible to perform a parametric analysis and obtain sensitivity ranges for λ for any given value of θ (and vice versa). A possible range for λ, within which the indifference region can be obtained by simple rescaling (i.e. in which there are no basis changes in all the linear programming problems involved in its determination), can then be obtained by taking the minimal range for all the values of θ that were used in the parametric analysis.

Standard parametric programming on θ will generate only those parameter values for which a basis change takes place. It is therefore likely that sensitivity analysis for λ using those values of θ will lead to very small intervals of possible change. Unfortunately, it cannot be stated in general whether choosing a value of θ, which lies within the possible interval for that parameter instead of at the boundary, will actually increase the interval for λ. The possible sensitivity ranges for λ also do not necessarily depend linearly on θ, since for different values of θ, different non-basic variables of the optimization problem might become the limiting column, which would enter the basis in case of a change of λ. Since it is not possible to maximize possible ranges for λ by this technique and computational experiments have shown that ranges for λ, obtained at boundary points for θ, are very narrow, sensitivity ranges for λ are not used in the prototype implementation presented in the next section.

4 The MCView Experimental Software System

In this section, we briefly describe an experimental software system implementing the approach outlined above, which will eventually be expanded into a general graphical front-end to various multi-criteria decision methods for discrete problems. The MCView system runs on personal computers under the MS-DOS operating system. It is written in TopSpeed Modula-2 version 3.0, using the graphics library contained in the TopSpeed system. It consists of about 3,500 lines of Modula-2 code (including a general linear programming module), leading to an executable file size of about 130 K bytes.

All interaction with MCView is initiated from the main screen shown in figure 1. This screen contains four major elements. The central element is the projection-based graphical problem representation, showing both the alternatives and the criteria. The main command menu and a status line are at the top of the screen. The right side of the screen contains a control bar for manipulating the projection parameter λ.

One major problem in screen design for projection-based graphical representations is the identification of alternatives and criteria, which are represented just by points on screen. If large, descriptive labels are used within the graphical representation, the graphics will soon become overloaded and major advantages of the graphical format would be lost. On the other hand, short labels (or just numbers as e.g. in GAIA, Mareschal, Brans, 1988) require additional cognitive effort on the part of the user to recognize alternatives. In MCView, the identification of data points is controlled by the user by means of a mouse interface: When the user moves the mouse-pointer close to the graphical point representing an alternative or criterion, basic information on that object is displayed in the status line at the top of the screen. By clicking on the data point, the user can obtain further information, which is displayed in a window within the problem representation (figure 2). The same window is also used to edit problem data (like attribute values of alternatives) and to specify whether the indifference region should be displayed for an alternative (figure 3). Projection parameter λ can be changed by clicking onto the slide bar on the right side of the screen and changing the size of the bar with the mouse.

During this operation, the projection of alternatives and criteria is constantly updated to reflect the current value of λ, allowing the user to choose a display that best suits his needs. Indifference regions, however, are not shown in this phase since the computing time for their determination would prevent rapid graphical interaction.

The main menu at the top of the screen, which in some instances leads to further pull-down menus, allows the user to perform tasks of file manipulation like loading and saving data as well as to create and edit new alternatives and criteria, for which no graphical

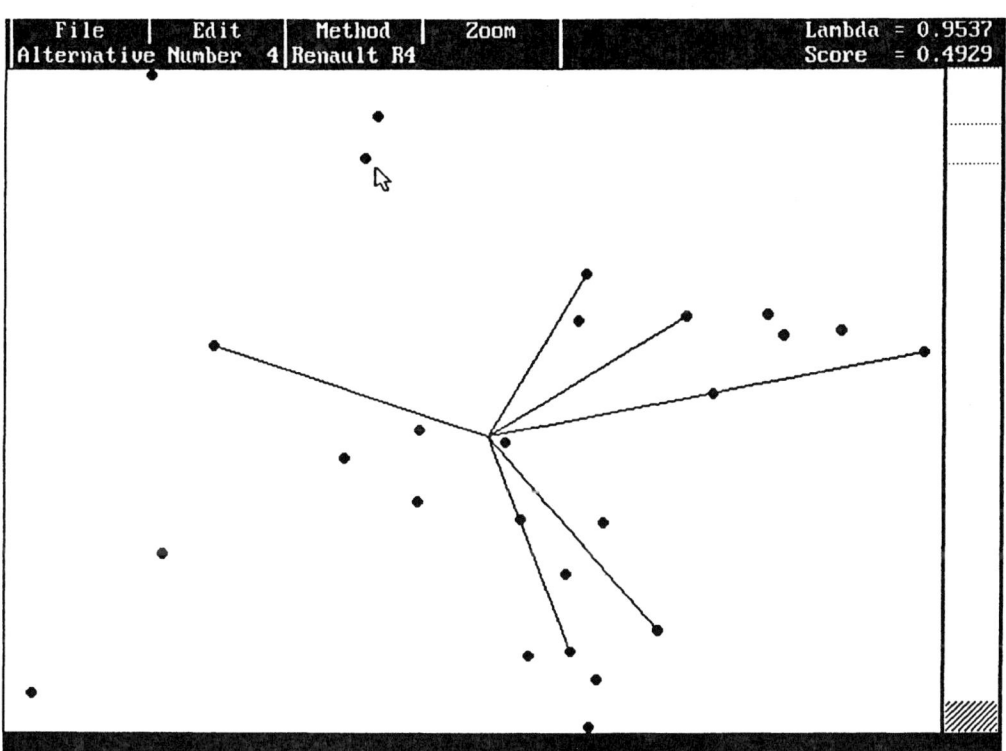

Figure 1: Main screen of MCView

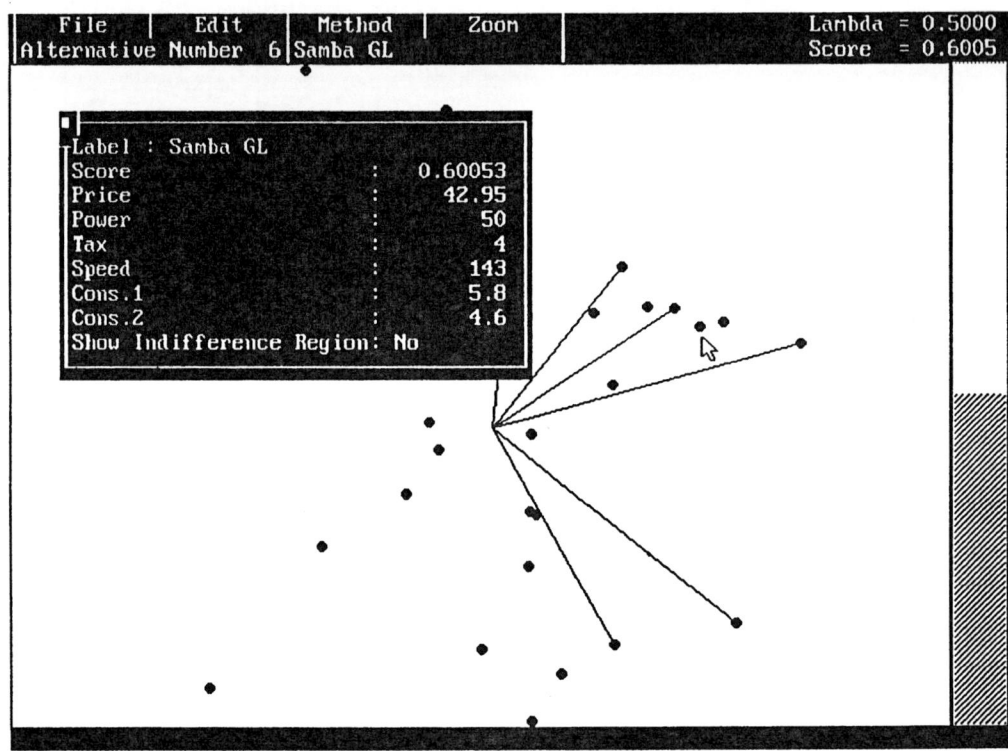

Figure 2: Data Edit Window

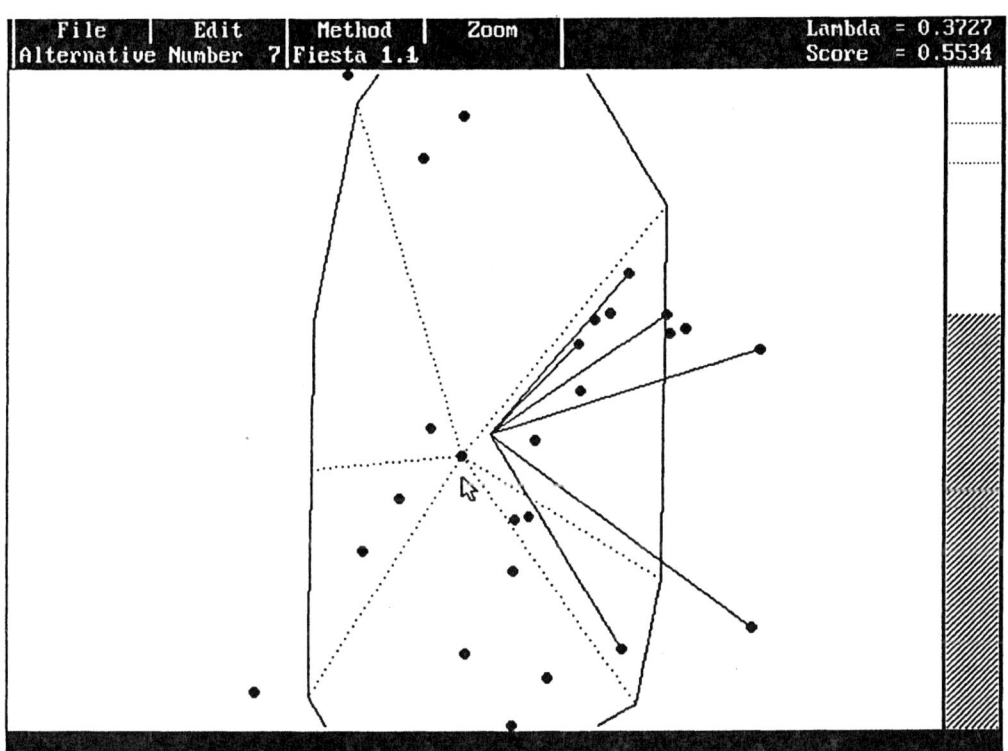

Figure 3: Indifference Regions

representation exists yet. In future versions of the system, it will also be used to select different evaluation methods to generate the ranking of alternatives.

5 Conclusions and Further Development

The MCView system, which was described in this paper both from a theoretical and implementation point of view, provides a comprehensive and intuitive graphical user interface for MCDM problems. The current version, which is still at an early stage of development, will serve as a starting point for further development in theoretical and empirical research. Further theoretical development is needed concerning the approximation of nonlinear preference evaluations for use with projection techniques. Alternative strategies for this task cannot be judged solely on a theoretical basis, but have to be tested empirically in controlled experiments, in which future versions of MCView will also play an important role.

References

Chernoff, H. (1973). The Use of Faces to Represent Points in k-Dimensional Space Graphically. *Journal of the American Statistical Association* 68, 361–368.

Kasanen, E., Stermark, R., and Zeleny, M. (1989). Gestalt Systems of Holistic Graphics: New Management Support View of MCDM. In: A.G. Lockett and G. Islei (eds.): *Improving Decision Making in Organisations*, Lecture Notes in Economics and Mathematical Systems, vol. 335, Springer, Berlin, 143–156.

Korhonen, P. (1987). VIG - A Visual Interactive Support System for Multiple Criteria Decision Making. *Belgian Journal of Operations Research, Statistics and Computer Science* 27, 3–15.

Korhonen, P. (1991). Using Harmonious Houses for Visual Pairwise Comparison of Multiple Criteria Alternatives. *Decision Support Systems* 7, 47–54.

Kreglewski, T., Paczynski, J., Granat, J. and Wierzbicki, A.P. (1988). IAC-DIDAS-N: A Dynamic Interactive Decision Analysis and Support System for Multicriteria Analysis of Nonlinear Models with Nonlinear Model Generator Supporting Model Analysis. IIASA Working Paper WP-88-112, Laxenburg, Austria.

Lehert, Ph. and de Wasch, A. (1983). Representation of Best Buys for a Heterogeneous Population. In: P. Hansen, (ed.): *Essays and Surveys on Multiple Criteria Decision Making*, Springer, Berlin, 221–228.

Lewandowski, A. and Granat, J. (1991). Dynamic BIPLOT as an Interaction Interface for Aspiration-Based Decision Support Systems. In: P. Korhonen, A. Lewandowski and J. Wallenius (eds.): *Multiple Criteria Decision Support*, Lecture Notes in Economics and Mathematical Systems, vol. 356, Springer, Berlin, 229–241.

Mareschal, B. and Brans, J.-P. (1988). Geometrical representations for MCDA. *European Journal of Operational Research* 34, 69–77.

Vetschera, R. (1991). A Preference-Preserving Projection Technique for MCDM. Working Paper I-256, Faculty of Economics and Statistics, University of Konstanz.

Parametric Programming Approaches to Local Approximation of the Efficient Frontier

Janusz Granat

Institute of Automatic Control
Warsaw University of Technology, Poland

Abstract

In this paper the set of the efficient solutions in criteria space is characterized locally by directional derivatives, which can be treated as a local linear approximation of the efficient frontier. The properties of the marginal function in nonlinear programming are applied. The results are used for building a prototype of a graphic interface for a decision support system.

1 Introduction

There is a broad class of the decision support systems, where the decision making problem is formulated as the multiobjective optimization problem. It is well-known, that usually there is a set of efficient solutions of such problem. This set can be scanned by using interactive procedures. For this purpose, the aspiration-led methodology is widely used (e.g. Wierzbicki, 1986; Steuer and Choo, 1983; Lewandowski and Wierzbicki, 1990). A point in the criteria space consisting of aspiration levels, specified by the user, reflects some reasonable values of the objectives. The system proposes to the user one or several alternative decisions, that are best attuned to this guiding information. In order to obtain the proposed decision, the system minimizes a scalarizing function, where the specified aspiration levels are treated as parameters. If the decision calculated by the system is not satisfactory to the user, he can modify his specification of the aspiration levels and the process is repeated until the user accepts the decision proposed by the system. After calculating an efficient decision, the user usually would like to know what are reasonable directions for modifying aspiration levels and how the solution will be changed after such modification. There are several approaches which can help in answering such questions. In a group of them, the system uses additional information which can be obtained after calculation the efficient solutions and which locally characterizes set of the efficient solutions. A survey of such methods will be presented in the next section. In the approach presented in this paper the set of efficient solutions is characterized locally by differentiability properties of the efficient solution point with respect to the perturbation of aspiration levels. The theoretical background will be presented in Section 3; Section 4 describes numerical aspects of calculating directional derivatives. The results from these sections will be applied to build "Past-Present-Future" graphic interface.

2 A Survey of local characterization approaches for the efficient frontier

2.1 Satisficing trade-off method (Nakayama and Sawaragi)

The following scalarization function is considered (Nakayama and Sawaragi, 1984):

$$\min_{x \in X} \max_{1 \le i \le p} w_i(f_i^* - f_i(x)) \tag{2.1}$$

$$w_i = \frac{1}{f_i^* - \bar{f}_i}$$

where f^* - ideal point, $\bar{f}$ aspiration point

During the trade-off analysis, the user classifies the criteria into three groups: criteria which he wants to improve, relax or accept as they are. The index set of each class is represented respectively by I_I^k, I_R^k, I_A^k. Let $f^k := f(x_k)$ be a solution of the problem (2.1) in k iteration. The user specifies a new aspiration level $\bar{f}_j^{k+1} = f^k + \Delta f_j^k$ for $j \in I_I^k$. The changes of the values of objectives to be relaxed are equal:

$$\Delta f_j = \frac{-1}{N \lambda_j w_j} \sum_{i \in I_I} \lambda_i w_i \Delta f_i \tag{2.2}$$

λ_i - is the Lagrange multipliers associated with the constraint in the problem (which is equivalent to the problem (2.1)) :

$$\min_{x,z} z \tag{2.3}$$

$$w_i(f_i^* - f_i(x)) \le z$$

$$x \in X$$

Recently Nakayama (Sawaragi et al., 1985) has presented a method for calculating exact trade-offs for linear problems by using parametric optimization techniques.

2.2 Trade-off information depending on Lagrange multipliers of the hyperplane problem

The hyperplane problem (Sakawa and Yano, 1990) is defined as follows:

$$HP(Q, t_{-k}) \qquad \min z \tag{2.4}$$

$$s.t. \quad Q\, F(f(x)) \le D(z, t_{-k})$$

Q is a $p \times p$, $F(\cdot)$ and $D(\cdot)$ are k dimensional vector functions (further conditions for Q, F, D are specified in Sakawa and Yano, 1990).

Minimization of all known scalarizing functions can be formulated as a problem of the form (2.4). The trade-off rates between the objective functions at the optimal solution to the $HP(Q, t_{-k}^*)$ on the Pareto surface in the objective space can be represented by:

$$\frac{\partial f_k(x)}{\partial f_i(x)} = -\frac{\sum_{i=1}^k \lambda_j^* q_{ji} \partial F_i(f_i(x^*))/\partial f_i}{\sum_{j=1}^k \lambda_j^* q_{jk} \partial F_k(f_k(x^*))/\partial f_k} \tag{2.5}$$

2.3 The efficient spanning directions

M. Halme (Halme, 1990) has presented a procedure for finding the minimum set of vectors that characterizes all efficient feasible directions emanating from the current efficient solution for linear multiobjective problems. Such vectors can be found without additional optimization; they can be calculated from the optimal simplex tableau corresponding to the current efficient solutions.

The results of Halme are associated with the methods developed by Korhonen (e.g Korhonen and Laakso, 1986).

2.4 Trade-off analysis by using methods of statistical data analysis

If we have calculated some efficient solution points it is possible to analyze approximate trade-off by methods of statistical data analysis. Although these are only approximative methods, in many cases we can obtain valuable information.

An interactive descriptive graphical approach has been developed by W. Y. Ng, (1991). He considers the matrix $\phi_{q \times p}$ with columns corresponding to efficient points and rows corresponding to objectives. He classifies the relationships between any two objectives into three kinds: the objectives which are in correspondence, in conflict and unrelated. The objectives are grouped into groups with members, that corresponds to each other, and the identification of conflicting and unrelated groups is performed. The correlation measure between objectives is a Spearkman's rank correlation coefficient. A weighted undirected graph is presented graphically to the user. Each node represent a group. Two nodes are connected by an arc, whenever the groups they represent are in conflict. Each arc is associated with the value of the median linkage as a measure of conflict.

Lewandowski and Granat, (1989) have presented a technique for data analysis based on the BIPLOT. A $n \times p$ matrix Y has been considered, with columns corresponding to objectives and rows corresponding to solutions. matrix Y can be approximated by matrix $Y_{[2]}$ of rank 2. Such matrix can be presented graphically by vectors emanating from one point of the graph and points. Vectors represents the objectives and the points. The angle between vectors represents the correlation between objectives and the distance between points represents the Machlanobis distance between solutions.

3 Theoretical background

3.1 Problem formulation

The multiobjective programming problem to be considered in this paper is defined as:

$$\min_{\underline{x} \in X} \underline{f}(\underline{x}) \tag{3.6}$$

where

$$\underline{f}(\underline{x}) = (f_1(\underline{x}), f_2(\underline{x}), \ldots, f_p(\underline{x}))^T$$

$$X = \{x \in R^n : \underline{g}(\underline{x}) = (g_1(\underline{x}), g_2(\underline{x}), \ldots, g_m(\underline{x}))^T \leq \underline{0}\}$$

$\underline{f}(\underline{x})$ - is the vector of objectives
X - is the set of admissible decisions
$\underline{x}$ - is the vector of decisions variables

$f_i(i = 1, \ldots, p), g_j(j = 1, \ldots, m)$ are continuously differentable.
    The achievement function is defined as:

$$s(q, \bar{q}) = \max_{1 \leq i \leq p} \frac{\bar{q}_i - q_i}{s_i} + \frac{\epsilon}{p} \sum_{i=1}^{p} \frac{\bar{q}_i - q_i}{s_i} \tag{3.7}$$

where

    $\bar{q}_i$ - aspiration level,
    $q_i = f_i(\underline{x})$,
    ϵ - small real number,
    s_i - scaling coefficient. For simplicity let $s_i = 1$.

An efficient solution can be obtained by solving the problem:

$$\min_{\underline{x} \in X} s(\underline{f}(\underline{x}), \bar{q}) \tag{3.8}$$

Solution of the problem (3.8) in the objective space is the function of changes of the aspiration point:

$$\hat{q}_i(\bar{q}) = f_i(\hat{x}(\bar{q})) \tag{3.9}$$

where

$$\hat{x}(\bar{q}) \in Arg \min_{\underline{x} \in X} s(\underline{f}(\underline{x}), \bar{q}), \quad i = 1, \ldots, p.$$

    We will try to find *the directional derivative (if it exist) of $\hat{q}_i(\bar{q})$ with respect to the perturbation of the aspiration level:*

$$D\hat{q}_i(\hat{\bar{q}}, d) = \lim_{t \downarrow 0} \frac{\hat{q}_i(\hat{\bar{q}} + td) - \hat{q}_i(\hat{\bar{q}})}{t} \tag{3.10}$$

d - vector of direction of changes of aspiration point.

To solve this problem we will apply parametric programming approach. Sensitivity analysis based on parametric programming is well developed. The standard theorem of Fiacco and McCormick, (1968) determining the perturbation of the optimal point, but under rather strong assumptions. Jittorntrum, (1984) has investigated the optimal solution of the perturbed problem without the strict complementarity assumption and suggested a method for computing directional derivative. These results can be useful for smooth scalarizing functions. Under weaker assumptions differential properties of the marginal function can be also investigated. (see e.g. Gauvin and Dubeau, 1982; Gauvin and Tolle, 1977; Rockafellar, 1984). We will use these results to solve the problem defined by (3.10), because the function $\hat{q}_i(\bar{q})$ can be nondifferentiable.

3.2   Properties of the marginal function

Sometimes the problem (3.8) is reformulated such that aspiration levels exist in the constraints, so let us consider the following parametric programming problem:

$$\min \xi(q, \bar{q}) \tag{3.11}$$

$$q = \underline{f}(\underline{x})$$

$$S(\bar{q}) = \{\underline{x} \in R^n : (\eta_1(\underline{x}, \bar{q}), \ldots, \eta_k(\underline{x}, \bar{q}))^T \leq \underline{0},$$

$$(\nu_{k+1}(\underline{x}, \bar{q}), \ldots, \nu_l(\underline{x}, \bar{q}))^T = \underline{0}\}$$

The marginal function is the optimal value of the problem (3.11) for each value of $\bar{q}$:

$$v(\bar{q}) = \inf_{x \in S(\bar{q})} \xi(q, \bar{q})$$

The function $v(\bar{q})$ have one sided directional derivative in the ordinary sense if the limits:

$$Dv(\bar{q}, d) = \lim_{t \downarrow 0} \frac{v(\bar{q} + td) - v(\bar{q})}{t}$$

exist, end in the Hadamard sense if the limit is defined as:

$$Dv(\bar{q}, d) = \lim_{d' \to d, \, t \downarrow 0} \frac{v(\bar{q} + td') - v(\bar{q})}{t}$$

The lower end upper Dini directional derivatives of $v(\bar{q})$ are respectively:

$$Dv(\bar{q}, d) = \liminf_{t \downarrow 0} \frac{v(\bar{q} + td) - v(\bar{q})}{t}$$

$$Dv(\bar{q}, d) = \limsup_{t \downarrow 0} \frac{v(\bar{q} + td) - v(\bar{q})}{t}$$

Mangasarian-Fromovitz (M-F) constraint qualification:
A feasible point $\hat{\underline{x}} \in S(\hat{q})$ is said to be (M-F) regular if:

1. There exist a direction $d \in R^n$ such that

$$\nabla_x \eta_i(\hat{\underline{x}}, \bar{q})d < 0, \quad i \in \{i : \eta_i(\hat{\underline{x}}, \bar{q}) = 0\}$$

$$\nabla_x \nu_i(\hat{\underline{x}}, \bar{q})d = 0$$

2. The Jacobian matrix $[\nabla_x \nu(\hat{x}, \bar{q})]$ has full row rank r (r-number of equality constraints).

Linear indepence (L-I) regularity condition:
the gradients $\{\nabla_x \eta_i(\hat{x}, \bar{q}), \ i \in \{i : \eta_i(\hat{x}, \bar{q}) = 0\}, \ \nabla_x \nu_j(\hat{x}, \bar{q}), \ j = 1, \ldots, k\}$
are linearly independent.

In the case, when Mangasarian-Fromowitz constraint qualification is satisfied Gauvin and Dubeau, (1982) have provides a formula for lower and upper Dini directional derivative and for directional derivative in ordinary sense when linear indepence constraint qualification is satisfied. Rockafellar, (1984) has obtained conditions for the existence of the directional derivative in the Hadamard sense.

The Lagrangian function corresponding to (3.11):

$$L(\underline{x}, \bar{q}, \lambda, \mu) = \xi(q, \bar{q}) + \lambda^T \eta(\underline{x}, \bar{q}) + \mu^T \nu(\underline{x}, \bar{q})$$

$K(\hat{x}, \bar{q})$ is the set of the Kuhn-Tucker vectors associated with an optimal point $\hat{x} \in S(\bar{q})$.

Theorem 3.1 (Gauvin and Dubeau, 1982) *If the regularity condition (M-F) is satisfied at some optimal point $\hat{x} \in S(\bar{q})$, then for and direction $d \in R^p$,*

$$D_+ v(\bar{q}, d) \geq \min_{(\lambda, \mu) \in K(\hat{x}, \bar{q})} \{\nabla_{\bar{q}} L(\hat{x}, \bar{q}, \lambda, \mu) d\}$$

Theorem 3.2 (Gauvin and Dubeau, 1982) *Suppose $S(\bar{q})$ is nonempty, $S(\bar{q})$ is uniformly compact near $\bar{q}$ and the (M-F) regularity condition hold at every $\hat{x} \in S(\bar{q})$ then for any direction $d \in R^p$, there is a $\hat{x}$ such that*

$$D^+ v(\bar{q}, d) \leq \max_{(\lambda, \mu) \in K(\hat{x}, \bar{q})} \{\nabla_{\bar{q}} L(\hat{x}, \bar{q}, \lambda, \mu) d\}$$

Corollary 3.1 (Gauvin and Dubeau, 1982) [1] *If $S(\bar{q})$ is nonempty, $S(\bar{q})$ is uniformly compact near $\bar{q}$ and the (M-F) regularity condition hold at every $\hat{x} \in S(\bar{q})$ then for any direction $d \in R^p$, we have*

$$\sup_{\hat{x} \in S(\bar{q})} \min_{(\lambda, \mu) \in K(\hat{x}, \bar{q})} \{\nabla_{\bar{q}} L(\hat{x}, \bar{q}, \lambda, \mu) d\} \leq$$

$$\leq D_+ v(\bar{q}, d) \leq D^+ v(\bar{q}, d) \leq$$

$$\leq \max_{\hat{x} \in S(\bar{q})} \max_{(\lambda, \mu) \in K(\hat{x}, \bar{q})} \{\nabla_{\bar{q}} L(\hat{x}, \bar{q}, \lambda, \mu) d\}$$

and if linear-indepence regularity condition holds (instead of (M-F)) then

$$Dv(\bar{q}, d) = \max_{\hat{x} \in S(\bar{q})} \{\nabla_{\bar{q}} L(\hat{x}, \bar{q}, \lambda, \mu) d\} \tag{3.12}$$

From computational point of view the following result can be useful (under assumption that the objective function and constraint functions are twice continuously differentable, see Rockafellar, 1984)

$$\max_{\hat{x} \in S(\bar{q})} \min_{(\lambda, \mu) \in M(\hat{x}, \bar{q})} \{\nabla_{\bar{q}} L(\hat{x}, \bar{q}, \lambda, \mu) d\} \tag{3.13}$$

---

[1] Gauvin and Dubeau have assumed maximization of the objective function

$M(\hat{x}, \bar{q})$ is the set of multipliers satisfying the second order necessary condition

$$w^T \nabla_x^2 L(\hat{x}, \bar{q}, \lambda, \mu)w \leq \underline{0}, \quad w \in R^n$$

$$\nabla_x \eta_i(\hat{x}, \bar{q})w = 0$$

for all $i \in 1, \ldots, k$ with

$$\eta_i(\hat{x}, \bar{q}) = 0$$
$$\nabla_x \nu_i(\hat{x}, \bar{q})w = 0 i = k+1, \ldots, l$$

3.3 Differentiability properties of the efficient solution point with respect to the perturbation of the aspiration levels

Let I be a set of indices such that:

$$I = \{i : \hat{\bar{q}}_i - \hat{q}_i + \frac{\epsilon}{p}\sum_{i=1}^{p}(\hat{\bar{q}}_i - \hat{q}_i) = s(\hat{q}, \hat{\bar{q}})\}$$

where $\hat{q}$ is a vector of efficient solution in the objective space.
Let take $i^* \in I$ and define the problem:

$$\min\{\bar{q}_{i^*} - q_{i^*} + \frac{\epsilon}{p}\sum_{i=1}^{p}(\hat{\bar{q}}_i - \hat{q}_i) = s(\hat{q}, \hat{\bar{q}})\} \tag{3.14}$$
$$\bar{q}_i - q_{i^*} + \frac{\epsilon}{p}\sum_{i=1}^{p}(\hat{\bar{q}}_i - \hat{q}_i) = s(\hat{q}, \hat{\bar{q}})$$
$$x \in X$$

Based on the optimality conditions for the problem (3.7) and (3.14) the following lemma can be proved:

Lemma 3.1 *If $s(q, \bar{q})$ has the form (3.7) then the solution of the problem (3.8) is equivalent to the solution of the problem (3.14)*

Lemma 3.2 *If directional derivative in ordinal sense of the marginal function of the problem (3.14) exist, then*

$$Dv_{i^*}(\bar{q}, d) = Dv'_{i^*}(\bar{q}, d) - Dq_{i^*}(\bar{q}, d) - \frac{\epsilon}{p}\sum_{j=1}^{p}D\hat{q}_{i^*}(\bar{q}, d)$$

where

$$v_{i^*} = \bar{q}_{i^*} + \frac{\epsilon}{p}\sum_{j=1}^{p}\bar{q}_j$$

By Lemma 1 and Lemma 2, the following theorem can be formulated:

Theorem 3.3 *If directional derivatives in ordinal sense of the marginal function of the problem (3.14) for all $i \in I$ exist, then*

$$\underline{D\hat{q}}'(\bar{q}, d) = M^{-1}(\underline{Dv}'(\bar{q}, d) - \underline{Dv}(\bar{q}, d))$$

where:

$$M = \begin{bmatrix} 1+\frac{\epsilon}{p} & \frac{\epsilon}{p} & \cdots & \frac{\epsilon}{p} \\ \frac{\epsilon}{p} & 1+\frac{\epsilon}{p} & \cdots & \frac{\epsilon}{p} \\ \vdots & \vdots & \ddots & \vdots \\ \frac{\epsilon}{p} & \frac{\epsilon}{p} & \cdots & 1+\frac{\epsilon}{p} \end{bmatrix} \qquad \underline{D\hat{q}}'(\bar{q},d) = \begin{bmatrix} D\hat{q}_{k1}(\bar{q},d) \\ D\hat{q}_{k2}(\bar{q},d) \\ \vdots \\ D\hat{q}_{kj}(\bar{q},d) \end{bmatrix}$$

$$\underline{Dv}'(\bar{q},d) = \begin{bmatrix} Dv'_{k1}(\bar{q},d) \\ Dv'_{k2}(\bar{q},d) \\ \vdots \\ Dv'_{kj}(\bar{q},d) \end{bmatrix} \qquad \underline{Dv}(\bar{q},d) = \begin{bmatrix} Dv_{k1}(\bar{q},d) \\ Dv_{k2}(\bar{q},d) \\ \vdots \\ Dv_{kj}(\bar{q},d) \end{bmatrix}$$

Matrix M has a dimension and $|I| \times |I|$, and $k1, \ldots, kj \in I$

4 Numerical aspects of calculating directional derivatives

The efficient solutions are calculated by modified version of DIDAS-N system. The NOA1 solver (developed by Kiwiel and Stachurski, 1988) for nondifferentiable optimization is used, which was included to DIDAS-N by T. Kreglewski.

The Lagrangian function corresponding to problem (3.14):

$$L_{i^\bullet}(\hat{x},\bar{q},\lambda,\mu) - [\bar{q}_{i^\bullet} - q_{i^\bullet} + \frac{\epsilon}{p}\sum_{j=1}^{p}(\bar{q}_j - q_j)] + \lambda^T g(\hat{x}) + \sum_{j\in I-\{i^\bullet\}} \mu_j[(\bar{q}_{i^\bullet} - q_{i^\bullet}) - (\bar{q}_j - q_j)]$$

and

$$\nabla_{\bar{q}} L_{i^\bullet} = \begin{bmatrix} \vdots \\ 0 \\ \vdots \\ \frac{\epsilon}{p} - \mu_j \\ \vdots \\ (1+\frac{\epsilon}{p}) + \sum_{j\in I-\{i^\bullet\}}\mu_j \\ \vdots \end{bmatrix} \begin{matrix} \vdots \\ i \notin I \\ \vdots \\ i \neq i^* \\ \vdots \\ i = i^* \\ \vdots \end{matrix}$$

In order to calculate directional derivative the following linear optimization problems are solved (by solver, which is used in DIDAS-L system, written by W. Ogryczak)

$$\min_{\mu,\lambda}(\max_{\mu,\lambda})\nabla_{\bar{q}}L_{i^\bullet} \cdot d \qquad (4.15)$$

$$\mu_1\nabla r_1(\hat{q},\bar{q}) + \ldots + \mu_j\nabla r_j(\hat{q},\bar{q}) + \lambda^T\nabla g(\hat{x}) = 0$$

$$\mu_1 + \ldots + \mu_j = 1$$

$$\mu \geq 0, \quad \lambda \geq 0$$

where $r_j = \bar{q}_j - q_j + \frac{\epsilon}{p}\sum_{i=1}^{p}(\bar{q}_i - q_i)$

5 "PAST-PRESENT-FUTURE" graphic interface and illustrative example

Based on the theory presented in the previous sections the following a scheme of the interaction with the user in aspiration-led decision support system can be applied:

1. Specification of the aspiration levels.

2. Quick scanning of the surroundings of the current efficient solution. The user specify the new aspiration levels and then computation of corresponding approximation of the efficient solution based on directional derivative is performed. The values of the directional derivatives are also presented to the user. This step can provide information to the user about tendency of changing solution, when the aspiration level has been changed in any direction. This process can be repeated until the reasonable direction of changing of the aspiration level is found.

3. Computation of the efficient solution by optimization. In this step the directional scan of the efficient solution can be performed in the direction, which was found in the previous step.

4. If the values of objectives are satisfactory we can finish process of interaction, if not we can go to step 1.

Table 1 presents the five results of the process of interaction. (In step 2 only one direction has been chosen) for the following simple nonlinear programming problem:

$$\max x_1$$
$$\max x_2$$
$$\max x_3$$

subject to
$$0 \leq x_1^2 + x_2^2 + x_3^2 \leq 1$$
$$0 \leq x_1 \leq 1$$
$$0 \leq x_2 \leq 1$$
$$0 \leq x_3 \leq 1$$

During the process of interaction the graphic user interface can be used. We developed a prototype of such interface, which is presented on Figure 1.

		obj1	obj2	obj3
r1	aspiration level	0.718	0.718	0.718
	aprox. solution			
	solution	0.577	0.577	0.577
	-	-	-	-
	direction			
	dir. derivative			
	-	-	-	-
r2	aspiration level	0.835	0.777	0.718
	aprox. solution	0.643	0.577	0.512
	solution	0.634	0.575	0.517
	-	-	-	
	direction	1.000	0.501	0.000
	dir. derivative	0.500	0.001	-0.500
	-	-	-	-
r3	aspiration level	0.990	0.835	0.718
	aprox. solution	0.712	0.567	0.447
	solution	0.709	0.554	0.437
	-	-	-	
	direction	1.000	0.375	0.000
	dir. derivative	0.474	-0.049	-0.425
	-	-	-	-
r4	aspiration level	0.990	0.835	0.777
	aprox. solution	0.685	0.529	0.486
	solution	0.693	0.538	0.480
	-	-	-	
	direction	0.000	0.000	1.000
	dir. derivative	-0.417	-0.417	0.834
	-	-	-	-
r5	aspiration level	0.951	0.893	0.777
	aprox. solution	0.645	0.595	0.471
	solution	0.650	0.592	0.476
	-	-	-	
	direction	-0.667	1.000	0.000
	dir. derivative	-0.688	0.816	-0.128
	-	-	-	-

Table 1

The graphic user interface consist of three windows.

The first window, which we called "PAST" presents the past process of interaction. The values of objectives are presented for each iteration.

The second window, which we called "PRESENT" presents a three bars for each objective. The first bar presents the last solution for the last aspiration level, the second

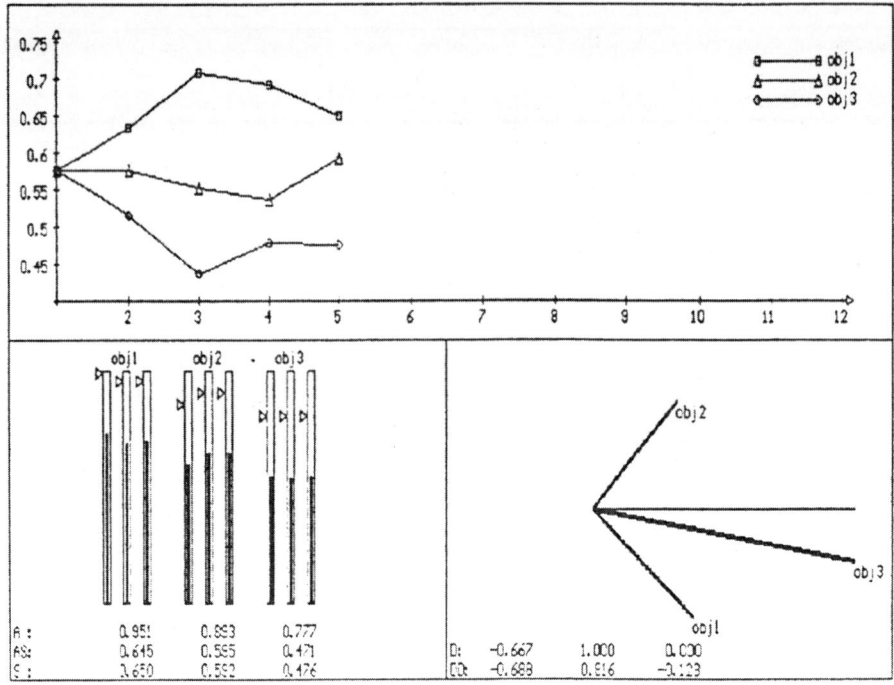

Figure 1: The prototype of graphic user interface

bar presents a new aspiration level, which can be changed dynamically on the screen during the process of interaction and the approximation of the solution based
    on directional derivatives. Comparing the aspiration level marked on the first bar and the second the user have the information about direction of changes of aspiration level. The third bar presents the new efficient solution, which was obtained as a result of optimization.Additionally the numerical values of aspiration levels (A), approximation of the solution (AS) and the solution (S) are presented to the user.

The third window, which we called "**FUTURE**" presents the prediction of changes of objectives when the aspiration point is changed in the specified direction based on values of directional derivatives. Additionally the numerical values of direction (D) and the directional derivatives are presented (DD) to the user.

6 Conclusions

We can find local liner approximation in terms of directional derivatives of the efficient solution by solving liner problems. The function $\hat{q}_i(\bar{q})$ can be nondifferentiable. Such information is useful for the user for finding quickly the tendency of changing of the efficient solution in its surroundings. The time of solving linear problems comparing with solving nonlinear problems especially by nondifferential solvers are much faster. It is very important during the process of interaction with the user when the system should give answer to the user as quick as possible. Presented graphic user interface is only a prototype and should be improved. Moreover the theory can be developed for broader class of scalarizing function, and for such a function more precise formulas can be obtained for directional derivative.

Acknowledgement The author express his gratitude to prof. A. Ruszczynski and prof. A. Wierzbicki for their help and useful suggestions.

References

Gauvin, J. and F. Dubeau (1982). Differential Properties of the Marginal Function in Mathematical Programming. *Mathematical Programming Study*, 19, pp. 101-109.

Gauvin, J. and J. W. Tolle (1977). Differential Stability in Nonlinear Programming. *SIAM Journal on Control an Optimization*, 15, pp. 294-311.

Fiacco, A.V. and G.P. McCormick (1968). Nonlinear Programming: Sequential unconstrained minimization techniques, Wiley, New York.

Halme, M. (1990). Finding Efficient Solutions in Interactive Multiple Objective Linear Programming. Helsinki School of Economics.

Jittorntrum, K. (1984). Solution Point Differentiability Without Strict Complementarity in Nonlinear Programming. *Mathematical Programming* Vol. 21, pp. 127-138.

Kiwiel, K. and A. Stachurski (1988). NOA1: A FORTRAN Package of Nondifferentiable Optimization Algorithms, Methodological and User's Guide, WP-88-116, IIASA, Laxenburg, Austria.

Korhonen, P.J. and J. Laakso (1986). A visual interactive method for solving the multiple criteria problem, *European Journal of Operational Research*, 24, pp. 277-278.

Kreglewski, T., Granat J. and A. P. Wierzbicki (1991). IAC - DIDAS - N: A Dynamic Interactive Decision Analysis and Support System for Multicriteria Analysis of Nonlinear Models , CP-91-010, IIASA, Laxenburg, Austria.

Lewandowski, A. and J. Granat (1989). Dynamic BIPLOT as the Interaction Interface for Aspiration Based Decision Support Systems. In P. Korhonen, A. Lewandowski, J. Wallenius, editors: Multiple Criteria Decision Support, Proceedings, Helsinki. Springer Verlag, Berlin.

Lewandowski, A. and A. P. Wierzbicki (eds) (1990). Aspiration Based Decision Support Systems, Springer. Springer Verlag, Berlin.

Nakayama, H. (1990). Trade-off Analysis using Parametric Optimization Techniques, Research Report 90-1, Dep. Appl. Math., Konan University, Japan.

Nakayama, H. and Y. Sawaragi (1984). Satisfacing Trade-off Method for Multiobjective Programming. In M. Grauer and Wierzbicki A.P. eds.: Interactive Decision Analysis. Proceedings Laxenburg. Springer Verlag, Berlin.

Ng, W.-Y. (1991). An interactive descriptive graphical approach to data analysis for trade-off decisions in multi-objective programming, *Information and Decision Technologies*, 17, pp. 133-149.

Rockafellar, R. T, (1984). Directional Differentiability of the Optimal Value Function in a Nonlinear Programming Problem. *Mathematical Programming Study*, 21, pp. 213-226.

Rogowski, T., Sobczyk J. and A. P. Wierzbicki (1988). IAC-DIDAS-L Dynamic Interactive Decision Analysis and support system linear version, WP-88-110. IIASA, Laxenburg, Austria.

Sakawa, M. and H. Yano (1990). Trade-off rates in the Hyperplane Method for Multiobjective Optimization Problems, *European Journal of Operational Research*, 44, pp. 105-118.

Sawaragi, Y., Nakayama H. and T. Tanino (1985). Theory of Multiobjective Optimization, Academic Press.

Steuer, R. E. (1986). Multiple Criteria Optimization: Theory, Computation and Application, John Wiley and Sons.

Steuer, R. E. and E.U. Choo (1983). An Interactive Weighted Tchebycheff Procedure for Multiple Objective Programming. *Mathematical Programming*, Vol. 26, No. 1, pp. 326-344.

Wierzbicki, A. P. (1986). On the Completeness and Constructiveness of Parametric Characterizations to Vector Optimization Problems. *OR Spectrum*, Vol. 8, No. 2, pp. 73-87.

Methods of Dynamic Multi–Criteria Problems Solutions and Their Applications

V. A. Gorelik

Computing Center of the Russian Academy of Sciences
Moscow, Russia

Abstract

Multicriteria dynamic optimization is an important scientific subject today. Such questions as the scalarization of multicriteria problems, the optimality conditions in phase and control spaces and the multicriteria dynamic programming technique are considered. Here we cannot review numerous works and results in this field; the reader is referred to a review by A. P. Wierzbicki (1991). This paper is dedicated to a special class of parametric scalarizing functions which is used for solving various dynamic multicriteria problems.

1 Formulations of multicriteria dynamic problems in the continuous time case

Consider a control process influencing a set of dynamic systems, described by differential equations

$$\dot{x}^i(t) = f_i(x^i(t), u(t)t), \quad x^i(0) = x_0^i. \tag{1}$$

Each control $u(\cdot)$ generates a bunch of trajectories $X(t) = (x^1(t), \ldots, x^n(t))$, $t \in [0, T]$, where $x^i(t)$ is a trajectory of i-th system, i.e. $x^i(\cdot)$ is a solution of vector-equation (1) with number i at given $u(\cdot)$, $i = \overline{1, n}$.

The quality of functioning of i-th system at the moment t is estimated by the function $f_i^0(x^i(t), u(t), t)$. The efficiency of control $u(\cdot)$ is determined by a quality of functioning of the whole set of systems. The concept of optimal control can be determined in various ways in such a situation; accordingly, we have various formulations of the problem. One of the possible ways is connected with an introduction of the integral type criterion for each system

$$F_i(u(\cdot)) = \int_0^T f_i^0(x^i(t), u(t)t)dt, \tag{2}$$

determined on the solutions of the set of equations (1). As a result we get a vector of criteria $\bar{F}(u(\cdot)) = (F_1(u(\cdot)), \ldots, (F_n(u(\cdot)))$. If we use the most widely used concept of vector optimality – the Pareto-optimality – then a feasible control $u(\cdot)$ is optimal if inequalities $F_i(u(\cdot)) \geq F_i(u_0(\cdot))$, $i = \overline{1,n}$, for any feasible control $u(\cdot)$ imply equalities $F_i(u(\cdot)) = F_i(u_0(\cdot))$, $i = \overline{1,n}$.

There are two questions connected with this definition. The first is a technical one but rather difficult from a mathematical point of view. The question is how to find Pareto-optimal decisions?

The second question is a principal one: how to compare different Pareto-optimal decisions? These questions are traditional for multicriteria problems but here they are complicated by dynamics.

The answer to both questions can be connected with an introduction of a general criterion which is formed from the vector of criteria by the operation of minimization :

$$F(u(\cdot)) = \min_{1 \leq i \leq n} [\lambda_i \int_0^T f_i^0(x^i(t), u(t), t)dt]. \tag{3}$$

It can be shown that if $F_i(u(\cdot)) > 0$, $i = \overline{1,n}$, for any feasible control, then each Pareto-optimal control can be found as a solution of the problem of maximization of $F(u(\cdot))$, determined on the solutions of the set of equations (1) with some $\lambda_i > 0$, $i = \overline{1,n}$. If a decision maker can intepret and determine λ_i as coefficients of importance of consecutive criteria (or dynamic systems) then a corresponding Pareto-optimal decision will be chosen.

If we suppose additionally that the choice of a control determines not only trajectories but also the set of criteria (or a composition of the system set), then the problem of the Pareto-optimal decision making can be transformed into the problem of maximization

$$\tilde{F}(u(\cdot)) = \min_{i \in R(u(\cdot))} [\lambda_i \int_0^T f_i^0(x^i(t), \quad u(t), t)dt \tag{4}$$

subject to (1), where $R(u(\cdot))$ is a set-valued mapping. The problems of such a type (even more general than (3) and (4)) were considered by Gorelik (1983), Gorelik and Kononenko (1982). Penalty function apparatus was developed for these problems, using a new idea of penalty constants congruence. and necessary conditions of optimality were obtained.

Consider another way of the defintion of optimality. If we are interested in Pareto-optimality at each moment t, then we can connect it with the general local criterion

$$f^0(x,(t),u(t),t) = \min_{1 \leq i \leq n} [\lambda_i(t) f_i^0(x^i(t), u(t), t)] \tag{5}$$

under condition that functions $f_i^0(x^i(t), u(t), t)$ are positive. Naturally, we can't achieve uniform (at each moment) Pareto-optimality in the general case, but it is possible to introduce a concept of integral Pareto-optimality of control determined as a solution of the problem of maximizing the integral functional

$$I(u(\cdot)) = \int_0^T \min_{1 \leq i \leq n} [\lambda_i(t) f_i^0(x^i(t), u(t), t)] dt \tag{6}$$

subject to (1), where $\lambda_i(t) > 0$, $i = \overline{1,n}$.

The problem (6), (1) is connected with nondifferentiable optimization. We also later show that this problem can be approximated by the usual variational problem by means of a penalty function approach which gives a possibility to construct computing methods and to obtain necessary conditions of optimality in the form of generalized maximum principle. These conditions are used for an analysis of a two-sector model of economic system control.

The above problem can be generalized in the case when it is necessary to choose a set of essential criteria at each moment at time. Then we have the problem of maximizing

$$\tilde{I}(u(\cdot)) = \int_0^T \min_{i \in \tilde{R}(u(t))} [\lambda_i(t) f_i^0(x^i(t), u(t), t)] dt \tag{7}$$

subject to (1), where $\tilde{R}(u(t))$ is a set-valued mapping. Analogous results are obtained for the problem (7), (1) using the same scheme and additionally penalty constants congruence, as it was done by Gorelik (1983). There also exist generalizations of these results for the case of additional phase and terminal constraints for all problems considered. The problem of multicriteria control for one dynamic system (when a quality of one trajectory is estimated by several criteria) and the problem of control over a set of dynamic systems with the same criterion of functioning can be interpreted as special cases. Some possible applications of such formulations are described in Gorelik, Gorelov, Kononenko (1991).

2   Approximation of the initial problem by the usual variational problem

Consider the maximization problem (6) subject to the differential constraints (1) over the class of measurable vector -functions $u(\cdot)$ with each component square-integrable on $[0,T]$ and with the values of $u(\cdot)$ in a convex compact $U \subset E_m$. Thus the set of feasible controls is

$$\mathcal{U} = \{u(\cdot) \in L_2[0,T] \big| u(t) \in U \quad \forall t \in [0,T]\}. \tag{8}$$

Suppose that the conditions of the existence and uniqueness of an absolutely continuous solution of vector-equation (1) on interval $[0,T]$ for each control $u(\cdot)$ and each index i are fulfilled. Denote by $W_2^1[0,T]$ the space of absolutely continuous functions with derivatives from $L_2[0,T]$ and introduce $x(\cdot) = (x^1(\cdot), \ldots, x^n(\cdot))$, $x_0 = (x_0^1, \ldots, x_0^n)$, and a set of functions

$$\mathcal{X} = \{x(\cdot) \in W_2^1[0,T] \big| x(t) \in E_k, \ x(0) = x_0\} \tag{9}$$

as well as a penalty functional

$$\Phi_i(x^i(\cdot), u(\cdot)) = -\int_0^T |\dot{x}^i(t) - f_i(x^i(t), u(t), t)|^2 dt. \tag{10}$$

Consider the problem

$$\hat{W}(c) = \sup_{u(\cdot) \in \mathcal{U}, x(\cdot) \in \mathcal{X}, v(\cdot) \in V} W(x(\cdot), u(\cdot), v(\cdot)c), \tag{11}$$

where

$$W(x(\cdot), u(\cdot)v(\cdot)c) =$$

$$\int_0^T \{v(t) - c\sum_{i=1}^n [\min(0, \lambda_i(t) f_i^0(x^i(t), u(t), t) - v(t))]^2\} dt + c\sum_{i=1}^n \Phi_i(x^i(\cdot), u(\cdot)), \tag{12}$$

where V is a space of continuous scalar functions on $[0,T]$ and C is a positive constant.

Theorem 2.1 *If functions $f_i(x, u, t)$ are continuous with respect to (x, u), measurable with respect to t and satisfy the conditions*

$$| < x, f_i(x, u, t) > | \leq d_i(1 + |x|^2),$$

functions $f_i^0(x, u, t)$ are upper bounded on $[0, T] \times E_{k_i} \times U$, U is a convex compact in E_m, functions $f_i(x, u, t)$ and $f_i^0(x, u, t)$ satisfy the Lipshitz condition with respect to x, functions $\lambda_i(t)$ are measurable, functionals (2) and (10) are weakly upper semicontinuous with respect to $u(\cdot)$ for each function $x^i(\cdot) \in W_2^1[0, T]$ (it is fulfilled, for example, when functions $f_i(x, u, t)$ and $f_i^0(x, u, t)$ are linear with respect to u), then the initial problem (6), (1) and the auxiliary problem (11), (12) have solutions (the latter one at any $C > 0$) such that the following limits are obtained:

$$I_0 = \lim_{c \to \infty} \hat{W}(c) = \lim_{c \to \infty} \int_0^T v_0(t, c) dt =$$

$$= \lim_{c \to \infty} \int_0^T \min_{1 \leq i \leq n} [\lambda_i(t) f_i^0(\hat{x}^i(t, c), u_0(t, c), t] dt,$$

where I_0 is the maximum value of the functional (6) on $\mathcal{U}$ subject to (1), $v_0(t, c), u_0(t, c)$ are solutions of the problem (11), (12), $\hat{x}^i(t, C)$ is solution of the system (1) at $u(t) = u_0(t, C)$.

Thus an approximate solution of the initial nondifferentiable problem (6),(1) can be found as a solution of the problem (11),(12) for sufficiently great C which is an usual variational problem if $f_i(x, u, t)$ and $f_i^0(x, u, t)$ are differentiable functions.

3 Necessary conditions of optimality

The auxiliary problem (11), (12) can be used not only for an approximate solution of the initial problem but also for obtaining necessary conditions of optimality. The following reasoning is used for this purpose: variations of the auxiliary problem functional (12) over all variables must be equal to zero as the conditions of maximum for the auxiliary problem (11); these equalities give us necessary conditions for the initial problem by passing to the limit at $C \to 0$. As the result we can obtain the following modification of maximum principle.

Theorem 3.1 *If all conditions of Theorem 1 are valid and additionally $f_i(x, u, t)$, $f_i^0(x, u, t)$ are continuously differentiable functions with respect to x, then there exist nonnegative* *measurable functions $p_i(\cdot)$, $i = \overline{1,n}$, $\sum_{i=1}^{n} p_i = 1$ (whereas $p_i(t) = 0$ if $i \notin J(t) = Arg$ $\min_{1 \le i \le n} \lambda_i(t) f_i^0 (x_*^i(t), u^*(t), t)$, where $u^*(\cdot)$ is the optimal control for the problem (6), (1) and $x_*^i(\cdot)$ is the solution of the system (1) at $u(t) = u^*(t)$) and such vector-functions $\Psi_i^*(t)$ which are solutions of equations*

$$\dot{\Psi}_i(t) = \frac{\partial}{\partial x} f_i(x_*^i(t), u^*(t), t) \Psi_i(t) - \lambda_i(t) p_i(t) \frac{\partial}{\partial x} f_i^0(x_*^i(t), u^*(t), t), \Psi_i(T) = 0,$$

that the Hamiltonian function

$$H(x_*(t), u, t, \Psi^*(t), p(t)) = \sum_{i=1}^{n} [f_i(x_*^i(t), u, t) \Psi_i^*(t) + \lambda_i(t) p_i(t) f_i^0(x_*^i(t), u^*(t), t)].$$

attains maximum over $u \in U$ at the optimal control $u^*(t)$ at almost all $t \in [0, T]$.

These conditions of optimality are transformed into the usual maximum principle if the minimum in (6) attained at single index, i.e., the set $J(t)$ consists of one element. If $J(t)$ consists of several elements, then the conditions are more complicated because of additional unknown variables $p_i(t)$, but the same number of additional equations as the unknown variables $p_i(t)$ appear for respective indexes from the conditions of minimum in (6).

The details of this approach and corresponding algorithms are described in Gorelik, Tarakanov (1988, 1989), Tarakanov (1988).

4 Multicriteria dynamic problems in discrete time case

Consider a multistep control process

$$x_k = f_k(x_{k-1}, u_k), \quad u_k \in U_k, \quad k = \overline{1, m},$$

$$x = (x_0, \ldots, x_m), u = (u_1, \ldots, u_m), \tag{13}$$

with n criteria

$$I_i(x, u) = \sum_{k=1}^{m} F_{ik}(x_{k-1}, u_k) + F_{im+1}(x_m), i = \overline{1, n}. \tag{14}$$

Pareto-optimal control for the problem (13),(14) can be found as a solution of the problem of maximizing

$$I_*(x,u) = \min_{1 \leq i \leq n} \alpha_i \left[\sum_{k=1}^{m} F_{ik}(x_{k-1}, u_k) + F_{im+1}(x_m) \right] \tag{15}$$

subject to restrictions (13).

Integral Pareto-optimality of control may be determined as a solution of the problem of maximization

$$I^*(x,u) = \sum_{k=1}^{m} \min_{1 \leq i \leq n} [\alpha_{ik} F_{ik}(x_{k-1}, u_k)] + \min_{1 \leq i \leq n} [\alpha_{im+1} F_{im+1}(x_m)] \tag{16}$$

subject to restrictions (13).

Necessary conditions of optimality for the problems (15) and (16) in the form of a generalized maximum principle can be obtained by means of the nondifferentiable optimization apparatus or the penalty function apparatus using an approximation by differentiable optimization problems and passing to the limit in necessary conditions for auxiliary problems. The second way gives us also a possibility to expand optimality conditions for discrete time problems on an unconvax case in the form of modified discrete maximum principle. The Hamiltonian functions for general criteria (15) and (16) are linear transformations of Hamiltonian functions for particular criteria, so we can state the relation between local and integral concepts of vector optimality (Gorelik, Alutina (1989,1991)).

It is interesting that the Bellman's principle of optimality is valid in the initial phase space for the problem (16), (13), but not for the problem (15), (13).

Bellman's function for the problem (16), (13) may be determined by equations

$$\omega(x_{k-1}) = \max_{u_k \in U_k} [\min_{1 \leq i \leq n} \alpha_{ik} F_{ik}(x_{k-1}, u_k) + \omega_k(f_k(y_{k-1}, u_k))], \ k = \overline{1,m},$$

$$\omega_m(x_m) = \min_{1 \leq i \leq n} \alpha_{im+1} F_{im+1}(x_m).$$

Then the maximum of the criterion $I^*(x,u)$ is equal to $\omega_0(x_0)$ and the integral Pareto-optimal solution satisfies the conditions:

$$\min_{1 \leq i \leq n} \alpha_{ik} F_{ik}(x^*_{k-1}, u^*_k) + \omega_k(f_k(x^*_{k-1}, u^*_k)) = \omega_{k-1}(x^*_{k-1}), \ k = \overline{1,m}.$$

For the problem (15), (13), the principle of optimality becomes valid under the usual expansion of the phase space. Then Bellman's function may be determined by recurrent equations

$$\varphi_{k-1}(x_{k-1}, y_{k-1}) = \max_{u_k \in U_k} \varphi_k(f_k(x_{k-1}, u_k), y_{k-1} + F_k(x_{k-1}, u_k)), \quad k = \overline{1, m},$$

$$\varphi_m(x_m, y_m) = \min_{1 \leq i \leq n} [\alpha_i(y_m^i + F_{im+1}(x_m))],$$

where

$$F_k(x_{k-1} u_k) = (F_{1k}(x_{k-1}, u_k), \ldots, F_{mk}(x_{k-1}, u_k)),$$

$$y_k = y_{k-1} + F_k(x_{k-1}, u_k), k = \overline{1, m}, \quad y_0 = 0.$$

Maximum value of the criterion $I_*(x, u)$ is equal $\varphi_0(x_0, 0)$ and the Pareto-optimal control can be found by the usual dynamic programming method with the function φ_k.

5   A control problem for a two-sector economic system

Consider a two-sector model of economic system which includes productive and unproductive spheres. These spheres are described by amounts of funds whose changes are connected with an investment of an additional homogeneous resource produced in the productive sphere, and an amortization. The control consists in determining proportions of the resource distribution between spheres. Achievements in each sphere are evaluated by an amount of funds at every time moment. The productive function is at first linear, i.e. an amount of a produced resource at every moment is proportional to an amount of productive funds. The general criterion has the form (6), where the only weighting coefficient is constant.

Here we consider the case of continuous planning interval $[0, T]$. Denote the values of funds in productive and unproductive spheres at the moment t by $x(t)$ and $y(t)$ respectively, where x and y are scalar variables. A homogeneous product (or resource) produced at the moment t in an amount $\alpha x(t)$ is distributed between productive and unproductive spheres in proportions $u(t)$ and $1 - u(t)$. A control $u(t)$ is supposed to be piece-wise continuous and satisfing the condition $0 \leq u(t) \leq 1$. Initial conditions x_0, y_0, amortization coefficients μ, δ and a weighting coefficient λ are given $(0 < \mu < \alpha, 0 < \delta < \alpha, x_0 > 0, y_0 > 0, \lambda > 0)$.

The problem consists in maximizing the functional

$$I = \int_0^T \min\{\lambda x(t), y(t)\}dt \tag{17}$$

subject to restrictions

$$\dot{x}(t) = [\alpha u(t) - \mu]x(t), \quad x(0) = x_0,$$

$$\dot{y}(t) = [1 - u(t)]\alpha x(t) - \delta y(T), y(0) = y_0,$$

$$0 \le u(t) \le 1, \quad 0 \le t \le T. \tag{18}$$

A phase plane (x, y) of the problem (17), (18) can be divided into three regions:

$$S^- = \{(x,y)|\lambda x < y\}, \quad S^+ = \{(x,y)|\lambda x > y\}, \quad S = \{(x,y)|\lambda x = y\}.$$

Necessary conditions of optimality in S^- and S^+ transform into the usual maximum principle, in S they are more complicated but this dividing line can be studied separately.

It is convenient to apply the following transformation of variables

$$W = e^{\mu t}x, \quad Z = e^{\mu t}(x + y).$$

Then the problem (17),(18) takes the form

$$\int_0^T \min\{\lambda W, Z - W\}e^{-\mu t}dt \to \max,$$

$$\dot{W} = \alpha u W, \dot{Z} = (\alpha + \delta - \mu)W + (\mu - \delta)Z. \tag{19}$$

The Hamiltonian function takes the form

$$H = \Psi_1 \alpha u W + \Psi_2[(\alpha + \delta - \mu)W + (\mu + \delta)Z] + [p_1 \lambda W + p_2(Z - W)]e^{-\mu t},$$

so that the optimal control u^* is connected with an adjoint variable Ψ_1 on the whole phase plane by a relation

$$u^* = \begin{cases} 1, & \Psi_1 > 0, \\ 0, & \Psi_1 < 0, \\ \text{arbitrary}, & \Psi_1 = 0. \end{cases}$$

The adjoint system for (19) is in S^- and S^+ respectively

$$\begin{cases} \dot{\Psi}_1 = -\Psi_1 \alpha u - \Psi_2(\alpha + \delta - \mu) - \lambda e^{-\mu t} \\ \dot{\Psi}_2 = -\Psi_2(\mu - \delta), \end{cases} \qquad \begin{cases} \dot{\Psi}_1 = -\Psi_1 \alpha u - \Psi_2(\alpha + \delta - \mu) + e^{\mu t} \\ \dot{\Psi}_2 = -\Psi_2(\mu - \delta) - e^{\mu t}. \end{cases}$$

These systems show that the function Ψ_1 cannot be equal to zero on an interval in S^- and S^+, hence for the pieces of the optimal trajectory which lie wholly in S^- or S^+, the optimal control is piece-wise constant and $u = 0$ or $u = 1$.

For the motion along S, it is necessary that:

$$u = u_0 = (1 + \lambda)^{-1}[1 + \lambda \alpha^{-1}(\mu - \delta)]$$

(this control is feasible if $\alpha > \lambda(\mu - \delta)$). Thus the optimal control can be equal to $0, 1, u_0$.

More detailed comparison of trajectories which are suspected to be optimal gives a possibility to construct the optimal control for all values of parameters and the initial conditions. This investigation shows that the problem (17),(18) has interesting qualitative properties which can be described briefly as follows.If the value T is not large then the optimal trajectory starting in S^- or S^+, tends to S and then moves along it till the end. When T becomes larger the optimal trajectory crosses S^- from S^+ into S, then returns to S and moves along it. While T becomes larger the duration of the motion in S^+ increases and finally the optimal trajectory reaches S only at the last moment and there is no the motion along S at all. So there is some kind of "antiturnpike" effect in this problem.

Consider now a variant of this model with the terminal criterion

$$\tilde{I} = \min\{\lambda x(T), y(T)\}$$

and a nonlinear productive function

$$\dot{x}(t) = u(t)f(x(t)) - \mu x(t), \quad x(0) = x_0,$$

$$\dot{y}(t) = (1 - u(t))f(x(t)) - \delta y(t), \quad y(0) = y_0.$$

Necessary conditions of optimality are also obtained for this case. An analysis shows (see Gorelik, Fomina (1991)) that under some additional assumptions there exists a special surface $(u(t) = u^*, x(t) = x^*)$ such that if the planning interval T is large then the optimal trajectory moves almost all time along it: hence the system has a kind of turnpike property.

References

Gorelik, V. A., Kononenko, A. F. (1982). Game-theoretical Decision Making Models in Ecological-economical Systems, Radio i Svyas Publishing Company.

Gorelik, V. A. (1983). Maximin Problems over Connected Sets in Banach Spaces. Kibernetika, N1.

Gorelik, V. A., Tarakanov, A. F. (1988). Minimization of Nondifferentiable Functionals Maximum Type with Phase Constraints/Decomposition and Optimization in Complex Systems, Computing Center of the USSR Academy of Sciences.

Tarakanov, A. F. (1988). The Method of Consequent Approach for Minimax Control Problem with Phase Constraints /Ibid.

Gorelik, V. A., Tarakanov, A. F. (1989). Penalty Method for Nondifferentiable Minimax Control Problems with Connected Constraints. Kibernetika, N4.

Gorelik, V. A., Alutina, E. F. (1989). Discrete Dynamic Multicriteria Problems/Simulation and Optimization of Complex Dynamic Processes, Computing Center of the USSR Academy of Sciences.

Gorelik, V. A., Alutina, E. F. (1991). Discrete Maximum Principle for Dynamic Multicriteria Problems/Decomposition and Optimization in Complex Systems, Computing Center of the USSR Academy of Sciences.

Gorelik, V. A., Fomina, T. P. (1991). Turnpike Properties for Two-sector Model/Ibid.

Gorelik, V. A., Gorelov, M. A., Kononenko, A.F. (1991). Analysis of Conflict Situations in Control Systems, Radio i Svyas Publishing Company.

Wierzbicki, A. P. (1991) Dynamic Aspects of Multiobjective Optimization. In A. Lewandowski, V. Volkovich (eds): Multiobjective Problems of Mathematical Programming, Proceedings, Yalta 1988, Springer-Verlag Lecture Notes in Economics and Mathematical Systems, Berlin 1991.

Quantitative Pareto Analysis and the Principle of Background Computations

Ignacy Kaliszewski

Systems Research Institute
Polish Academy of Sciences, Poland

Abstract

It is shown that by appropriate interpretation of results on scalarizing vector optimization problems the scope of Pareto analysis can be significantly broadened. As a result, new and valuable information on mutual relations between efficient decisions is available for decision makers in (interactive) decision making processes.

1 Introduction

In this presentation we shall be dealing primarily with Vector Optimization (VO) problems. An attempt to exploit those problems beyond the present state–of–the–art is driven by the will to show that they can be something more than just efficient decision generators for decision making processes. In the literature of the subject there is a number of papers in which authors investigate properties of efficient decisions with respect to their neighbourhoods (the local approach) or with respect to the whole set of efficient decisions (the global approach). Much is known about local properties of efficient decisions (Sawaragi et al., 1985, cf also Granat, 1991 for a short review of the existing results and an original development) whereas the global approach, though some strong results have been obtained (Korhonen, Laakso, 1986; Halme, 1990; Wierzbicki, 1990) and they will be quoted when appropriate, has attracted less attention.

By applying the concept of *cone separability* (Wierzbicki, 1977) and a related technique called *Cone Separation Technique* (CST) (Kaliszewski, 1991) Decision Maker (DM) can be provided with valuable information about various mutual relations between admissible (efficient) decisions. Traditional use of VO problems gives DM an important *qualitative* information about decisions, namely, it splits decisions into classes of efficient and nonefficient decisions. It provides also some *quantitative* information about efficient decisions, such as values of objective functions, a distance from a certain ideal (may be fictitious) decision, and also about the whole set of efficient decisions such as maximal and minimal values of separate criteria attained over the admissible set of decisions. What we are going to present is a coherent methodology, referred to as *Quantitative Pareto Analysis* (QPA) (Kaliszewski, 1991), which provides, in addition to the efficiency status and the information listed above, also:

- a simple way to impose a certain hierarchical structure over the set of efficient decisions,

- a way for visualizing decision making processes by offering a method for fast approximations of sets of efficient outcomes (Pareto sets),

- bounds on trade-offs,

- approximate sensitivity analysis of efficient decisions with respect to perturbations of utility functions,

- approximate sensitivity analysis of efficient decisions with respect to perturbations of objective functions.

All the additional information comes from interpretations of specific numerical characterizations of efficient decisions related to the notions of proper efficiency and *substantial efficiency*. Those characterizations and their interpretations are cast into a specific language of CST which provide a unified framework for the development and the presentation of results.

QPA can be applied in any interactive decision making method to enhance the quality of decisions. DM is free to decide on each iteration which elements of QPA are to be performed. The analysis is rather demanding in computing capacity and computation time. However, in the decision phase (man acts) of any man–machine interactions, machine (computer) is idle for an amount of time which is often significant. This idle time can be used to initiate and proceed elements of Qualitative Pareto Analysis in *the background* (and, in parallel, if computing requirements cannot be met by sequential computations), even if some results of the analysis may be later not used.

The paper presents briefly elements of QPA and discusses performing Qualitative Pareto Analysis in the background of interactive decision making processes.

2   Quantitative Pareto Analysis

Here we describe briefly the non-standard elements of Quantitative Pareto Analysis. We assume that the underlying model for decision making problems is the following vector optimization problem:
(VO)

$$V \max f(x) \quad \text{s.t.} \quad x \in X_0,$$

where $X_0 \subseteq X$, $f : X \to R^k$, $f = (f_1, f_2, ..., f_k)$, $f_i : X_0 \to R$, $i = 1, ..., k$, the set $Z = f(X_0)$ (the outcome set in an outcome space) is compact (bounded and closed). Boundedness of Z ensures the existence of an element y^* such that $Z \subseteq \{y^*\} - \mathrm{int}\, R_+^k$. We shall make use of standard definitions of efficiency and proper efficiency of decisions (elements of X_0) and outcomes (elements of Z).

2.1 Imposing a hierarchical structure over the set of efficient outcomes/decisions

Geoffrion's classical definition of proper efficiency assigns to any properly efficient decision x and also to its outcome $y = f(x)$ a finite number M which is a bound on trade-off coefficients. We shall consider here properly efficient outcomes with an a priori given, least upper bound M_0, defined in the following way.

Given a properly efficient outcome $\bar{y}$, M_0 is a least upper bound on the ratios $(y_i - \bar{y}_i)/(\bar{y}_j - y_j)$ for each i and for at least one j, whenever $y_i - \bar{y}_i > 0$, $\bar{y}_j - y_j > 0$, $y \in Z$. The following theorem relates properly efficient outcomes with the corresponding bounds M_0.

Theorem 2.1. (Kaliszewski, 1991). An outcome $\bar{y} \in Z$ is properly efficient with a least bound M_0 if and only if there exists $\lambda = \lambda(M_0) > 0$ and $\rho = \rho(M_0) > 0$ such that $\bar{y}$ solves the following problem:
(P^∞)

$$\min_{y \in Z} \max_i \lambda_i((y_i^\star - y_i) + \rho e^k(y^\star - y)),$$

where $e^k = (1, ..., 1) \in R^k$.

For each properly efficient outcome $y \in Z$ there exists a $\lambda > 0$ such that y solves P^∞ uniquely for every positive ρ satisfying $M_0 \le ((k-1)\rho)^{-1}$. For each outcome $y \in Z$ solving the above problem the inequality $M_0 \le (1 + (k-1)\rho)\rho^{-1}$ holds.

The above necessary and sufficient condition is implicit in general results of Wierzbicki, 1977 (cf also Choo, Atkins, 1983; Kaliszewski, 1987).

Solving the problem P^∞ is equivalent to searching for an outcome $\bar{y} \in Z$ such that $(\{\bar{y}\} + K_\rho) \cap Z = \{\bar{y}\}$, where $K_\rho = \{y \in R^k \mid y_i + \rho e^k y \ge 0\}$. The cone K_ρ is a basic construct of CST and a main tool for QPA.

By selecting a sequence of positive parameters ρ such that $\rho_j > \rho_{j+1}$ and solving for each ρ_j the problem P^∞ for all λ we get a family of subsets of properly efficient outcomes $Z_1, Z_2, ...$ For each outcome of Z_i, $M_0 \le (1 + (k-1)\rho_i) \, \rho_i^{-1}$. As shown by Lemma 2.1, there is a natural hierarchy over the sets of efficient decisions and outcomes.

Lemma 2.1. Subsets Z_i are nested, ie $Z_1 \subseteq Z_2 \subseteq ....$

2.2 Approximations of Pareto sets

If the set Z is convex, then there exist several methods for geometrical approximation of Pareto sets for bicriteria vector optimization problems (ie. for problems with $k = 2$, cf Solanki, Cohon, 1989). A generalization of those methods is the *sandwich algorithm* for approximating values of convex functions of one variable, presented in (Burkard et al., 1987; Fruhwirth et al., 1988). We are not aware of any algorithm for approximating Pareto set in the general (ie nonconvex) case except that we are going to present now. Assume $k = 2$. Each efficient outcome $\bar{y}$ generated by P^∞ determines in R^2 a set $\bar{y} + K_\rho$, where K_ρ is a cone defined in Section 2.1. The set $\bar{y} + K_\rho$ is a certain "conical" approximation of Z. The union of such approximations determined for a number of efficient decisions is an approximation of Z; the more efficient outcomes we use the better approximation. To improve the approximation we can formulate the problem of determining the maximal value of the parameter ρ for which $\bar{y}$ solves P^∞. The same principle of Pareto set approximations by cones K_ρ applies to any $k > 2$.

2.3 Bounds on trade–offs

Suppose that a properly efficient decision $\bar{y}$ has been found by solving the problem P^∞. By this (cf Theorem 2.1) an upper bound on M_0 is established. Since $\bar{y}$ is properly efficient, by the definition of proper efficiency for each outcome $y \in Z$ and for each index i, an increase in y_i, $y_i - \bar{y}_i$, relative to a decrease in y_j, $\bar{y}_j - y_j$, is not greater than M_0 for at least one j.

In general, there may be pairs of indices for which the corresponding increase to decrease ratio over Z is greater than M_0. This does not happen, however, when $k = 2$ since then for each outcome $y \in Z$ and for each index i, an increase in y_i, $y_i - \bar{y}_i$, relative to a decrease in y_j, $\bar{y}_j - y_j$, is not greater than M_0, $i, j = 1, 2, i \neq j$.

A trade–off $T_{i,j}(\bar{y}, \hat{y})$ between outcomes $\bar{y}$ and $\hat{y}$ involving i and j is defined as $(\hat{y}_i - \bar{y}_i)/(\hat{y}_j - \bar{y}_j)$, $i, j = 1, ..., k$, $i \neq j$. The notion of trade–off is widely used in multiple criteria decision making methods and therefore it may be of interest to observe that for $k = 2$ some partial information on trade–offs in the set Z is provided by the number M_0.

Assume $k = 2$. Let $\bar{y}$ be properly efficient outcome of Z. Let

$$A = \{y \in Z \mid y_1 > \bar{y}_1, y_2 < \bar{y}_2\},$$

$$B = \{y \in Z \mid y_1 < \bar{y}_1, y_2 > \bar{y}_2\}.$$

All efficient outcomes y of Z, $y \neq \bar{y}$, belong either to A or to B. It follows directly from the definition of proper efficiency that for any outcome y, $y \neq \bar{y}$,

$$|T_{12}(\bar{y}, y)| \leq M_0 \text{ and } |T_{21}(\bar{y}, y)| \geq M_0^{-1} \text{ if } y \in A,$$

$$|T_{21}(\bar{y}, y)| \leq M_0 \text{ and } |T_{12}(\bar{y}, y)| \geq M_0^{-1} \text{ if } y \in B. \qquad (2.1)$$

By its nature, trade–off information is pointwise, ie it concerns a pair of outcomes. The above bounds on trade-offs, however, are global, ie they are valid for any outcome of $A \cup B$.

By Theorem 2.1, for $k = 2$, $M_0 \leq 1 + \rho^{-1}$. Hence, the above upper bound on trade–offs is essentially that obtained in Wierzbicki, (1990). Moreover, a significantly stronger result was proved there. Namely, it was shown that the upper bound $1 + \rho^{-1}$ is still valid for $k > 2$, ie that

$$|T_{ij}(\bar{y}, y)| \leq 1 + \rho^{-1},$$

if $y \in \{y \in Z \mid \bar{y}_j > y_j, \bar{y}_l \leq y_l, l = 1, ..., k, l \neq j\}$ for $i, j = 1, ..., k$, $i \neq j$.

It has been shown in Kaliszewski, (1991), that for $k > 2$ relations analogous to (2.1) exist between trade-offs and a certain number S_0 whose interpretation for $k = 2$ coincides with that of M_0. As M_0 is related to the notion of proper efficiency, S_0 is related to the notion of substantial efficiency. It should be noted, however, that the bound on M_0 is established at no cost if an efficient outcome is generated by solving the problem P^∞. In contrast, to calculate a bound on S_0 consistency of $k(k-1)/2$ systems of conditions have to be verified (cf. Kaliszewski, 1991).

For $k = 2$ and linear vector optimization problems exact values of trade–offs can be easily determined by the method of Halme (Halme, 1990).

2.4 Approximate sensitivity analysis with respect to perturbations of utility functions

Assume that in R^k a utility function u, $u : R^k \rightarrow \Re$, is given. Suppose that $\bar{y}$ maximizes $u(\bar{y})$ over Z. The outcome $\bar{y}$ is efficient. Suppose it is properly efficient. The function u may be perturbed in many ways and it is of importance to know which perturbations preserve the optimality status of $\bar{y}$.

The outcome $\bar{y}$ is insensitive to a perturbation of u resulting in a new function u' if

$$\{y \in R^k \mid u'(y) > u'(\bar{y})\} \cap Z = \{\emptyset\}.$$

With the cone K_ρ defined in Section 2.1 a weaker condition for insensitivity, namely

$$\{y \in R^k \mid u'(y) \geq u'(\bar{y})\} \subseteq \{y \in R^k \mid y - \bar{y} \in K_\rho\}, \tag{2.2}$$

forms a base for what may be called a CST sensitivity analysis. Its potential strength lies in its theoretical simplicity. Observe that in some situations the condition (2.2) is very convenient since it is independent of Z. This can simplify significantly verification of insensitivity, especially if the analytical description of Z is complex. However, the usefulness of the above condition greatly depends on the cone K_ρ. For example, if a utility function u' is linear, then for any finite ρ the condition (2.2) is not satisfied. On the other hand, the analysis proposed here is perfectly suited for utility functions whose isoquants are in a sense close to displaced cones.

2.5 Approximate sensitivity analysis with respect to perturbations of objective functions

In the same spirit as in the previous section we can formulate sufficient conditions for admissible perturbations of objective functions in VO problems. Here by admissibility of perturbations we mean perturbations of f_i, $i = 1, ..., k$, which do not change the efficiency status of a selected efficient decision.

Consider a VO problem. Let $\bar{x}$ be a properly efficient decision to this problem. Suppose that a cone K_ρ is known.

Let us consider vector optimization problem VO' which differs from VO in the mapping function, ie

(VO')

$$V \max f'(x), \quad x \in X_0,$$

where f', $f' = (f_1', f_2', ..., f_k')$, $f_i' : X_0 \rightarrow R$, $i = 1, ..., k$.

Now we formulate a sufficient condition for $\bar{x}$ to be an efficient decision of VO', or equivalently, for $f'(\bar{x})$ to be an efficient outcome of VO'.

Theorem 2.2 (Kaliszewski, 1991). Assume $f(x) > 0$, $f'(x) > 0$. A sufficient condition for a decision $\bar{x}$, $\bar{x} \in X_0$, efficient with respect to f, to be efficient also with respect to f' is that

$$\min_i (f_i(x) + \rho e^k f(x) - \beta_i f_i'(x)) \geq 0, \quad x \in W, \tag{2.3}$$

where $\beta_i = (f_i(\bar{x}) + \rho e^k f(\bar{x})) f_i'(\bar{x})^{-1}$, and W is an arbitrary set such that $X_0 \subseteq W$.

An advantage of the condition (2.3) over a brute–force approach, ie just solving *VO'* to check whether $\bar{x}$ is an efficient decision, is the possibility to work with relaxations of X_0 (the sets $X_0 \subseteq W$) but of course for $W = X_0$ the condition (2.3) is the strongest possible. It is of importance for numerical tractability of the condition (2.2) in cases where analytical description of X_0 is complex. The advantage is even more appealing if there is a family of functions depending on a parameter, eg $f'(x, \delta)$, $\delta \in [0, \bar{\delta}]$, $f'(x, 0) = f(x)$, and we want approximate the range of δ for which $\bar{x}$ is an efficient decision to *VO'* with $f'(x, \delta)$ as an mapping function.

The generality of this approach follows from the generality of CST - assumptions about *VO* problems considered are rather mild. The condition (2.3) corresponds to a series of standard mathematical programming problems.

3 Computational tractability of Quantitative Pareto Analysis

3.1 Computing capacity requirements for QPA

A basic vehicle for the Quantitative Pareto Analysis is the cone K_ρ. All the outcomes of QPA (except trade–off information for $k > 2$) is based upon knowing this cone explicitly. For each (properly) efficient outcome generated by solving P^∞ with given $\rho > 0$, the cone K_ρ is the same, namely $K_\rho = \{y \in R^k \mid y_i + \rho e^k y \geq 0\}$. However, all the results of QPA remain valid and become stronger with cones K_ρ' such that:

$$K_\rho \subseteq K_\rho' \text{ and } (\{\bar{y}\} + K_\rho) \cap Z = \{\bar{y}\} \text{ entails } (\{\bar{y}\} + K_\rho') \cap Z = \{\bar{y}\},$$

where $\bar{y}$ is an outcome of Z. Then, it is essential for the strength of QPA to formulate and solve (exactly or approximately) the problem of determining the maximal cone $K_{\hat{\rho}}$ (or equivalently: determining $\hat{\rho}$ - the maximal value of ρ) such that $(\{\bar{y}\} + K_{\hat{\rho}}) \cap Z = \{\bar{y}\}$. It should be noted that the amount of computations to solve this problem may equal or exceed the amount of computations to generate an efficient outcome $\bar{y}$.

As it has been already mentioned in Section 2.3 the amount of computations to perform full trade–off analysis for $k > 2$ may grow exponentially with the size of *VO* problems.

3.2 QPA and interactive decision making schemes

At the first glance it seems that the amount of computations required by QPA in its full extent is prohibitively large. Especially, trade–off analysis may be very demanding. We can always advocate the use of supercomputers but this is rather impractical in the decision making context because of still limited access to such installations and high costs of their services. Therefore we should look for a more reasonable approach.

It is quite obvious that QPA is well suited for interactive decision making schemes. In such schemes DM expresses progressively his preferences over the set of efficient outcomes and in consequence over the set of decisions (the weighting approach – Chankong, Haimes, 1983; Yu, 1985; Steuer, 1986; Zionts, Wallenius, 1976; or the aspiration levels approach by Wierzbicki, 1980). One interaction (iteration) consists of the decision phase (DM expresses his partial preferences), and the computing phase (the computer determines an efficient decision suiting best DM's preferences). If required, computer can also perform any part of QPA. The question of QPA computational tractability should be raised here since by obvious reasons in interactive processes the time of individual computer phases must be kept within reasonable limits. It happens, however, that even for medium size problems the time consumed for determining one efficient decision is significant and in such situations practical usefulness of QPA may be questionable. We can propose two remedies to heal this.

3.3 Background computations

The first remedy we propose is the idea of *background computations*. In all interactive decision making support algorithms during the decision phase for the most of the time the computer is idle. Even if DM uses the computer to store, sort, retrieve, or compare previously derived decisions, this usually consumes a negligible part of computer capacity. The remaining part of computer capacity can be used to start QPA in advance regardless whether DM will later make a request for its results or not. For example, an approximation of Pareto set can be progressively updated and invoked when required. Other elements of QPA can be started immediately after DM is provided with a new efficient decision. Usually, it takes some time for DM to decide about his next move and during this period, which is otherwise lost, some elements of QPA can be significantly advanced if not completed.

It is important that all the computations started by the computer (precisely speaking - by a support algorithm) on its own initiative should not harass DM in his process of decision making (information presented on the screen should not be affected) and therefore all the related computations must be done in "the background". This, however, calls for a capability of software to create submodules of a program, called *treads*, which can be processed concurrently. Mechanisms of this type are present in several algorithmic languages as Ada, Modula, and various "parallel" extensions of C, Fortran, and Pascal. Concurrency means that threads are processed interchangeably, where processor after some time spent on processing a thread suspends it and starts (or resumes) processing a subsequent thread. Usually threads are structured by some priority rules.

3.4 Multiprocessing on networks of transputers

If it happens that computer capability is still limited for implementing QPA in its full extent, then the next possible step is to make use of multiprocessor computers. In computers of this sort threads can be physically distributed among several processors. This, if done skillfully, results in a speed-up of computations with the theoretical bound on speed-up equal to the number of processors. Though some academic and even commercial multiprocessor computers are available, again limited access and high cost make them hardly advisable in the decision making context. One must remember that most implementations of support algorithms for decision making have been done with desk-top minicomputers.

Fortunately, quite recently a technology has emerged which seems to be perfectly suited to the needs of decision making and solves, at least to some extent, the problem of computational tractability of QPA. It features a family of processors, called transputers (Whitby-Stevens, Hodgkins, 1990), each with four links, which can be easily connected via links with other transputers into a network of processors. Moreover, the whole network can be connected via an idle link to any computer turning it into a multiprocessor computer of significant capability. A preliminary application of PC based transputer networks to *VO* problems has been already successfully completed (Kaliszewski, 1990).

3.5 A pilot DSS implementing QPA on a network of transputers

A pilot DSS implementing elements of QPA is currently tested in the Mathematical Programming Department of the Systems Research Institute. At present only linear multiple criteria decision making problems can be approached by this methodology. The next step will be to extend the system to linear integer multiple criteria programming problems.

A hardware platform for the system is a network of up to six transputers hosted by a PC computer.

References

Burkard R. E., Hamacher H. W., Rothe G. (1987). Approximation of convex functions and Applications in mathematical programming. Report 89-1987, Institut für Mathematik, Technische Universität Graz.

Choo E. U., Atkins D. R. (1983). Proper efficiency in nonconvex programming. Mathematics of Operations Research, vol 8, 467-470.

Chankong V., Haimes Y. Y. (1983). Multiobjective Decision Making. Theory and Methodology. Elsevier, New York.

Fruhwirth B., Burkard R. E., Rote G. (1988). Approximation of convex curves with applications to the bicriterial minimum cost flow problem. Report no 119, Institut für Mathematik, Technische Universität Graz.

Geoffrion A.M. (1968). Proper efficiency and the theory of vector maximization. Journal of Mathematical Analysis and its Applications, vol 22, 618-630.

Granat J. (1991). Parametric programming approaches to local approximation of the efficient frontier. This volume.

Halme M. (1990). Finding efficient solutions in interactive multiple objective linear programming. Discussion Paper, Helsinki School of Economics, Helsinki.

Kaliszewski I. (1987). A modified weighted Tchebycheff metric for multiple objective programming. Computers and Operations Research, vol 14, 315-323.

Kaliszewski I. (1988). Substantially efficient solutions of vector optimization problems. Systems Research Institute Technical Report ZPM30/88, Warszawa.

Kaliszewski I. (1990). Determination of maximal elements in finite sets on a network of transputers. Systems Research Institute Technical Report ZPM2/90, Warszawa.

Kaliszewski I. (1991). Quantitative Pareto Analysis by Cone Separation Technique (book in preparation).

Korhonen P. J., Laakso J. (1986). A visual interactive method for solving the multiple criteria problem. European Journal of Operations Research, vol 24, 277-287.

Sawaragi Y., Nakayama H., Tanino T. (1985). Theory of Multiobjective Optimization. Academic Press, New York.

Solanki R. S., Cohon, J. L. (1989). Approximating the noninferior set in linear biobjective programs using multiparametric decomposition. European Journal of Operation Research, vol 41, 355-366.

Steuer R. E. (1986). Multiple Criteria Optimization: Theory, Computation and Application. John Wiley & Sons, New York.

Wierzbicki A. P. (1977). Basic properties of scalarizing functionals for multiobjective optimization. Mathematische Operationsforschung und Statistik, Ser. Optimization 8, No 1.

Wierzbicki A. P. (1980). The use of reference objectives in multiobjective optimization. In: Multiple Criteria Decision Making; Theory and Applications, G. Fandel, T. Gal (eds), Lecture Notes in Economics and Mathematical Systems, vol 177, Springer Verlag, Berlin, 468-486.

Wierzbicki A. P. (1990). Multiple criteria solutions in noncooperative game theory, Part III: Theoretical Foundations. Discussion Parer No 288, Kyoto Institute of Economic Research, Kyoto, Kyoto University, Kyoto.

Whitby-Stevens C. (1990). Transputers – past, present and future. IEEE Micro, 16-19, 76-82.

Yu P. L., (1985). Multiple Criteria Decision Making: Concepts, Techniques and Extensions. Plenum Press, New York.

Zionts S., Wallenius J. (1976). Am interactive programming method for solving the multiple criteria problem. Management Science, vol. 22, 652-663.

Applications of Linear Approximation Structures to the Description of Linear Preferences Structures

Jacinto Gonzalez Pachon and Sixto Rios-Insua
Department of Artificial Intelligence
Politechnical University of Madrid, Spain

Abstract

The general preference structure concept does not constitute a good practical tool to represent the different attitudes in multiobjective decision making problems. In this paper we restrict ourselves to those preference structures which can be analytically represented by means of families of functions. The linear approximation structure concept is used to describe a particular case of preference structure, one that can be described by means of a family of linear functions.

1 Introduction

In multiobjective decision making under certainty where it is assumed that the decision maker (DM) controls all the external factors, he may reveal, for a pair of consequences or alternatives y_1 , y_2 , four types of attitudes:

a) y_1 is more preferred than y_2 ($y_1 \succ y_2$)

b) y_1 is less preferred than y_2 ($y_1 \prec y_2$)

c) y_1 is indifferent to y_2 ($y_1 \sim y_2$)

d) Between y_1 and y_2 , doubt arises about which one is preferred (y_1 ? y_2)

In this paper the last attitude, denoted as doubt, will be considered as a part of the information revealed by the DM. Thus, we will follow the approach considered in Chien, Yu and Zhang (1990) and Pachon (1990).

Doubt may be revealed, as we have previously mentioned, in an intentional way by the DM. However, it may appear in an unintentional way, revealing judgments about his/her attitudes which do not follow a certain "logical reasoning". In the next paragraphs, we will see what we mean with this last assertion.

2 The preference structure

Our aim is to model the DM's attitudes by means of binary relations. Their properties will form the foundation to the "logical reasoning" revealing the unintentional attitude denoted as doubt.

From now on, we will denote by $Y \subseteq \mathbf{R}^n$ the objective or consequence space in a multiobjective decision making problem.

Definition 2.1 *A preference structure on the set Y is a pair of binary relations on Y, denoted by $(\mathcal{R}_1, \mathcal{R}_2)$, that fulfills the following axioms*

E1: $\mathcal{R}_1$ *is asymmetric and transitive.* $\mathcal{R}_1$ *is called* preference *on Y.*

E2: $\mathcal{R}_2$ *is an equivalence relation.* $\mathcal{R}_2$ *is called* indifference *on Y.*

E3: $\mathcal{R}_1$ *and $\mathcal{R}_2$ are disjoint $(\mathcal{R}_1 \cap \mathcal{R}_1 = \emptyset)$.*

Furthermore, $\mathcal{R}_1$ and $\mathcal{R}_2$ are related by means of the next coherence axioms

E4: *If $(y_1, y_2) \in \mathcal{R}_1$ and $(y_2, y_3) \in \mathcal{R}_2 \Rightarrow (y_1, y_3) \in \mathcal{R}_1$ for every $y_1, y_2, y_3 \in Y$*

E5: *If $(y_1, y_2) \in \mathcal{R}_2$ and $(y_2, y_3) \in \mathcal{R}_1 \Rightarrow (y_1, y_3) \in \mathcal{R}_1$ for every $y_1, y_2, y_3 \in Y$*

This definition, introduced in Pachon (1990), acts as a model for rationality of the DM's attitudes in a decision process. However, this concept only regards two of the four attitudes mentioned in the introduction. If we wish to work with a concept in which appears the four attitudes revealed by a DM in his/her judgments about paired comparisons appear, we would have to consider the next definition.

Definition 2.2 *Let $(\mathcal{R}_1, \mathcal{R}_2)$ be a preference structure on Y. A quaternion of binary relations associate to $(\mathcal{R}_1, \mathcal{R}_2)$, is the set of binary relations $(\mathcal{R}_1, \mathcal{R}_1^s, \mathcal{R}_2, \mathcal{R}_{12}^c)$, where $\mathcal{R}_1^s$ is the reflection of $\mathcal{R}_1$ on $Y \times Y$ with respect to the diagonal Δ, and $\mathcal{R}_{12}^c$, is the complement of $\mathcal{R}_1 \cup \mathcal{R}_1^s \cup \mathcal{R}_2$ on $Y \times Y$. The latter is called* doubt *or indecision.*

The reason why the preference structure definition only refers to the preference and indifference properties, and not to doubt, comes from the fact that this last one can be easily derived from the others. This is obtained in the following propositions proved in Pachon (1990).

Proposition 2.3 *The relation which represents doubt, $\mathcal{R}_{12}^c$, is irreflexive and symmetric.*

Proposition 2.4 *Given the preference structure $(\mathcal{R}_1, \mathcal{R}_2)$ on $Y \subseteq \mathbf{R}^n$, we have*

a) *If $(y_1, y_2) \in \mathcal{R}_2$ and $(y_2, y_3) \in \mathcal{R}_{12}^c \Rightarrow (y_1, y_3) \in \mathcal{R}_{12}^c$*

b) *If $(y_1, y_2) \in \mathcal{R}_{12}^c$ and $(y_2, y_3) \in \mathcal{R}_2 \Rightarrow (y_1, y_3) \in \mathcal{R}_{12}^c$*

There is a difference between the attitudes which the DM reveals to the analyst $(\prec, \succ, \sim, ?)$ and the binary relations which model these relations by means of the concept of a quatern associated to a preference structure. Such a difference appears in $\mathcal{R}_{12}^c$, which contains, besides the doubt (?) revealed by the DM, those pairs where axioms **E1**– **E5** are not fulfilled (unintentional doubt). The way from (?) to $\mathcal{R}_{12}^c$ is given in the following general result.

Theorem 2.5 *Given two binary relations $\mathcal{R}$ and $\mathcal{S}$ on Y such that the diagonal Δ on $Y \times Y$ is contained in $\mathcal{S}$, and let $\mathcal{C}_i$ be a maximal subset of $\mathcal{R} \cup \mathcal{S} \subseteq Y \times Y$ which fulfills the following properties:*

1) $\mathcal{R}_{1i}^* = \mathcal{R} \cap \mathcal{C}_i$ *is asymmetric and transitive*

2) $\mathcal{R}_{2i}^* = \mathcal{S} \cap \mathcal{C}_i$ *is an equivalence relation*

3) *On $\mathcal{C}_i$ the relations $\mathcal{R}_{1i}^*$ and $\mathcal{R}_{2i}^*$ fulfill axioms **E4** and **E5***

   Then, $(\mathcal{R}_{1i}^* \backslash \mathcal{R}_{2i}^*, \mathcal{R}_{2i}^*)$ is a preference structure.

This preference structure will be called *preference structure associated to $\mathcal{R}$, $\mathcal{S}$ and $\mathcal{C}_i$* , and denoted by $(\mathcal{R}_1, \mathcal{R}_2)_{\mathcal{R}, \mathcal{S}, \mathcal{C}_i}$.

The proof can be found in Pachon and Rios-Insua (1991a)

Observe that $\mathcal{C}_i$ arises when we remove some pairs of consequences of an information chain on preferences in which the DM reveals incoherences with regard to the axioms. However, it seems logical to remove the complete information chains, since we do not have an exact criterion to find out where the incoherences appear. This leads us to define a unique preference structure associated to $\mathcal{R}$ and $\mathcal{S}$, and to asses the set

$$C = \bigcap_{i \in I} C_i$$

(the intersection of all the maximal sets defined in the above theorem). This structure will be denoted by $(\mathcal{R}_1, \mathcal{R}_2)_{\mathcal{R}, \mathcal{S}}$. (see Pachon and Rios-Insua, 1991a).

Remark 2.6 *Thus, the DM, from paired comparisons of consequences, reveals his/her attitudes to the analyst. The analyst screens such information (by means of theorem 2.5) and models it as a preference structure. However, this last stage should not be communicated to the DM, because taking cognisance of incoherences via the analyst would lead to a bias in the judgments which would make it impossible to take the original idea about the problem into account.*

3 Preference structure represented by a family of functions: $\succ_V$– Preference

Although the preference structure concept rigorously models the different attitudes, it does not constitute by its own a good tool to represent such attitudes and to search for the efficient solutions, which is an essential purpose in multiobjective problems.

   Therefore, it seems reasonable to restrict attention to those preference structures which can be analytically represented by means of a family of functions.

Definition 3.1 *Let* $(\mathcal{R}_1, \mathcal{R}_2)$ *be a preference structure on* $Y \subseteq \mathbf{R}^n$ *and* V *a family of functions of class* $\mathcal{C}^k(\mathbf{R}^n)$. *We say that* $(\mathcal{R}_1, \mathcal{R}_2)$ *is a* $\succ_V-$*preference of class* k *(if* $k = 1$ *it will be called* $\succ_V-$ *preference) if* $(\mathcal{R}_1, \mathcal{R}_2) = (\mathcal{R}_1, \mathcal{R}_2)_{\mathcal{R},\mathcal{S}}$ *where*

$$\mathbf{y}_1 \ \mathcal{R} \ \mathbf{y}_2 \Leftrightarrow v(\mathbf{y}_1) \geq v(\mathbf{y}_2) \quad \forall v \in V$$
$$\mathbf{y}_1 \ \mathcal{S} \ \mathbf{y}_2 \Leftrightarrow v(\mathbf{y}_1) = v(\mathbf{y}_2) \quad \forall v \in V$$

Proposition 3.2 *Let* $(\mathcal{R}_1, \mathcal{R}_2)$ *be a* $\succ_V-$ *preference on* $Y \subseteq \mathbf{R}^n$, *then*

$$(\mathbf{y}_1, \mathbf{y}_2) \in \mathcal{R}_1 \Leftrightarrow [v(\mathbf{y}_1) \geq v(\mathbf{y}_2) \ \forall v \in V \ \ and \ \ \exists v' \in V \ \ such \ that \ \ v'(\mathbf{y}_1) > v'(\mathbf{y}_2)]$$
$$(\mathbf{y}_1, \mathbf{y}_2) \in \mathcal{R}_1^{\bullet} \Leftrightarrow [v(\mathbf{y}_1) \leq v(\mathbf{y}_2) \ \forall v \in V \ \ and \ \ \exists v' \in V \ \ such \ that \ \ v'(\mathbf{y}_1) < v'(\mathbf{y}_2)]$$
$$(\mathbf{y}_1, \mathbf{y}_2) \in \mathcal{R}_2 \Leftrightarrow v(\mathbf{y}_1) = v(\mathbf{y}_2) \ \forall v \in V$$
$$(\mathbf{y}_1, \mathbf{y}_2) \in \mathcal{R}_{12}^c \Leftrightarrow [\exists v, v' \in V \ \ such \ that \ \ v(\mathbf{y}_1) > v(\mathbf{y}_2) \ \ and \ \ v'(\mathbf{y}_1) < v'(\mathbf{y}_2)]$$

This last result, which can easily be proved from definitions 2.2 and 3.1, provides the analytical representation of the preference structure considered above.

4 Linear approximation structures

Following the idea of making the concepts operative, we may now observe that, in many situations, the set of consequences Y is continuous or finite but with a large number of elements. In these cases, the preference structure concept merely appears as an abstraction. Then, a first stage to solve a problem would be a set discretization or a filtering, depending on the case, of the set of consequences which would lead to a reduced and finite set denoted by Y_o .

Different algorithms to obtain the set Y_o appear in Steuer (chapter 11, 1986). However in practice, it is not possible to go from a preference structure on Y_o to the original one on Y. So, we need an approximation concept to the preference structure to obtain conclusions about all Y from Y_o . To introduce such a concept we will use the notions of globally and locally preferred, dominated, indifferent and doubt cones, introduced in Chien, Yu and Zhang (1990) and later considered in Pachon and Rios-Insua (1991b).

Definition 4.1 *Let* $(\mathcal{R}_1, \mathcal{R}_2)$ *be a preference structure on* $Y \subseteq \mathbf{R}^n$,

1. **d** *is a globally preferred, dominated, indifferent or a global* doubt direction *for* $\mathbf{y}_0$, *if for every* $\alpha > 0$, $(\mathbf{y}_0 + \alpha\mathbf{d}, \mathbf{y}_0)$ *belongs to* $\mathcal{R}_1$, $\mathcal{R}_1^{\bullet}$, $\mathcal{R}_2$ *or* $\mathcal{R}_{12}^c$, *respectively*.

   The collection of all globally preferred, dominated, indifferent and doubt directions for $\mathbf{y}_0$ *are, respectively, called the* global preferred, dominated, indifferent *and the* global doubt cone *for* $\mathbf{y}_0$, *and will be denoted by* $P(\mathbf{y}_0)$, $D(\mathbf{y}_0)$, $I(\mathbf{y}_0)$ *and* $DD(\mathbf{y}_0)$, *according to the case.*

2. **d** *is a* locally preferred, dominated, indifferent *or a local* doubt direction *for* $\mathbf{y_0}$, *if there is an* $\alpha_0 > 0$ *($\alpha_0 \in \mathbf{R}^n$ and fixed) such that, whenever* $0 < \alpha < \alpha_0$, $(\mathbf{y_0} + \alpha\mathbf{d}, \mathbf{y_0})$ *belongs to* $\mathcal{R}_1$, $\mathcal{R}_1^s$, $\mathcal{R}_2$ *or* $\mathcal{R}_{12}^c$, *respectively.*

Analogously, we will have the locally preferred, dominated, indifferent *and the local* doubt cone *for* $\mathbf{y_0}$, *which will be denoted by* $LP(\mathbf{y_0})$, $LD(\mathbf{y_0})$, $LI(\mathbf{y_0})$ *and* $LDD(\mathbf{y_0})$, *respectively.*

Let us now see what we understand by linear approximation to a preference structure

Definition 4.2 *Let* $(\mathcal{R}_1, \mathcal{R}_2)$ *be a preference structure on* $Y \subseteq \mathbf{R}^n$. *The* lower linear approximation structure *is the quaternion of binary relations* (L_1, L_2, L_3, L_4) *on* $\mathbf{R}^n$ *defined by*

$$L_i = \bigcup_{\substack{\mathbf{y_0} \in Y \\ \mathbf{d} \in K_i(\mathbf{y_0}) \\ \alpha > 0}} \{(\mathbf{y_0} + \alpha\mathbf{d}, \mathbf{y_0})\} \quad i = 1, 2, 3, 4$$

where $K_i(\mathbf{y_0})$, $i = 1, 2, 3, 4$, *is equal to* $P(\mathbf{y_0})$, $D(\mathbf{y_0})$, $I(\mathbf{y_0})$, $DD(\mathbf{y_0})$, *respectively.*

If $K_i(\mathbf{y_0})$, $i = 1, 2, 3, 4$, is equal to $LP(\mathbf{y_0})$, $LD(\mathbf{y_0})$, $LI(\mathbf{y_0})$, $LDD(\mathbf{y_0})$, we will have the *locally linear approximation structure*, denoted by $(L_1^*, L_2^*, L_3^*, L_4^*)$. If the cones $K_i(\mathbf{y_0})$, $i = 1, 2, 3, 4$, were defined by those directions $\mathbf{d} \in \mathbf{R}^n$ such that exists an $\alpha' > 0$ verifying that $(\mathbf{y_0} + \alpha'\mathbf{d}, \mathbf{y_0})$ belongs to $\mathcal{R}_1$, $\mathcal{R}_1^s$, $\mathcal{R}_2$ or $\mathcal{R}_{12}^c$, respectively, we would obtain the *upper linear approximation structure*, denoted by (L_1', L_2', L_3', L_4').

In these definitions, we have used the term "linear". This is due to the fact that

$$(\mathbf{y_1}, \mathbf{y_2}) \in \mathcal{R} \Rightarrow (\mathbf{y_2} + \alpha(\mathbf{y_1} - \mathbf{y_2}), \mathbf{y_2}) \in \mathcal{R} \quad \forall \alpha > 0 \quad [\mathbf{P}]$$

Let us now consider a result, which has been proved in Pachon and Rios-Insua (1991b) and which characterizes the lower and upper approximation preference structures.

Theorem 4.3 *Let* $(\mathcal{R}_1, \mathcal{R}_2)$ *be a preference structure on* $\mathbf{R}^n$ *and* $(\mathcal{S}, <<)$ *the ordered set of quaternions of binary relations on* Y, *where the order is defined by*

$$(A, B, C, D) << (A', B', C', D') \Leftrightarrow A \subseteq A', B \subseteq B', C \subseteq C', D \subseteq D'$$

Let us consider the sets

$$\mathcal{C} = \{(\mathcal{S}_1, \mathcal{S}_2, \mathcal{S}_3, \mathcal{S}_4) \in \mathcal{S} : \mathcal{S}_i, \quad i = 1, 2, 3, 4 \quad satisfies \ [\mathbf{P}] \ and$$
$$(\mathcal{S}_1, \mathcal{S}_2, \mathcal{S}_3, \mathcal{S}_4) << (\mathcal{R}_1, \mathcal{R}_1^s, \mathcal{R}_2, \mathcal{R}_{12}^c)\}$$

$$\mathcal{C}' = \{(\mathcal{S}_1, \mathcal{S}_2, \mathcal{S}_3, \mathcal{S}_4) \in \mathcal{S} : \mathcal{S}_i, \quad i = 1, 2, 3, 4 \quad satisfies \ [\mathbf{P}] \ and$$
$$(\mathcal{R}_1, \mathcal{R}_1^s, \mathcal{R}_2, \mathcal{R}_{12}^c) << (\mathcal{S}_1, \mathcal{S}_2, \mathcal{S}_3, \mathcal{S}_4)\}$$

then

$$(L_1, L_2, L_3, L_4) = \max \mathcal{C} \quad and \quad (L_1', L_2', L_3', L_4') = \min \mathcal{C}'$$

5 Characterization of linear $\succ_V-$ Preference by the linear approximation structures

Let $(\mathcal{R}_1, \mathcal{R}_2)$ be a $\succ_V-$ preference on $Y \subseteq \mathbf{R}^n$. We say that $(\mathcal{R}_1, \mathcal{R}_2)$ is a linear $\succ_V-$ preference, if the elements of the family V are linear scalar functions on $\mathbf{R}^n$.

Our purpose is, given the linear approximations to $\succ_V-$ preference, to deduce from them whether we have a linear $\succ_V-$ preference or not.

Lemma 5.1 *The quaternion of binary relations associated to a linear $\succ_V-$ preference $(\mathcal{R}_1, \mathcal{R}_2)$ fulfills the linearity property* [P].

Proof. Let us first show that $\mathcal{R}_1$ fulfills property [P]. Let $(y_1, y_2) \in \mathcal{R}_1$, because $(\mathcal{R}_1, \mathcal{R}_2)$ is a $\succ_V-$preference and in view of proposition 3.2, we have that

$$v(y_1) \geq v(y_2) \; \forall v \in V \text{ and } \exists v' \text{ such that } v'(y_1) > v'(y_2) \; .$$

Because V consists of linear functions, we obtain

$$\forall v \in V \; v(y_1 - y_2) \geq 0 \text{ and } \exists v' \text{ such that } v'(y_1 - y_2) > 0 \; .$$

Given $\alpha > 0$, we have

$$v(y_2) + \alpha v(y_1 - y_2) \geq v(y_2), \; \forall v \in V \text{ and } \exists v' \text{ such that}$$

$$v'(y_2) + \alpha v'(y_1 - y_2) > v'(y_2)$$

and by the linearity of the elements in V we have

$$\forall \alpha > 0 \; v(y_2 + \alpha(y_1 - y_2)) \geq v(y_2), \; \forall v \in V$$

and $\exists v' \in V$ such that
$$v'(y_2 + \alpha(y_1 - y_2)) > v'(y_2) \; .$$

and by proposition 3.2

$$\forall \alpha > 0 \; (y_2 + \alpha(y_1 - y_2), y_2) \in \mathcal{R}_1 \; ,$$

and thus $\mathcal{R}_1$ fulfills [P].

The proof for $\mathcal{R}_1^s$ is derived from the above. Let us show the property for $\mathcal{R}_2$. Let us consider $(y_1, y_2) \in \mathcal{R}_2$. By proposition 3.2 and from the linearity in V, we will have that

$$v(y_1) = v(y_2) \; \forall v \in V \quad \text{and hence} \quad v(y_1 - y_2) = 0 \; \forall v \in \mathbf{V}$$

From this

$$v(y_2 + \alpha(y_1 - y_2)) = v(y_2) + \alpha v(y_1 - y_2) = v(y_2), \; \forall v \in V \; .$$

which is equivalent to

$$\forall \alpha > 0 \quad (\mathbf{y}_2 + \alpha(\mathbf{y}_1 - \mathbf{y}_2), \mathbf{y}_2) \in \mathcal{R}_2 \ ,$$

and $\mathcal{R}_2$ verifies [P]. Finally, because $\mathcal{R}_1$, $\mathcal{R}_1^s$ and $\mathcal{R}_2$ fulfill [P], its complement $\mathcal{R}_{12}^c$ fulfills [P], and the lemma has been proven. $\square$

The next theorem provides a characterization of the linear $\succ_V -$ preferences from their linear approximation structures

Theorem 5.2 *Let* $(\mathcal{R}_1, \mathcal{R}_2)$ *be a preference structure on* $\mathbf{R}^n$. $(\mathcal{R}_1, \mathcal{R}_2)$ *is a linear* $\succ_V -$*preference if and only if*

1. $L_1 = L_1'$ *and* $L_3 = L_3'$

2. $L_i, \ i = 1, 3$ *is compatible with addition on* $\mathbf{R}^n$ *, that is*

$$\forall \mathbf{y} \in \mathbf{R}^n \ \ if \ \ (\mathbf{y}_1, \mathbf{y}_2) \in L_i \Rightarrow (\mathbf{y}_1 + \mathbf{y}, \mathbf{y}_2 + \mathbf{y}) \in L_i, \ \ i = 1, 3$$

Proof. "$\Longrightarrow$". If $(\mathcal{R}_1, \mathcal{R}_2)$ is a linear $\succ_V -$ preference we will have, by lemma 5.1, that $(\mathcal{R}_1, \mathcal{R}_1^s, \mathcal{R}_2, \mathcal{R}_{12}^c)$ fulfills [P]. Therefore, by theorem 4.3

$$L_1 = \mathcal{R}_1 = L_1' \ \ and \ \ L_3 = \mathcal{R}_2 = L_3'$$

which proves the first affirmation. Let us show the second one. Suppose that $(\mathbf{y}_1, \mathbf{y}_2) \in L_1$. Because $L_1 \subset \mathcal{R}_1$ and $(\mathcal{R}_1, \mathcal{R}_2)$ is a $\succ_V -$ preference, we have

$$v(\mathbf{y}_1) \geq v(\mathbf{y}_2) \ \forall v \in V \ \ and \ \ \exists v' \in V \ \ such \ that \ \ v'(\mathbf{y}_1) > v'(\mathbf{y}_2)$$

Since V consist of linear functions, we have

$$v(\mathbf{y}_1 - \mathbf{y}_2) \geq 0 \ \forall v \in V \ \ and \ \ \exists v' \in V \ \ such \ that \ \ v'(\mathbf{y}_1 - \mathbf{y}_2) > 0$$

Adding and subtracting $\mathbf{y} \in \mathbf{R}^n$, we obtain that

$$v((\mathbf{y}_1 + \mathbf{y}) - (\mathbf{y}_2 + \mathbf{y})) \geq 0 \ \forall v \in V \ \ and \ \ \exists v' \in V \ \ such \ that$$
$$v'((\mathbf{y}_1 + \mathbf{y}) - (\mathbf{y}_2 + \mathbf{y})) > 0$$

Thus, from the linearity of the elements on V and proposition 3.2,

$$((\mathbf{y}_1 + \mathbf{y}), (\mathbf{y}_2 + \mathbf{y})) \in \mathcal{R}_1$$

which implies in view of 1. and theorem 4.3, that

$$((\mathbf{y}_1 + \mathbf{y}), (\mathbf{y}_2 + \mathbf{y})) \in L_1 \ \ since \ \ \mathcal{R}_1 = L_1$$

The proof for L_3 is analogous.

"⟸". By 1. and theorem 4.3, we obtain that $\mathcal{R}_1 = L_1$ and $\mathcal{R}_3 = L_3$. From 2. we will prove that the globally preferred and indifferent cones for any $\mathbf{y}_0 \in \mathbf{R}^n$ are all constants, that is

$$P(\mathbf{y}_0) = P \text{ and } I(\mathbf{y}_0) = I \ \forall \mathbf{y}_0 \in \mathbf{R}^n$$

Then, we will show that $\forall \mathbf{y}_1, \mathbf{y}_2 \in \mathbf{R}^n$ one has $P(\mathbf{y}_1) = P(\mathbf{y}_2)$. For the globally indifferent cones the reasoning is analogous.

Let $\mathbf{d} \in P(\mathbf{y}_1)$. By the definition of L_1, we have $(\mathbf{y}_1 + \alpha\mathbf{d}, \mathbf{y}_1) \in L_1 \ \forall \alpha > 0$ and by 2., we have

$$\forall \mathbf{y} \in \mathbf{R}^n \ \forall \alpha > 0 \ (\mathbf{y}_1 + \mathbf{y} + \alpha\mathbf{d}, \mathbf{y}_1 + \mathbf{y}) \in L_1$$

Again from the definition of L_1, we will have that $\mathbf{d} \in P(\mathbf{y}_1 + \mathbf{y}) \ \forall \mathbf{y} \in \mathbf{R}^n$. Thus $P(\mathbf{y}_1) \subseteq P(\mathbf{y}_1 + \mathbf{y}) \ \forall \mathbf{y} \in \mathbf{R}^n$. Taking $\mathbf{y} = \mathbf{y}_2 - \mathbf{y}_1$ and exploiting the symmetry in $\mathbf{y}_1$ and $\mathbf{y}_2$, we obtain the desired equality, since $P(\mathbf{y}_1) \subseteq P(\mathbf{y}_2)$ and $P(\mathbf{y}_2) \subseteq P(\mathbf{y}_1)$. Now, taking into account the definitions of L_1 and L_3 , and the equalities $L_1 = \mathcal{R}_1$ and $L_3 = \mathcal{R}_3$, we have that

$$(\mathbf{y}_1, \mathbf{y}_2) \in \mathcal{R}_1 \ \Leftrightarrow \ \mathbf{y}_1 - \mathbf{y}_2 \in P$$
$$(\mathbf{y}_1, \mathbf{y}_2) \in \mathcal{R}_2 \ \Leftrightarrow \ \mathbf{y}_1 - \mathbf{y}_2 \in I$$

Both cones P and I are disjoint, since, if $\mathbf{y}_1 \in P \cap I$, then $\mathbf{y}_1 \in P$, which means that $(\mathbf{y}_1, 0) \in \mathcal{R}_1$. Furthermore, $\mathbf{y}_1 \in I$ and then $(\mathbf{y}_1, 0) \in \mathcal{R}_2$. But this is a contradiction because $(\mathcal{R}_1, \mathcal{R}_2)$ is a preference structure and $\mathcal{R}_1 \cap \mathcal{R}_2 = \emptyset$. For this reason, if we denote $K = P \cup I$, we have

$$(\mathbf{y}_1, \mathbf{y}_2) \in \mathcal{R}_1 \ \Leftrightarrow \ \mathbf{y}_1 - \mathbf{y}_2 \in K \ \text{and} \ \mathbf{y}_2 - \mathbf{y}_1 \notin K$$
$$(\mathbf{y}_1, \mathbf{y}_2) \in \mathcal{R}_2 \ \Leftrightarrow \ \mathbf{y}_1 - \mathbf{y}_2 \in K \ \text{and} \ \mathbf{y}_2 - \mathbf{y}_1 \in K$$

Let us consider the polar set of K, that is

$$K^* = \{\mathbf{x} \in \mathbf{R}^n \ / \ \mathbf{x} \cdot \mathbf{k} \geq 0 \ \forall \mathbf{k} \in K\}$$

This set consists of linear functions on $\mathbf{R}^n$ with positive values on all the elements of K, that is

$$K^* = \{l : \mathbf{R}^n \longrightarrow \mathbf{R} \ /l(\mathbf{k}) \geq 0 \ \forall \mathbf{k} \in K\}$$

Furthermore, since it is on $\mathbf{R}^n$, we have that $(K^*)^* = K$. This leads us to write the preference structure as follows

$$(\mathbf{y}_1, \mathbf{y}_2) \in \mathcal{R}_1 \Leftrightarrow l(\mathbf{y}_1) \geq l(\mathbf{y}_2) \ \forall l \in K^* \ \text{and} \ \exists l' \in K^* \ \text{such that}$$

$$l'(\mathbf{y}_1) > l'(\mathbf{y}_2)$$

$$(\mathbf{y}_1, \mathbf{y}_2) \in \mathcal{R}_2 \Leftrightarrow l(\mathbf{y}_1) = l(\mathbf{y}_2) \ \forall l \in K^*$$

Then $(\mathcal{R}_1, \mathcal{R}_2)$ is a linear $\succ_V-$ preference where $V = K^*$. $\square$

6 Conclusions

The linear approximation structures to a preference structure are easy concepts to obtain in practice, from the interaction between the analyst and the DM. This ease is complemented by the usefulness of this tool to describe preference structures represented by a family of functions. In this paper we have only considered the representation by means of families of linear functions. However, an open problem in this context, would be to describe, from these approximations, families of functions under more complex analytical conditions.

Acknowledgements

This work has been supported by the Direccion General de Investigacion Cientifica y Tecnica (DGICYT) of the Ministry of Education under Grant PS89-0027.

References

Chien, I.S., P.L. Yu and D. Zhang (1990). Indefinite Preference Structures and Decision Analysis. *J. Optim. Theory Appl.* **64**, 71-86.

Pachon, J. G. (1990). Aproximaciones de Estructuras de Preferencia en Problemas de Decision Multiobjetivo. Ph. D. Thesis, Universidad Politecnica de Madrid.

Pachon, J.G and S. Rios-Insua (1991a). Preference Structures: Modelling the Attitudes in Multiobjective Decision Making. (submitted to *Operations Research Letters*).

Pachon, J.G and S. Rios-Insua (1991b). Linear Approximation Structures to Preference Structures. Working Paper 1/1991, Dept. of A.I., Politechnical University of Madrid.

Steuer, R. (1986). Multiple Criteria Optimization: Theory, Computation and Application. Wiley, New York.

Yu, P.L. (1985). Multiple Criteria Decision Making. Plenum Press, New York.

Smooth Relations in Multiple Criteria Programming

V. I. Borzenko, M. V. Polyashuk

Institute of Control Sciences

Moscow, Russia

Abstract

This paper addresses some methodological aspects of multiple criteria (MC) programming. Special attention is paid to smooth (binary) relations as a model of decision maker's (DM) preference structure (PS). The theoretical results are presented that allow correct utilization of mighty means of mathematical programming in much wider a context than that of smooth value function.

Also the classification of MC-problems from the viewpoint of PS modelling is proposed. In the framework of this classification a number of approximational MC-methods for solving problems with "smooth-relation" preferences are discussed.

1 Introduction

In this paper we discuss some aspects of multiple criteria decision making (MCDM), i.e. decision making which considers several criteria that estimate the decisions quality. To put it more accurately, the paper deals with multiple criteria (MC) programming, i.e. theory and methods of solving MCDM-problems with the help of computer.

Computer-aided decision making methods are used on condition that decision maker (DM) is not able to reliably choose by himself the best solution(s) from a given presentation. Such a situation (that is quite typical for the decision making practice) may stem from the size of presentation which may be infinite (e.g. a domain in some decisions space) but they also may be traced to the complexity of DM's preference structure (PS), where several conflicting criteria characterize the quality of solutions. In all such cases some formal choice mechanism (e.g. value function, binary relation, choice function) is to be used as a model of the real DM's preferences for supporting the decision process. Obviously, the adequacy of modelling real DM's preferences is one of the most important requirements to MC programming, for the accuracy of modelling to a great extent determines the quality of the resultant decision; on the other hand, mistakes that are made while choosing MC-model (and MC-method) sometimes cannot be corrected afterwards.

Analysis of various approaches to solving MC-problems (Berezovskiy et al. (1987)) has led us to the following conception of PS modelling: the formal object associated with the notion of PS is a choice function in the criteria space R^n, i.e. triple (A, P, C) where $A \subseteq R^n$ is referred to as a universal set, $P \subseteq 2^A$ is a family of presentations,

$C : P \to 2^A : X \mapsto C(X) \subseteq X$ is a choice operator (this operator C is often referred to as choice function, as in the case of mappings). To solve a given MC-problem means to extrapolate operator C from a certain set of simple presentations to all presentations required. This is done by a model of DM's preferences which is a parametric family of choice functions $\{C_\alpha\}$ defined on the same universal set A.

The following principles of PS modelling can be formulated:

- all the assumptions on general properties of DM's preferences must be explicitly stated and properly justified;

- in the framework of these assumptions the model must allow approximating DM's PS with any required accuracy ($\{C_\alpha\} \to C$);

- any question to DM may be stated in terms of his preferences only, and never in terms of a model;

- DM may answer "I don't know" whenever he really does not.

The above conception and principles of "approximational approach to solving MC-problems" (Borzenko, 1989) allows the use of real-space topology and metrics for formalizing the notion of approximation accuracy as far as PS modelling is concerned. These principles, natural and simple as they are, are not taken into consideration in quite a few of MCDM-methods. For example, a number of publications (Berezovskiy et al., 1981; Wierzbicki, 1979) soundly criticize the assumption that the DM's preferences correspond to a smooth value function for this approach hardly accounts of the existence of incomparable (for DM) solutions. Moreover, many of MC-methods consist of some more or less arbitrary combinations of certain heuristic steps and the interpretation of DM's messages in terms of the chosen model is often unsound. The above facts have motivated constructing and investigating new MC-models.

2 Smooth relation as a model of decision makers's preferences

Many publications, as well as the existing practice of MCDM, confirm that in quite a few of situations it may be assumed that DM's preferences correspond to a binary relation defined in the criteria space. Actually, very often the notions of "optimal" and "best" solutions coincide, especially if only one solution is required to be chosen. We believe that one of the binary MC-models that enable us to adequately use the information on the DM's preferences is the model of smooth relation.

DEFINITION. Consider a relation $\mathcal{R} \subseteq A \times A$, $A \subseteq R^n$. Relation $\mathcal{R}$ is referred to as smooth if its boundary (frontier) in $A \times A$ is a smooth surface (submanifold).

Simply saying, the above definition means that smooth relation as a subset in $R^n \times R^n$ is defined by a smooth constraint. Unlike the smooth value function case which requires $\mathcal{R}$ to be a full order and to be continuous, i.e. an open set in $A \times A$ (see the classical result of Debreu), here we strengthen the second requirement but we do not require that every two alternatives can be compared.

Let us now consider the most important theoretical results concerning geometrical characteristics of smooth relations.

THEOREM. If $\mathcal{R}$ is a smooth relation and $\mathcal{R} \supseteq Par^n$ - Pareto relation, then its (upper) section $\mathcal{R}_x = \{y \in A : x\mathcal{R}y\}$ is also bounded by a smooth surface (in A) for every point $x \in A$. Tangent (hyper)planes at point x for the upper and lower sections of $\mathcal{R}$ coincide.

DEFINITION. The tangent plane of the section of relation $\mathcal{R}$ at point x ($T_x\mathcal{R}_x$) is referred to as indifference plane of $\mathcal{R}$ at x: $I_\mathcal{R}(x)$.

Now such well-known notions as indifference (hyper)plane, substitution coefficients, etc., that have been considered before only in the context of smooth utility/value functions, can be extended to the smooth function case. However, unlike the utility-function case, the assumption of smooth relation fully recognizes incomparable points and seems in general much more realistic. Nevertheless, it has been demonstrated (Borzenko, 1989) that for a smooth relation $\mathcal{R}$ the integral of its indifference planes field $I_\mathcal{R}$ exists and is isomorphic (isomorphism of relations) for transitive closure $Tr\mathcal{R}$.

THEOREM. Let $\mathcal{R}$ be smooth. Then $I_\mathcal{R}$ is integrable if $\mathcal{R}$ is acyclic, the integral being an ordinal function (denote it as $f_\mathcal{R}$) homomorphic for $\mathcal{R}$ and isomorphic for $Tr\mathcal{R}$:

$$x\mathcal{R}y \Longrightarrow f_\mathcal{R}(x) < f_\mathcal{R}(y);$$

$$x(Tr\mathcal{R})y \Leftrightarrow f_\mathcal{R}(x) < f_\mathcal{R}(y).$$

The two following theorems justify the use of binary relations in many situations when the choice differs from just obtaining maximum with respect to some relation; these results also allow to find out whether the presentation contraction procedure can be correctly used.

THEOREM. Let a choice function C satisfy H-condition (Izerman et al. , 1982): for any presentations X, X' $X' \subseteq X \longrightarrow \bar{C}(X') \subseteq \bar{C}(X)$, and M-condition for a given presentation X_0: for any presentation X $C(X_0) \subseteq X \subseteq X_0 \longrightarrow |C(X)| \le |C(X_0)|$. Then for any $C' \supseteq C$ and any family of presentations $P' \subseteq P \subseteq 2^{X_0}$ all the C'-rejected points may be ignored:

$$C(X_0 \setminus \bigcup_{Y \in P'} \bar{C}'(Y)) = C(X_0)$$

THEOREM. Let $\mathcal{R}_i \uparrow \mathcal{R}$. Then $Max_{\mathcal{R}_i} \downarrow Max_\mathcal{R}$. (Here the topology is pointwise, the condition $\mathcal{R}_i \subseteq \mathcal{R}$ is essential for convergence.)

The above results substantially expand the application area of the powerful tools of mathematical programming, so far restricted in this context to the case of somewhat dubious value-function model.

3   Classification of multiple criteria problems

Besides a general approach to solving MC-problems (like the approximational approach), on the one hand, and theoretical results allowing building MC-methods, on the other, in response to actual needs of the MCDM developments it is necessary to have a constructive tool for choosing an adequate model and method for a given MC-problem. This can be done by means of classifying the models of MCDM from the viewpoint of adequate PS modelling.

In this paper we present new classification of MC-problems (Polyashuk, 1990) which is built from the viewpoint of PS modelling and is guided by the following principles:
- the classification is intended for the DM-computer dialog determining the type of the problem and proposing some adequate methods according to its dimension and the resources allocated for its solution (DM's time, computer time, memory size, etc.) ;
- two MC-problems are attributed to different classes if their solution require different methods;
- the notion of an adequate method is understood in the approximational sense;
- no separate classification is proposed for MC-methods, but for each problem type some approximational methods are proposed;
- the classification is open in that new parameters of classification may be added, and new values of the existing parameters may be introduced.

Before listing the parameters of our classification, recall the main components of each MC-problem (in the deterministic case):
- decisions space R^m, the space of the parameters values of the alternatives (solutions);
- universal set $A' \subseteq R^m$;
- one or more presentations $X \in P \in 2^A$;
- criteria space R^n, the space of quality characteristics of the solutions;
- criteria mapping $F : A' \to R^n$, attributing criteria values to all the feasible solutions.
- DM's preference structure defined in the criteria space;
- a priori given characteristics of the problem.

The further detalization of these main elements of the MC problem structure gives us the following list of classification parameters (the parameters values are omitted):

1. Parameters characterizing MC-problem in general (general parameters).
1.1. Possibility of a dialog with DM after fixing the presentation.
1.2. Existence of a feedback.
1.3. A number of DMs with conflicting vector criteria.
2. Parameters characterizing the decisions space R^m (decisions space parameters, DSP).
2.1. DSP-type.
2.2. Cardinality of presentations.
2.3. Regularity of presentations.
2.4. Presentations mode.
2.5. Feasibility type.
3. Parameters characterizing criteria space R^n and criteria mapping.
3.1. Way to determine criteria mapping.
3.2. Criteria type.
3.3. Existence of criteria constraints.
3.4. Localization of optima.
4. Parameters characterizing the DM's preference structure.
4.1. Type of model which adequately describes PS.
4.2. PS invariance characteristics.
4.3. Type of comparative criteria importance.

4   Approximational methods for smooth–relation problems

Now, after we have at our disposal the above classification of MC-problem, it becomes clear that the main restrictions for using the smooth-relation model are related to the values of parameter 2.1 (DSP-type) and parameter 4.1 (the type of a model). It is necessary that all the criteria are (pseudo-)continuous and DM's preferences correspond to a binary relation. These assumptions seem quite natural and hold true in a vast majority of actual situations. In any case, this context is much wider than that of continuous value function.

Using the general principles of approximational approach to preferences modelling, along with the above theoretical results, we (together with our colleagues) succeeded to construct dialog procedures for solving following smooth-relation problems:

1. The problem of unique choice with a "quick" (not time-consuming) criteria mapping.

2. The problem of unique choice with a "slow" (time-consuming) criteria mapping.

3. The problem of unique choice from a finite presentation.

4. The problem of unique choice from a small presentation.

5. The problem of "automatic" choice.

6. The problem of compromise choice.

7. The problem of choice from a sequential presentation.

Almost all of the enumerated smooth-relation methods contain the following standard modules:

- algorithm for determining substitution coefficients (their upper approximation) in a point x in the criteria space R^n;

- algorithm for building the indifference hyperplane in $x \in R^n$;

- algorithm for determining the best direction in the decisions space R^m.

Some of these methods are, as a matter of fact, approximational versions of well-known methods (Frank et al., 1956; Geoffrion, 1970), while the others are quite original. Now we shall describe one of the original methods that were constructed in the context of smooth relation. This method (more exactly, two modifications of one algorithm) is intended for solving the problem of choice from a small presentation by means of the two points comparison in the criteria space. Simple as it is, such an idea has not been considered, as far as we know, even in the value-function context. This surprising fact may be attributable to what was mentioned in the foreword: the value-function model does not allow for incomparability, thus making the very setting of the comparison problem unnatural.

The main idea of the method consists in drawing a polygonal approximation of integral curves (for indifference hyperplane field) lying in a certain two-dimensional space.

Modification 1.

Step 1. In one of the two points $x \in R^n$ the approximate substitution coefficients $\tilde{\mu}_{ij}(x)$ are determined; the corresponding indifference hyperplane is constructed. Set $l = 1, x^0 = x$.

Step 2. In the other point of the pair $y \in R^n$ the indifference hyperplane and the normal to it n_y are constructed.

Step 3. The two-dimensional plane α^2 containing x^{l-1}, y and n_y is constructed.

Step 4. The intersection of α^2 and the indifference hyperplane in x^{l-1} defines the direction p^l; move to the point $x^l = x^{l-1} + \Delta p^l$, where Δ denotes the step-length.

Step 5. If our trajectory has not yet crossed n_y, then the indifference hyperplane in x^l is constructed. Set $l = l + 1$ and go to Step 3.

Step 6. If the trajectory intersects the positive ray of n_y, it is interpreted as a comparison in favour of the point x $(y\tilde{\mathcal{R}}x)$; otherwise - as a comparison in favour of y $(x\tilde{\mathcal{R}}y)$.

Modification 2.

Steps 1-3 coincide with those of the Modification 1.

Step 4. The convex closure L^l of the points x^l, $\{y : x^l\tilde{\mathcal{R}}y\}$ is constructed, where $\tilde{\mathcal{R}}$ is the (inner) approximation of the transitive closure of the real DM's preference relation $\mathcal{R}$: $\tilde{\mathcal{R}} \subseteq Tr\mathcal{R}$.

Step 5. Within the intersection of L^l and α^2 the ray p^l is constructed which provides the maximum angle to the positive ray of the normal n_y. Move to the point $x^l = x^{l-1}+\Delta p^l$.

Step 6. If our trajectory has not yet crossed n_y, then the indifference hyperplane in x^l is constructed. Set $l = l + 1$ and go to Step 3.

Step 7. If the trajectory intersects the negative ray of n_y, then $x\tilde{\mathcal{R}}y$ and, therefore, $x(Tr\mathcal{R})y$.

In the second modification we can guarantee that $x\tilde{\mathcal{R}}y \Longrightarrow x\mathcal{R}y$, but it may happen that x and y are $\tilde{\mathcal{R}}$-incomparable. On the contrary, in the first version there are no $\tilde{\mathcal{R}}$-incomparable points, but errors are possible.

THEOREM. Let $\mathcal{R}$ be a smooth acyclic relation, $\tilde{\mu}_{ij} \to \mu_{ij}$, $\Delta \to 0$, where $\mu_{ij}, i, j = 1, \ldots, n$, — substitution coefficients, Δ — step-length. Then the probability of errors (incomparability) tends to 0.

Thus, both versions of the described method allow DM after due efforts to perform comparison with any required accuracy. In other words, it means that the method is approximational. It has been also shown that the trajectories which are constucted in both modifications of the method possess all the necessary properties (they reach n_y in a limited number of steps, do not degenerate, etc.).

Besides approximational methods, the special dialog procedure was developed that enables DM to choose by himself an adequate MC-method for solving a real-life MC-problem. This procedure was built as a universal one, though it has been practically used in the smooth-relation case only.

It is clear that the whole manifold of MC-problems can not be covered by the description of some particular cases. Ideally, for each problem type, i.e. for each possible combination of the values of classification parameters, there must be a number of approximational methods for DM to choose from. But, in any case, the context of smooth binary relation is much wider than that of smooth value function.

References

Berezovskiy, B. A., Borzenko, V. I. and Kempner, L. M. (1981). Binary relations in multicriteria optimization Nauka Pbl., Moscow (in Russian).

Berezovskiy, B. A., Borzenko, V. I. and Polyashuk, M. V. (1987). Modelling DM's preference structure (models and methods of multicriteria optimization), Informatsyonnyie matierialy: Kibernetika, No.6, (in Russian).

Borzenko, V. I. (1989). Approximational Approach to Multicriteria Problems, Lecture Notes in Economics and Mathematical Systems, Springer-Verlag, vol.337.

Frank, M. and Wolf, P. (1956). An algorithm for quadratic programming. Nav. Res. Logist. Quat., vol. 3., No. 1,2.

Geoffrion A. M. (1979). Vector maximal decomposition programming. In: 7th Mathematical Programming Symp., The Hague.

Iserman, M. A. and Malishevskiy, A. V. (1982). General choice theory: some aspects. Avtomatika i Telemekhanika, No. 2.

Polyashuk, M. V. (1990). An interactive procedure for choosing adequate methods for multicriteria problems. Ph.D. Thesis, Moscow, Pbl. of the Institute of Control Sciences (in Russian).

Wierzbicki A. P. (1979). The use of reference levels in group assessment of solutions of multiobjective optimization. IIASA Working Paper, Austria: Laxenburg.

Pairwise Comparisons in Decision Support for Multi–Criteria Choice Problems

Janusz Majchrzak

Systems Research Institute

Polish Academy of Sciences, Poland

Abstract

Two threads: reduction of the number of alternatives and generation of the proposal for the DM's most preferred alternative are distinguished in the process of decision support. Presented approach is based on pairwise comparisons of alternatives.

1 Introduction

In the last few years a rapid evolution have been observed in the field of techniques and concepts of decision support systems for multicriteria choice problems. The usefulness of several former approaches has been contested and new directions of development have been proposed. New approaches emphasize the aspects of aiding, tutoring and flexibility rather than computational efficiency.

The formalism of multicriteria problems allows for a reduction of the set of feasible alternatives to the set of nondominated alternatives. In order to reduce further the set of alternatives under considerations one has to supply to the problem description some additional information about the decision maker (DM). In general, this knowledge comes from both of the following two sources:

- assumptions about the DM and his preferences,

- information supplied by the DM.

Explicit or implicit assumptions are usually extremely hard to verify in practice. Methods based on strong assumptions produce results which the DM quite often refuse to accept. The level of acceptance depends also on the forms of preferences exhibition and their accordance with the notions used by the DM in his intuitive analysis of the problem. The supply of additional information about his preferences may be a difficult and heavy task for the DM in some of the approaches.

The preferences may be exhibited by the DM in different forms. Below, some of the most common examples are listed.

- Intuitive decisions (usually with graphical displays of problems with 2 - 4 criteria and/or 3 - 10 alternatives).

- Explicit preference (utility) function.

- Criteria ordering (lexicographic).

- Weights of criteria.

- Aspiration and/or reservation levels of criteria values.

- Alternatives ranking.

- Pairwise comparisons of alternatives.

The presented approach is based on pairwise comparisons and we will focus on this subject but in practical systems several options have to be offered to the DM.

2   Motivations

Let us assume that a multicriteria choice problem is given, i.e. for all feasible alternatives criteria values have been evaluated and listed in a file.

Several simple tools can be used by the DM in the early stage of the problem analysis. He may reject alternatives which are clearly out of the region of his interest by setting bounds on criteria values, turn some of the criteria into constraints, ignore some of the criteria introducing a (group) lexicographic approach, obtain a representation (subset) of the nondominated set by introducing criteria values tolerances and thus ignoring small differences in criteria values, etc. All this actions lead to a reduction of the number of alternatives still under consideration.

Using some other tools the DM may ask the supporting system for a proposal of his most preferred alternative. A good example of a relevant technique is the reference point, goal, compromise solution or trade-off approach (see Majchrzak, J., 1987, for an example).

If the DM was not able to accept any of the alternatives proposed by any of the mentioned techniques he has to be asked to supply some more specific information about his preferences. Pairwise comparisons of alternatives is a technique frequently used here (see Koksalan, M., et al., 1984; Malakooti, B., 1988).

Before the presentation of our approach to the pairwise comparisons utilization some general assumptions will be discussed.

- The DM cannot be forced to compare any pair of alternatives selected by the system.

- The number of comparisons supplied by the DM may be very small.

- The DM selects by himself the alternatives for comparisons.

- Also nonfeasible alternatives can be compared.

Such a situation may occur if the DM is willing to compare only those few alternatives he knows from his practice and some of them may by currently not available.

Let us consider the following multicriteria decision making problem. A decision maker (a person or an institution) wants to buy a new car and has some difficulties in choosing from the variety of models available on the market. He is not an expert in cars and he knows just a few models: his old car and those possessed by his friends and relatives. So, all he is able to say about his preferences is a number of statements concerning cars he knows, like for example:

VW Golf is preferred to Opel Kadett,

Fiat Uno is preferred to Peugeot 205, etc.

He refuses to compare cars he doesn't know or to supply any other kind of information about them. The reference point approach might be adopted in this case, but what if the DM would not be satisfied with the result?

The task can be formulated as follows. A relatively small number of pairwise comparisons of alternatives is available. What can be said about the DM's preferences on the basis of this small amount of information and what can be said about the quality of that information ? Note that a statement: "a cheap good car is preferred to an expensive bad car" is a rather low quality information since, once price and performance have been established as criteria, this is an obvious statement. The DM should be informed about the quality of the alternatives evaluation he had made. Also his inconsistencies should be discovered.

Because the number of supplied comparisons may be small, we have to expect a small amount of information about DM's preferences. The method will not tend to determine the most preferred solution but to reduce the set of alternatives. It may be shown that in some instances even a single pairwise comparison may allow for a significant reduction, in some problems even up to a single alternative!

The information about the DM's preferences contained in a pairwise comparison may be of two different types. Let us assume that price and quality are two criteria for a selection of a car.

compromise type : a better but more expensive car is preferred to a worse but cheaper car.

non-compromise type : a good cheap car is preferred to an expensive bad car.

It will be shown that also the information of the second type may be utilized in our approach.

In the presented below method we will distinguish two threads (goals) of the decision support process: reduction of the number of alternatives under consideration and generation of a proposal (candidate) for the DM's most preferred alternative.

3 Basic ideas

Let F be the space of m criteria, $\Lambda = R^m_+$ be the domination cone and let $Q \subset F$ be the set of feasible alternatives. We will assume that there exist an underlying implicit quasiconvex utility function $U : \longmapsto R$ behind the DM's preferences. The DM need not recognize it existence; however, we will assume that whenever he decides that alternative $b \in Q$ is preferred to alternative $a \in Q$, it is equivalent to $U(b) > U(a)$.

The DM's utility function U is in general a nonlinear function of criteria. Identifying such function usually requires large amount of data and a significant computational effort. Therefore, keeping nonlinearity of U in mind, we shall restrict ourselves to a set of linear approximations of U only.

Suppose that k pairs of alternatives were compared by the DM:

$$b_i \text{ is preferred to } a_i , \ a_i, b_i \in Q , \ i = 1, ..., k$$

This set of data may be considered as a set W of k vectors in the criteria space F, pointing from a less preferred alternative a_i to a more preferred alternative b_i.

$$W = \{ w_i : w_i = [a_i, b_i] , \ a_i, b_i \in Q , \ i = 1, ..., k \}$$

Let us also consider the set V of normalized vectors $w_i \in W$:

$$V = \{ v_i : v_i = \frac{w_i}{\|w_i\|} \ i = 1, ..., k \}$$

Each of the vectors v_i represents a direction of improvement in the space of criteria of the function $U(f)$. Hence, the cone C spanned by vectors $v_i \in V$ is the cone of improvement for $U(f)$ and can be defined as:

$$C = \{ \sum_{i=1}^{i=k} \alpha_i v_i : \alpha_i \in R_+ , \ v_i \in V \}$$

The cone $C^\star$ is the corresponding polar cone and can be defined as:

$$C^\star = \{ y_i \in F : y, v_i \geq 0 , \ v_i \in V, \ i = 1, 2, ..., k \}$$

Both cones C and $C^\star$ can be expressed by their generators. The set of cone generators is the minimal subset of vectors belonging to that cone that still span the cone. The generators of cones C and $C^\star$ will be denoted by c and $c^\star$, respectively.

$$C = \{ \sum_j \alpha_j c_j , \ \alpha_j \in R_+ , \ c_j \in \tilde{C} \}$$

$$C^\star = \{ \sum_j \alpha_j c_j^\star , \ \alpha_j \in R_+ , \ c_j^\star \in \tilde{C}^\star \}$$

where $\tilde{C}$ and $\tilde{C}^\star$ are corresponding generator sets.

Let us return to the pairwise comparisons. Since we shall consider linear approximations of the utility function, for the sake of presentation simplicity, assume that U is linear. If the DM has decided that alternative $b \in Q$ is preferred to alternative $a \in Q$, then $U(b) > U(a)$. It is clear that $< v, u > > 0$, where $v = [a, b]$, and u is a vector normal to hyperplanes $U(f) = const$. Hence, the vector u is contained in cone $C^\star$.

From the above analysis it follows that an accurate determination of the vector u normal to the hyperplanes of U will be possible only in the case when the cone $C^\star$ is spanned by a single vector (namely u). In this case the DM's utility function (or rather actually its linear approximation only) has been obtained and we can easily calculate the DM's most preferred solution by minimizing U over the set Q.

In general, because of obvious reasons, the cone C will be smaller than a halfspace and its polar cone $C^\star$ will have a nonempty (relative) interior. In such a case, each of the vectors contained in $C^\star$ may appear to be the vector u. Fortunately, we can restrict ourselves to the generators $c^\star$ of the cone $C^\star$ only. Considering each $c_j^\star$ to be the vector u (minimizing linear function based on $c_j^\star$) one can obtain a set of $q_j \in Q$ being the linear approximation minimizers of DM's utility function. These elements q_j define a subset $S \subset Q$ of nondominated elements of Q in which the DM's most preferred alternative (minimizer of U) is contained.

As it can be seen now, our approach does not pretend to determine the DM's most preferred solution exactly. It will rather tend to find a domain in which it is contained. The more information about DM's preferences is contained in alternatives pairwise comparisons supplied, the smaller this domain will be. Besides, also a good candidate for the most preferred solution may be presented to the DM. It can be obtained a kind of average vector for the cone $C^\star$: a sum of $c_j^\star$, a sum of v_i, a gravity center of v_i, etc. The author's favorite method for the candidate selection is the calculation of the minimal (Euclidean) norm element from the convex hull spanned by the cone $C^\star$ generators $c_j^\star$. This technique based on the method of Wolfe (Wolf, P., 1975) appeared to be very useful in our approach, serving also for some other purposes. Let us denote the minimal norm element from the convex hull spanned by the set V of vectors v as

$$z = MNECH(C^\star)$$

The minimizer of the linear function based on vector z will be chosen as the candidate for the DM's most preferred solution.

4 Details

In this chapter, we shall discuss the basic cases that can occur for different sets of pairwise comparisons of alternatives supplied by the DM.

Case 1. Cone C is a halfspace of F and $\|z\| = 0$.

As it has been already mentioned, in this case the linear approximation of the DM's utility function is defined by the vector u normal to the halfspace spanned by C. The DM's most preferred solution may be found by the optimization of the linear function based on u.

Case 2. Cone C is not a halfspace of F and $\|z\| = 0$.

Since the DM's utility function is assumed to be quasiconvex, the set V of pairwise comparisons supplied by the DM is inconsistent. Conflicting elements should be selected from the set V and presented to the DM. They are those elements which spann a convex hull containing zero and hence cause $\|z\| = 0$. Their selection is automatic during the calculation of the element z.

Case 3. Cone C is contained in a halfspace of F, it contains the domination cone Λ and $\|z\| > 0$.

This is the basic case. After the set of linear functions based on vectors $c^\star$ optimization, a subset of nondominated elements of set Q will be obtained. This subset is defined by the set of linear approximation optimizers of the utility function. A candidate for the DM's most preferred solution will be found by optimizing over the set Q the linear function based on vector z. Notice that if the number of supplied pairwise comparisons is small (too small to spann a non-degenerate cone C), then generators of the domination cone Λ can be added to the set V.

Case 4. Cone C is contained in the domination cone Λ and $\|z\| \geq 0$.

This is the case of a low quality of information contained in pairwise comparisons of alternatives supplied by the DM (and corresponds to statements like: "a good cheap car is preferred to an expensive bad car"). The DM should be informed about this fact and perhaps he will be able to give some more restrictive statements. If he refuses for some reasons, we cannot proceed along the Case 3 line. However, instead of of considering the supplied information as being of a discriminative type we can treat it as an instructive type information. Each of the vectors $v \in V$ can be treated now as an approximation of the DM's improvement direction or his utility gradient approximation. Hence, we can proceed just like in Case 3, taking the cone C instead of C^* into consideration. Of course the DM should be aware of the new interpretation of the information he has supplied.

Cases 3 and 4 can be distinguished a priori by checking whether $C \supset \Lambda$ or $C \subset \Lambda$, respectively.

5   Concluding remarks

A method for the utilization of pairwise comparisons of alternatives has been presented within a framework of decision support in multicriteria choice problems. However, it can be used for the DM's preferences assessment in multicriteria problems of any type (linear, nonlinear).

An arbitrary small sample of pairwise comparisons is accepted. The DM is not forced to compare alternatives selected by the systems, he may compare both feasible and non-feasible alternatives of his choice. In the case of inconsistencies conflicting elements are separated and presented to the DM for conflict resolution. Pairwise comparisons may contain both compromise and non-compromise type of information about the DM's preferences.

The DM may be quite sure that a is preferred to b, but he may have some doubts whether c is preferred to d. A natural way of dealing with uncertainty of this kind is to group comparisons with different levels of certainty or confidence and analyze them separately and combine results.

If the DM is able to supply a large amount of results on alternatives evaluations, then a technique similar to one presented in (Koksalan, K., et al., 1984), should be used in order to eliminate dominated alternatives from further considerations. If it is not the case, the DISCRET package methodology should be applied. Actually, the presented approach is planned to be included into the DISCRET framework. At the moment it is on the stage of an independent experimental program.

Several practical aspects of the presented approach are still to be further investigated. Probably the most interesting question is how to provide some help for the DM if he is not able to choose pairs for comparisons, i.e. how to select a small sample of such alternatives that their evaluation by the DM may result in significant improvement of an approximation of DM's preferences.

References

Wolfe, P. (1975). Finding the Nearest Point in a Polytope. *Mathematical Programming*, Vol. 11, pp. 128-149.

Koksalan, M., Karwan, M. H. and S. Zionts (1985). An Improved Method for Solving Multiple Criteria Problems Involving Discrete Alternatives. *IEEE Transactions on Systems, Man and Cybernetics*, Vol. SMC-14, No. 1, pp. 24-34.

Majchrzak, J. (1987). Methodological Guide to the Decision Support System "DIS-CRET" for Discrete Alternatives Problems. In: Lewandowski, A., Wierzbicki, A. P. (eds) Theory, Software and Testing Examples for Decision Support Systems, WP-87-26, International Institute for Applied Systems Analysis, Laxenburg, Austria.

Rouse W. B. (1987). Model-based evaluation of an integrated support system concept. *Large Scale Systems*, Vol. 13, pp. 33-42.

Malakooti, B. (1988). A Decision Support System and a Heuristic Interactive Approach for Solving Discrete Multiple Criteria Problems. *IEEE Trans. Syst. Man. Cybernet.*, Vol. 18, pp. 273-285.

Interval Value Tradeoffs Methodology and Techniques of Multicriteria Decision Analysis

Vadim P. Berman, Gennady Ye. Naumow,
Vladislav V. Podinovskii
National Software Institute
Rostov–on–Don, Russia

1 Introduction

We consider here a problem of choice of the best alternative from a given set of alternatives which are characterized by multiple criteria. We can assume that the first step in such choice is a selection of the set of Pareto–optimal alternatives.

This set has one remarkable feature. If one of the two Pareto–optimal alternatives is better than the other by one criterion it is worse inevitably by some other criterion. Hence if we want to choose only one alternative it is necessary to point out what failing off by one criterion we are ready to donate for the improvement by the other one. This is obtained by the analysis of tradeoffs.

In the classical theory the tradeoffs must be pinpointed in the form of a rate of substitution (see, for example, Keeney and Raiffa, 1976). But it is very difficult for two reasons, which are:

- it is difficult for a decision maker (DM) to determine the exact value of the rate of substitution,

- the exact rate of substitution value varies when the criteria value changes.

These difficult can be surmounted by developing a new theory. It is just enough to point out to point out the interval containing the rate of substitution value of one criterion for another. This interval may be broad enough. The lower bound of the interval is the maximal value of increment by one criterion such that any smaller number is certainly not acceptable for you in the exchange for decrement of another criterion by one unit. Similarly the upper bound of this interval is the minimal value such that any greater number is certainly acceptable.

2 Interval approximation of DM's preferences

Let S be a given set of available alternatives (or strategies), and $f = (f_1, ..., f_n)$ be a vector criterion on S. Each alternative s is represented by its vector estimate $f(s)$ in the criteria space $\Re^n$. The criteria f_i will be denoted by their numbers i, $i = 1, ..., n$. For definiteness suppose that all criteria are positively oriented, i.e. their larger values are more preferable than the smaller ones. Negatively oriented criterion smaller values of which are more preferable than larger ones can be formally transformed easily into the positively oriented one, for example, by changing the sign.

The interval approach to the solution of a multicriterial problem presupposes a specification of interval estimations of substitution for some (not obligatory for all) pairs of criteria. To receive such estimation for a pair of criteria $< i, j >$ the DM has to compare the following pairs of vector estimates:

$$x = (x_1, ..., x_i, ..., x_j, ..., x_n),$$

$$y = (x_1, ..., x_i - 1, ..., x_j + \Delta, ..., x_n)$$

where all components $x_1, ..., x_n$ are fixed and $\Delta > 0$ is variable. As a result the positive numbers Δ_{ij}^- and Δ_{ij}^+, $\Delta_{ij}^- < \Delta_{ij}^+$ should be found, such that

a) if $\Delta \leq \Delta_{ij}^-$ then x is more preferable than y,

b) if $\Delta \geq \Delta_{ij}^+$ then y is more preferable than x,

c) if $\Delta_{ij}^- < \Delta < \Delta_{ij}^+$ then x and y are not comparable.

Then the DM increases the interval $(\Delta_{ij}^-, \Delta_{ij}^+)$ until the conditions a)–c) are valid in any point $x \in \Re^n$. Hence for a pair of criteria $< i, j >$ the DM specifies the interval $\mu_{ij} = [\mu_{ij}^-; \mu_{ij}^+]$ having the following property:
for every $x \in \Re^n$, $t > 0$

$$(x \parallel x_i - t, x_j + \mu_{ij}^+ t) \ P \ x \quad \text{and} \quad x \ P \ (x \parallel x_i - t, x_j + \mu_{ij}^- t) \tag{1}$$

Such interval is called the (constant) interval rate of substitution (IRS).

Axiomatics and structure of IRS–relation as well as properties of interval rates of substitution were given in Berman and Naumov, (1989). Interval evaluation of substitution of a fixed criterion for each other criteria was considered by Passy and Levanon, (1984), while the interval estimates were not constant in the criteria space.

It should be noted that there is a fundamental distinction between the interval approach and methods of separation of some part of the Pareto set with the help of a generalized criterion when the weight coefficients are restricted by intervals (Steuer, 1976).

A binary relation generated by μ_{ij} in accordance with property (1) will be denoted by $P(\mu_{ij})$. Let M be an information on DM's preferences formed by accumulated μ_{ij}. This information induces a relation P^M on $\Re^n$ which is the transitive closure of union of P^O and all $P(\mu_{ij})$:

$$P^M = Tr[P^O \cup (\cup_{\mu_{ij} \in M} P(\mu_{ij}))].$$

In accordance with definition $x P^M y$ is valid if and only if there exists a chain $x P^1 z^1$, $z^1 P^2 z^2, ..., z^{m-1} P^m y$, in which each $z^k \in \Re^n$, P^k is P^O or $P(\mu_{ij})$ for some $\mu_{ij} \in M$.

The information M is said to be consistent if P^M is irreflexive. Having a consistent M we can consider the intervals μ_{ij} as estimates of interval rates of substitution for some IRS–relation.

A construction of preference relation using an information on constraints of rates of substitution (not only for single criteria but also for sets of criteria) was suggested in the framework of the theory of criteria importance in Podinovskii, (1978), (see also Podinovskii, this volume). Methods for verifying the consistency of such information and for constructing the corresponding preference relation were presented in Menshikova and Podinovskii, (1988).

Using these results we can construct the preference relation P^M. For each μ_{ij} we introduce two row vectors $\alpha^-(\mu_{ij})$ and $\alpha^+(\mu_{ij})$ as follows:

$$\alpha_k^-(\mu_{ij}) = \begin{cases} -\mu_{ij}^- & k = j \\ 1 & k = i \\ 0 & k \neq i, j \end{cases}$$

$$\alpha_k^+(\mu_{ij}) = \begin{cases} \mu_{ij}^+ & k = j \\ -1 & k = i \\ 0 & k \neq i, j \end{cases}$$

Using all these row vectors we form $2q \times n$ – matrix A^M where q is a number of μ_{ij} in M. We shell use the following notation (for $x, y \in \Re^n$): $x \geq y \Leftrightarrow x_i \geq y_i$, $i = 1, ..., n$; $x > y \Leftrightarrow x_i > y_i$, $i = 1, ..., n$.

Consider the set $B^M = \{\beta \in \Re^n; \ \beta > 0, \ A^M\beta > 0\}$.

From the point of view of the theory of criteria importance M is the information on criteria importance of a special kind (proportional importance) and elements of B^M are vectors of importance coefficients.

Theorem 1. The information M is consistent if and only if B^M is non–empty.

Theorem 2. Given any $x, y \in \Re^n$ and $x \neq y$, xP^My is valid if and only if there exists a vector $u \in \Re^{2q}$ with nonnegative components such that the following vector inequality is satisfied:

$$x - y \ \geq \ uA^m$$

Theorem 3. xP^My if and only if $(x - y)\beta > 0$ for each $\beta \in B^M$.

Let $Y = f(S)$ be the set of vector criteria assessments of alternatives. The relation P^M determines a subset of nondominated points Y^M in Y.

As a results the Pareto set of choice S^O is reduced to a set of nondominated alternatives S^M (such that their vector criteria assessments belong to Y^M).

In practical situations, when the set S was finite and included several dozens of alternatives, we had only a few nondominated ones. Quite often there was only one nondominated alternative.

3 Decision support system MCITOS

The interval approach to the solution of multicriterial problem is now supported by a modern software. The authors have developed the system MCITOS (MultiCriteria Interval

Trade–Offs System). Source data for MCITOS are following:
set of criteria;
set of alternatives with criteria assessments for all criteria.

MCITOS asks DM about his preferences and computes all non–dominated alternatives out of the given set.

MCITOS can be used on any IBM PC–compatibles. Color graphics monitor is recommended.

The number of criteria and alternatives is limited only by accessible memory (approximately 1000 alternatives with 15 criteria).

System interface is based on windows, menu, spreadsheets and is user–friendly. All the data are in an easy form to examine and modify. Spreadsheet in the table of alternatives allows to solve easily such problems as "what if ...".

English and Russian versions of MCITOS are available. MCITOS can use files created by dBase, Clipper and FoxBase systems.

4 Applications

MCITOS was successfully used for solving several practical problems including a problem of choice of equipment for automatic stores, problems of constructing power gears and some problems of personal choice.

Let us consider an example of the application of MCITOS for the problem of the choice of the best Super VGA computer monitor. The list of alternatives and criteria was taken from "PC Magazine" (May 15, 1990) and "Your Computer" (August, 1989). Alternatives (monitors) were evaluated by 11 criteria:

1) cost (dollars);

2) picture size (inches);

3, 4, 5) monitor dimensions: length, width, height (inches);

6) monitor weight (kg);

7) compatibility with display adapters (the number of types of adapters monitor can be used with);

8, 9) screen resolution: horizontal and vertical (pixels);

10) video bandwidth (MHz);

11) dot inch (mm/line).

The criteria 1, 3,...,6 and 11 are negatively oriented, the criteria 2, 7,...,10 are positively oriented.

Following monitors were compared: 1) Mitsuba 710 VH; 2) Acer 7015 Multiscanning Color Monitor; 3) Cordata CMC–141M Multiscanning Color Monitor; 4) Dell Super VGA Color Monitor; 5) MAC Coputronic PMV14VC; 6) TW Casper 5156H; 7) GoldStar 1450

Plus VGA; 8) Tatung CM–1496X; 9) TVM SuperSync; 10) GoldStar 1460 3APlus VGA; 11) NEC Multisync 2A; 12) Relisys RE–5155; 13) Amdek AM738 Smartscan; 14) Seiko CM–14400; 15) Sony CPD–1302; 16) Mitsubishi FA3415ATK; 17) Idek Multiflat Digiana MF–5015; 18) NEC MultiSync 3D; 19) Nanao FlexScan 9060S; 20) Sony CPD–1304; 21) Electrohome ECM 131OU; 22) Microvitec 1019/SP; 23) Mitsubishi Diamond Scan 20C; 24) Panasonic Panasync C1391; 25) Princeton Ultra 14.

All alternatives are Pareto–optimal. The Table shows vector estimates of all alternatives.

The problem was analyzed with help of MCITOS. The following interval rates of substitution were estimated:

$$\mu_{21} = (100; 155), \quad \mu_{31} = (100; 200), \quad \mu_{41} = \mu_{51} = (100; 250),$$
$$\mu_{61} = (80; 120), \quad \mu_{71} = (320, 600), \quad \mu_{81} = \mu_{91} = (12; 18),$$
$$\mu_{10,1} = (120; 270), \quad \mu_{11,1} = (6000; 10000).$$

Alterna-	Criteria										
tives	1	2	3	4	5	6	7	8	9	10	11
1	495	14	14	14	14.25	37	3	1024	768	45	0.28
2	560	12.75	17.25	18	18	28.5	4	800	600	35	0.31
3	599	14	14.25	14.5	15.25	32.5	4	800	600	35	0.31
4	599	13.5	12.75	13.5	15	28.75	2	800	600	38	0.29
5	635	14	17.25	18	18	30	3	1024	768	45	0.28
6	680	13.25	13.75	14.25	14.25	25.25	4	1024	768	30	0.31
7	699	14	14	14	14.75	28.5	2	800	600	45	0.31
8	749	13	12.5	14.5	15.75	27.5	3	1024	768	45	0.31
9	795	14	12	14.5	15.75	29.75	3	1024	768	45	0.28
10	799	13	14	14	14.75	28.5	3	1024	768	45	0.28
11	799	13	13	13.75	15.25	28.5	2	800	600	38	0.31
12	799	14	14.5	14.75	15.25	31	4	800	600	30	0.31
13	835	14	13.25	14	14.5	30	3	1024	768	25	0.31
14	899	13	13	13.75	15.75	33	3	1024	768	35	0.25
15	995	13.25	11.75	14	15.5	32	5	900	560	25	0.26
16	1015	13.25	14	13.75	15.25	31	5	1024	768	40	0.28
17	1045	15	14.75	14.5	15.75	34.25	5	800	600	30	0.31
18	1049	13	13	13.75	15.25	30.75	5	1024	768	45	0.28
19	1053	13	14.25	16.25	14.5	32	5	1024	768	30	0.28
20	1095	14	11.75	14	15.5	32	3	1024	768	35	0.26
21	1259	13.5	13	14.25	15	29.25	4	800	800	30	0.31
22	2396	19	17.5	18.5	19.5	59	5	800	600	40	0.31
23	2679	19	17.75	19.75	21	66	5	1120	780	50	0.31
24	899	14	14.5	14	14.25	25.25	5	800	600	30	0.31
25	899	14	13	13.75	15.5	35	5	1024	768	45	0.28

Vector estimates of monitors

The information $M = \{\mu_{21}, \mu_{31}, ..., \mu_{11,1}\}$ is consistent. The only alternative whose vector estimate is nondominated is the 18th one. Hence the monitor 18 "NEC MultiSync 3D" was the most preferable for the DM.

5 Conclusion

The interval approach to the solution of multicriterial problem is well justified and based and based on reliable information. It makes it possible to obtain reliable results. The defect of this method is an incomplete comparability of alternatives. But this incompleteness reflects the inexactness of additional information on tradeoffs.

References

Berman, V. P., and G. Ye. Naumov (1989). Preference Relations with Interval Value Tradeoffs in Criterial Space. *Automation and Remote Control*, No.3, 1989, pp. 398–410.

Keeney, R. L., and H. Raiffa (1976). Decision with Multiple Objectives, New York, Wiley.

Menshikova, O. R., and V. V. Podinovskii (1988). Construction of Preference Relation and Core in Multicriteria Problems with Non–Homogeneous Importance–Ordered Criteria. *USSR Computational Mathematics and Mathematical Physics*, Vol. 28, No. 3, pp. 15–22.

Passy, U., and Y. Levanon (1984). Analysis of Multiobjective Decision Problems by the Indifference Band Approach. *Journal of Optimization Theory and Application*. Vol. 43, No. 2, pp. 205–235.

Podinovskii, V. V. (1978). Relative Importance of Criteria in Multiobjective Decision Problems. *Mashinostroenie Publishing House*, Moscow, pp. 48–92 (in Russian).

Steuer, R. E. (1976). Multiple Objective Linear Programming with Interval Criterion Weights. *Management Science*. Vol. 23, No. 3, pp. 305–316.

Problems with Importance–Ordered Criteria

Vladislav V. Podinovskii

Academy of Labour and Social Relations
Moscow, Russia

Abstract

An analysis presented in (Podinovskii, 1991) shows that the concept of criteria importance is a representation of a special kind of regularity of preference structures. This regularity provides increase or stability of preferences under specific ratio of increments of those components of vector estimates which correspond to the criteria compared by importance (with all other components fixed). We present here some basic results of the theory of symmetrical importance - which is one of the most developed sections of the general theory - and discuss possibilities of its applications.

1 Mathematical model and basic definitions

A multicriteria optimization problem under certainty is represented by a set of available alternatives (strategies)S and a vector criterion $f = (f_1, ..., f_m)$ defined on S. Each strategy s is characterized by its vector criteria assessment $f(s) \in X$, where $X = X_1 \times ... \times X_m \subseteq \Re^m$ is a non–empty set of all vector assessments.

The decision maker (DM) preferences are represented by a non-strict preference relation R which is partial quasiorder, or pre–order (it is a reflexive and transitive relation).

We shall denote by I and P respectively the indifference and (strict) preference relations corresponding to R:

$$I = R \cap R^{-1}; \quad P = R \setminus I.$$

The relation R is unknown and should be restored (completely or partially) from an information on DM's preferences. Such information might contain statements on the relative importance of criteria consisting of single judgments ω of the criteria importance.

We shell consider a multicriteria maximization problem for which the Pareto relation R^o is defined as follows: $x R^o y$ iff $x \geq y$, i.e. $x_i \geq y_i$, $i = 1, ..., m$. In fact, R^o is a part of the relation R and thus it is imbedded into R: $R^o \geq R$, i.e. $I^o \subseteq I$ and $P^o \subseteq P^\star$ both hold.

Let all criteria be homogeneous, i.e. have a common scale:

$$X_1 = X_2 = ... = X_m.$$

Definition 1. Assertion $\omega^\sim = < i_\sim j > = \ll$the criteria f_i and f_j are of equal importance$\gg$ means that every two vector assessments x, y such that

$$x_i = y_j, \quad x_j = y_i, \quad x_k = y_k, \quad k \in \{1, ..., m\} \setminus \{i, j\} \tag{1}$$

are indifferent ($xI^{\omega\sim}y$ and $yI^{\omega\sim}x$).

Definition 2. Assertion $\omega^{\succ} = < i \succ j >= \ll$the criterion f_i is more important than the criterion $f_j \gg$ means that, given any two vector assessments x, y that satisfy (1) with $x_i > x_j$, the first assessment is more preferable than the second one ($xP^{\omega^{\succ}}y$).

According to these definitions, every judgment ω about criteria importance induces a relation R^ω on X which is an indifference relation I^ω or a preference relation P^ω. Using all the importance information $\Omega = \{\omega_1, ..., \omega_k\}$, we can introduce on X a relation

$$R^\Omega = Cl(R^o \cup R^{\omega_1} \cup ... \cup R^{\omega_k})$$

where by ClR we denote a transitive closure of R.

Definition 3. The information Ω is said to be consistent iff $R^{\omega_j} \leq R^\Omega$, $j = 1, ..., k$, and $R^o \leq R^\Omega$.

Definition 4. Any positive numbers $\alpha_1, ..., \alpha_m$ which satisfy the conditions

$$< i \sim j > \in \Omega \Rightarrow \alpha_i = \alpha_j, \quad < i \succ j > \in \Omega \Rightarrow \alpha_i > \alpha_j, \qquad (2)$$

are called the criteria importance coefficients generated by Ω.

Let $A^\Omega \subset \Re_+^m$ designate the set of vectors of importance coefficients generated by Ω.

2 Principal results

Theorem 1. The information Ω is consistent iff there exist $\alpha_1, ..., \alpha_m$, i.e. $A^\Omega \neq \emptyset$.

In accordance with *Theorem 1* the information Ω is inconsistent if and only if for some $i_1, i_2, ..., i_n$ we have $< i_1 \overset{\succ^1}{\sim} i_2 > \in \Omega, ..., < i_{n-1} \overset{\succ^{n-1}}{\sim} i_2 > \in \Omega, < i_n \overset{\succ^n}{\sim} i_1 > \in \Omega$, where each $\overset{k}{\sim}$ is $\sim$ or $\succ$ but $\overset{k}{\sim}$ is $\succ$ for at least one k. For example, the information $\Omega = \{< 1 \succ 2 >, < 2 \succ 3 >, < 3 \succ 1 >\}$ is inconsistent.

In order to construct the relation R^Ω we introduce the sets $E^\Omega(y)$ and $B^\Omega(y)$ as follows:

$E^\Omega(y) = \{y\} \cup \{x \in X \mid$ there exists $\{z^1, ..., z^n\} \subseteq X$ such that $xR^1z^1, z^1R^2z^2, ..., z^nR^{n+1}y$

where each R^j is I^ω, $\omega \in \Omega\}$;

$B^\Omega(y) = \{y\} \cup \{x \in X \mid$ there exists $\{z^1, ..., z^n\} \subseteq X$ such that $xR^1z^1, z^1R^2z^2, ..., z^nR^{n+1}y$

where each R^j is I^ω or P^ω, $\omega \in \Omega\}$.

It is obvious that $E^\Omega(y) \subseteq B^\Omega(y)$ and $\mid E^\Omega(y) \mid \leq \mid B^\Omega(y) \mid \leq m!$.

Theorem 2. $xR^\Omega y$ iff $xR^o z$ for some $z \in B^\Omega(y)$. Given the consistent information Ω, $xI^\Omega y$ iff $x \in E^\Omega(y)$.

It is easy to show with the help of *Theorem 2* that :

if $< i \sim j > \in \Omega$ and $< j \sim k > \in \Omega$ then $R^\Omega = R^{\Omega'}$ where $\Omega' = \Omega \cup \{< i \sim k >\}$,

if $< i \sim j > \in \Omega$ and $< j \succ k > \in \Omega$ then $R^\Omega = R^{\Omega'}$ where $\Omega' = \Omega \cup \{< i \succ k >\}$,

if $< i \succ j > \in \Omega$ and $< j \sim k > \in \Omega$ then $R^\Omega = R^{\Omega'}$ where $\Omega' = \Omega \cup \{< i \succ k >\}$.

But if $< i \succ j > \in \Omega$ and $< j \succ k > \in \Omega$ then for $\Omega' = \Omega \cup \{< i \succ k >\}$ we have only $R^\Omega \subseteq R^{\Omega'}$.

Let us consider an example: $m = 3$, $x = (1,4,0)$, $y = (0,3,1)$, $\Omega = \{< 1 \succ 2 >, < 2 \succ 3 >\}$. Here $B^\Omega(y) = \{y, (3,0,1), (3,1,0)\}$ and $B^\Omega(x) = \{x, (4,1,0)\}$, so that x and y are incomparable with respect to R^Ω. However for $\Omega' = \Omega \cup \{< 1 \succ 3 >\}$ we have $(1,3,0) \in B^{\Omega'}(y)$ and, since $xR^o(1,3,0)$ we obtain $xR^{\Omega'}y$.

Now we can introduce the relations of equality $\sim^\Omega$ and superiority in importance $\succ^\Omega$ on the set $\{1, 2, ..., m\}$ of criteria indices as follows:

$$i \sim^\Omega j \quad \text{when } i = j \text{ or a chain}$$

$$\{< i \overset{1}{\succsim} i_1 >, \ < i_1 \overset{2}{\succsim} i_2 >, ..., < i_n \overset{n+1}{\succsim} j >\} \subseteq \Omega \tag{3}$$

exists where each $\overset{k}{\succsim}$ is $\sim$,

$i \succ^\Omega j$ when there exists chain of the form (3) where exactly one $\overset{k}{\succsim}$ is $\succ$ and the other $\overset{k}{\succsim}$ are $\sim$.

The relation $\sim^\Omega$ is an equivalence. If Ω is consistent then $\succ^\Omega$ is irreflexive and asymmetric. Let $\succ^\Omega_T$ be a transitive closure of $\succ^\Omega$: $i \succ^\Omega_T j$ when the chain (3) exists where each $\overset{k}{\succsim}$ is $\succ$ or $\sim$ but $\overset{k}{\succsim}$ is $\succ$ for at least one k. The information Ω is consistent if and only if $\succ^\Omega_T$ is a strict partial order.

Let us consider consistent information Ω^o when all criteria are ordered by importance. This information enables us to measure the importance of all criteria in a common ordinal scale in accordance with (2). Consequently, $\alpha^o_1, ..., \alpha^o_m$ may be called ordinal coefficients of importance.

Given $x, y \in X$, let $G(x,y)$ be the ordered set $\{x_1\} \cup ... \{x_m\} \cup \{y_1\} \cup ... \cup \{y_m\}$ of numbers $g_1 > g_2 > ... > g_{k(x,y)}$. For example let $m = 3$, $x = (1,3,4)$, $y = (3,4,7)$. Then $G = < 7,4,3,1 >$ and $k(x,y) = 4$. Further, let $\alpha^j(z) = (\alpha^j_1(z), ..., \alpha^j_m(z))$ where $\alpha^j_i(z)$ is α^o_i for $z_i \geq g_j$ and 0 otherwise. Finally, let $u_\downarrow = (u_{[1]}, ..., u_{[m]})$ be a permutation of u by decrease: $u_{[1]} \geq u_{[2]} \geq ... \geq u_{[m]}$.

Theorem 3. $xR^{\Omega^o}y$ iff $\alpha^j(x) \geq \alpha^j(y)$, $j = 1, ..., k(x,y)$, moreover, $xI^{\omega^o}y$ iff all nonstrict inequalities are satisfied as equalities.

Let us consider the special case $\Omega^o = Sym$ when all criteria are of equal importance.

Theorem 4. $xR^{Sym}y$ iff $x_\downarrow \geq y_\downarrow$, moreover $xI^{Sym}y$ iff $x_\downarrow = y_\downarrow$.

For a generalized criterion

$$f_s(x, \alpha) = \sum_{i=1}^{m} \alpha_i x^s_i, \quad s > 0, \ X \subset \Re^m_+.$$

with the importance coefficients $\alpha \in A^\Omega$ we have

$$xR^\Omega y \Rightarrow f_s(x, \alpha^\Omega) \geq f_s(y, \alpha^\Omega).$$

3 An Example

Let us consider an illustrative application of the symmetrical importance theory to the material selection problem from (Kirkwood and Sarin, 1985). Table 1 shows all the candidate materials and their assessments by four criteria with 3–level discrete scale. The pairwise ranking method using the linear generalized criterion ($s = 1$):

$$f_1(x, \alpha) = \alpha_1 x_1 + \alpha_2 x_2 + \alpha_3 x_3 + \alpha_4 x_4$$

and the information

$$\alpha_1 > \alpha_2 > \alpha_3 > \alpha_4 > 0$$

gives the partial ordering of materials (see Fig.1).

Using only the importance information

$$\Omega = \{<1 \succ 2>, <1 \succ 3>, <1 \succ 4>, <2 \succ 3>, <2 \succ 4>, <3 \succ 4>\}$$

without any generalized criterion in accordance with *Theorem 2* or *Theorem 4* we have the same partial ordering of materials with respect to R^Ω.

It is interesting to note that the partial ordering is the same one for

$$\Omega = \{<1 \succ 4>, <2 \succ 4>\}.$$

TABLE 1

Candidate materials and levels of evaluation measures for
materials to provide physical continuity

Material	Strength	Creep Characteristics	Thermal Expansion	Stability
Igneous & Metamorphic Rocks				
Basalt	3	3	2	2
Granite	3	3	2	2
Slate	2	3	2	3
Minerals				
Quartz	3	3	2	3
Native Elements				
Gold	3	1	1	3
Silver	3	1	1	2
Platinum	3	1	1	3
Copper	3	1	1	2
Processed Metals Lead	2	1	1	2
Metal alloys	3	1	1	2
Ceramics Mullite	3	3	2	3
Stcatite	3	3	2	3
Concretes	2	3	2	3

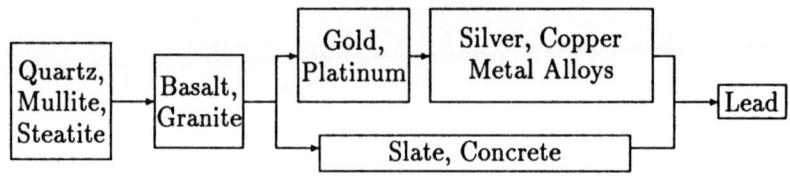

Fig. 1. The partial ordering of materials

4 Concluding remarks

Applications of this theory to real–life multicriterial problems presuppose a preliminary transformation of original criteria with various scales to homogeneous ones with a common scale. To do this one can use well–known formal methods of criteria transformation for constructing a generalized criterion. The simplest and widely–distributed methods are based upon the following formulae:

$$f_i/\alpha_i, \ (f_i - \alpha_i)/(b_i - \alpha_i),$$

where α_i is the minimum (or "standard") value of f_i and b_i is the maximum (or "target") value of f_i. But such methods are suitable for cardinal criteria only and do not take into account subjective levels (and intensity) of the decision maker's preferences on scales of criteria.

A more appropriate and rather simple way is as follows. The discrete scale with 3-6 levels (estimates) such as "very good", "good" etc. is introduced for all criteria. For convenience one can number these levels, for example according to an increase of preference. It is permissible since the methods for constructing the quasiorder R^Ω are adequate for the ordinal scale: criteria estimates are compared only by their values (see *Theorem 2*). For each qualitative criterion it is necessary to give a substantial verbal description of each level. The continual scales of qualitative criteria are to be divided into intervals so that all estimates from a unique interval correspond to the same level of discrete scale.

To obtain the information about the criteria importance from a decision maker it is necessary to proceed from the proper definitions 1 and 2. For example, according to (1) criteria f_i and f_j are equally important when for every vector assessment x the permutation of its components x_i and x_j produces a new vector, which is indifferent to x. It is clear that in practice to check the preservation of preferences one should begin with a few test vector estimates x, and only then one should suggest to the decision maker an appraisal of the correctness of a general assertion of preference preservation under permutations of i-th and j-th components in any vector estimate. Such approach is fully analogous to practical methods of verification of different independence conditions for multicriterial structures.

The approaches to obtaining and processing the information about criteria importance have been implemented in a decision support system "SIVKA".

This system was developed by a group including V. V. Podinovskii, V. A. Osipova , N. S. Alekseev and E. A. Gushin and used at the early stages of designing several technical systems.

References

Podinovskii, V.V. (1991). Criteria Importance Theory. In: A. Lewandowski and V. Volkovich, eds., Multiobjective Problems of Mathematical Programming, Berlin, Springer Verlag, Vol.351, pp.64-70.

Podinovskii, V.V. (1975). Multicriterial Problems with Uniform Equivalent Criteria. *USSR Computational Mathematics and Mathematical Physics*, Vol. 15, No.2, pp.47-60.

Podinovskii, V.V. (1976). Multicriterial Problem with Importance-Ordered Criteria. *Automation and Remote Control*, Vol. 37, No. 11, Part 2, pp. 1728-1736.

Podinovskii, V.V. (1978). Importance Coefficients of Criteria in Decision–Making Problems. *Automation and Remote Control*, Vol. 39, No. 11, Part 2, pp. 1514-1524.

Podinovskii, V.V. (1978). On Relative Importance of Criteria in Multiobjective Decision–Making Problems. In: Multiobjective Decision Problems , Mashinostroenie Publishing House, Moscow, pp. 48-92, (in Russian).

Podinovskii, V.V. (1979). An Axiomatic Solution of the Problem of Criteria Importance in Multicriterial Problems. In: Operation Research Theory: State of the Art, Nauka, Moskow, pp.117-1499, (in Russian).

Kirkwood, C.W. and Sarin, R.K. (1985). Ranking with Partial Information: A Method and an Application. *Operations Research*, Vol.33, No.1, pp.38-48.

Part B:
SDS Tools and Applications

Types of Decision Support Systems and Polish Contributions to Their Development

Andrzej P. Wierzbicki

Institute of Automatic Control
Warsaw University of Technology, Poland

Abstract

After a general review of basic concepts related to decision support, the paper gives a synthetic presentation of various results obtained in Poland in the field of multiple criteria decision support theory and methodology, while concentrating on subjects related to the reference point methods in this field. Finally, comments on new standards of formulation and modes of analysis of multiple-objective models are presented.

1 Introduction

Generally accepted features of modern decision support systems are that they should contain a type of *knowledge representation* in a given field – a model base, in addition to a data base – and help to rationalize and support an entire *decision process* as opposed to its selected phases, while taking into account some *representation of uncertainty* inherent in this decision process as well as other specific features of a given decision situation. All these aspects were subjects of research and contributions by several Polish research groups.

Knowledge representation, besides the *logical form* of a knowledge base in expert systems, can also have various *analytical forms.* Polish researchers have contributed significantly to the development of methodology of multi-objective analysis and decision support based on analytical models, in particular by developing the *reference point methods* in multi-objective optimization that are related to the aspiration–led methodology of interaction, the theory of conical separation of sets and of order-consistent achievement functions. This was followed by many related theoretical and methodological results, as well as by the development of many *prototype decision support* systems for various classes of analytical models.

The representation of uncertainty in decision support can use probabilistic models, set-valued models, fuzzy set theory, or rough set theory; the last one was developed in Poland, while there are also significant Polish contributions to the theory and applications of fuzzy sets as well as to stochastic optimization, in particular with its multi-objective aspects in decision support. Another contribution is related to multi-objective decision support in group decision situations with multiple criteria; this concerns cooperative aspects of bargaining theory as well as the development of noncooperative multiple criteria game theory.

However, one of the important lessons from various applications of decision support systems based on analytical models is the relevance of standards of model definition for selected classes of models and the need of the development of more *versatile software tools for multi-objective definition, simulation, and analysis of various classes of models*. These issues are addressed in the last part of the paper.

2 Types of Decision Support Systems

There are many possible definitions of decision support systems (DSS) and it is not clear whether it would be useful to try and work out a unique and sharp definition for a concept with such a rich meaning. Andriole (1989) writes *"Decision support ... consists of any and all data, information, expertise and activities that contribute to option selection."* This definition is too narrow because it concentrates on option selection, which is only a phase in a decision process. Thus, we should rephrase this broad definition as follows: a DSS is collection of computerized tools supporting a user in an interactive decision process.

However, there is also a consensus between DSS specialists that, in order to be useful, a DSS must contain expertise and knowledge pertinent to a given decision situation, encoded in the form of a model base or knowledge base, as well as various tools (algorithms) that help in using this knowledge. Thus, it is advisable to make this broad definition slightly more specific. *DSS is a computerized system that supports its users in a rational organization and conduct of a decision process (or some selected phases) and, besides a data base, it also contains a pertinent knowledge representation in the form of models of decision situations as well as appropriate algorithms for using these models.* This definition corresponds to the general scheme of the architecture of a DSS presented in Figure 1.

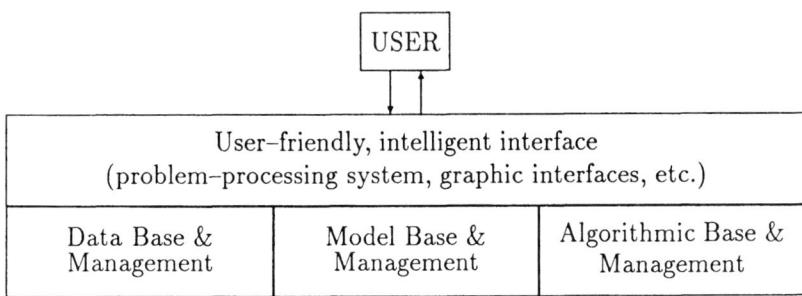

Fig. 1. A general scheme of the architecture of a DSS.

This scheme also stresses the first and possibly most important aspect of the definition of DSS: the *sovereign position of the user*. The user can be a *decision maker*, an *expert*, or an *analyst*; an ordinary *person* working in an office on a well–defined task that might benefit from decision support; or even a *group of persons* with varied expertise and decision authority. In any of these cases, a *DSS must not replace the user in his sovereign decision making*; it must help him in the organization of the decision process and in processing information customized for selected situations and phases of the process by drawing upon expertise and knowledge encoded in the models included in the system and by using various algorithms for the evaluation of options and alternative decisions.

This does not mean that repetitive, operational decisions might not be automated; but this is rather the field of automatic control of technical processes and even in this case the human decision maker usually has the authority to override the automatic equipment. In all other cases, the modern market demand just does not respond to high-technology products that limit the sovereignty of the user; this is the essence of *the high-tech – high-touch* megatrend as formulated by Naisbitt (1982) or the significant trend toward *user friendliness* in software development.

The principle of user sovereignty relates also to the recent recognition of the role of human intuition in decision making. While *Mind over Machine* by Dreyfus and Dreyfus (1986) is often perceived as a passionate criticism of the claims of artificial intelligence, it has nevertheless stressed and documented empirical evidence of an important fact: *master experts use subconscious, intuitive parts of their minds to make decisions.* Although *Mind over Machine* does not pursue this idea further, it opens the road to a rational investigation of subconscious decision making, with all its *Heureka, aha, gut feeling,* and other *enlightenment* phenomena. Until we truly learn to understand subconscious decision making, we should assume that the intuition and expertise of the DSS user is irreplaceable, and the value of a DSS should be judged by its capability to enhance human intuition. Finally, we must remember that the user, not the DSS, bears the responsibility for the decisions made.

The second important aspect of the above definition of a DSS is the concept of a decision process. Simon (1957) defined three main phases of this process: *intelligence, design,* and *choice*; later, Cooke and Slack (1984) analyzed these phases in much more detail. The phase of intelligence consists of observation, problem recognition, data gathering, and diagnosis. The phase of design comprises problem specification and model and option specification; Lewandowski et al. (1989) have added model fitting, identification, and verification; simulation and preliminary analysis; and model–based option generation to the design phase. The phase of choice consists of option evaluation and selection (Cooke and Slack) but might be augmented by multi-objective option analysis, sensitivity analysis, and post-optimal analysis, and include recourses to earlier phases (Lewandowski et al., 1989). Finally, a fourth major phase – *implementation* – should be also included, together with monitoring of results and possible adjustments of the decisions in a feedback loop.

All of these phases might involve various users – for example, analysts in the phases of intelligence and design, actual decision makers or their advisers in the phase of choice, finally operational officers in the phase of implementation. These phases require also various methodologies of analysis; it is then necessary to unify and modify these diversified methodologies for the purposes of decision support. This often raises new research questions; for example, even if there is much experience in methodologies that support choice, we must use this experience in modifying methodologies that support design: in fact, there is a need of further development of a general *methodology of modeling and simulation for decision support.* This includes the questions of sufficiently broad but specialized classes of models developed in the anticipation of adequate methods of supporting choice: the issues of user-friendliness of a model format and the easiness of problem instance specification; the impact of model errors on decision support; hard and soft constraints as well as multi-objective analysis methodology in model simulation; methods of preliminary scenario generation; etc.

However, the existing experience in modeling for decision support makes it possible to specify several basic distinctions. The first of them concerns *preferential* versus *substantive models.* We must be clear whether a model used for decision support intends to represent the preferences of the user (see Keeney and Raiffa, 1976) or an independent expertise and knowledge in a substantive area pertinent to the decision situation (if the answer is both, then the appropriate parts of the model should be separated). The preferential models might be explicit – say, based on utility or value theory – or implicit – expressed by various dominant structures or ranking procedures. It should be stressed that while some types of preferential models are usually necessary for decision support, one should be careful in using them: if we rely too much on explicit preferential models, we might violate the principle of sovereignty of the user by replacing him in the crucial phases of the decision process.

Therefore, substantive models actually constitute the essence of decision support: when they are formulated by expert specialists and analysts, we hope that they can help to enhance the intuition of the user. There are two basic classes of substantive models. In expert systems, artificial intelligence, etc., *logical models* are used; in fact, some specialists in this field would claim that the methodologies of so-called knowledge bases (the name used in AI for model bases) and inference engines (the name used in AI for algorithmic bases) provide enough tools for decision support of any type. While these methodologies are certainly powerful, there are nevertheless inherent limitations to knowledge representation by logical models.

One of the main conceptual advances of the 20th century is the realization that thinking in terms of direct, static cause-and-effect relations is insufficient to represent the complexity of the real world and that the concepts of feedback loops in dynamic models are essential (particularly in some aspects of representing uncertainty through chaotic models).

While there exist methods of modifying classical logic to include dynamic clauses or even loops, these phenomena can be represented much easier in the second major class of *analytical models.* More diversified and richer than logical models, analytical models can be in turn subdivided into discrete (discrete event systems, queuing theory models, Petri nets, etc.) and continuous (linear, nonlinear, dynamic models etc.). Finally, what might be most important for decision support, all analytical models can be treated either as single-objective, or multi- objective (there is no principal difficulty in including multiple objectives in logical models, but the related methodology is not fully developed as yet).

An essential aspect of all models for decision support (be they preferential or substantive, logical or analytical) is the way of representing uncertainty. The development of the theory of chaos (emerging even in simple deterministic dynamic models with strong feedback) has essentially changed our understanding of uncertainty: there is no sense in the dispute whether the universe is deterministic or indeterministic when a simple deterministic model can represent equally uncertain behavior as an indeterministic one. Thus, besides the classical probabilistic or stochastic models for representing uncertainty, many other types are used - fuzzy sets (starting with Zadeh, 1978, see also Seo and Sakawa, 1988, for relations of multi–attribute choice and fuzzy set theory), rough sets (Pawlak, 1991), set-valued models (Kurzhanski, 1986).

The way of representing uncertainty is essentially related to the algorithmic base of a DSS. Because algorithms for dealing with models of uncertainty are usually complicated, averaged, or "deterministic," models are often preferred. When they are insufficient (it can be justifiably argued that any realistic decision support must include some aspects of representing uncertainty), probabilistic or stochastic models are preferred because the related algorithms of dealing with them are relatively best developed. However, further development of algorithms for fuzzy set, rough set, and set-valued models is very intense and we can expect more applications of such models in the near future. We must nevertheless note that such concentration on the development of algorithms has its dangers: various algorithms contained in a DSS should be treated as tools supporting the user, never as goals.

Moreover, it should be stressed that including even well-developed algorithms in the algorithmic base of a DSS presents new challenges. In a DSS, algorithms run in the background and the user typically refuses to be bothered by possible algorithmic failures. Therefore, new *robust* variants of known *algorithms* usually must be developed before their inclusion in a DSS; they must run effectively for a broad class of models with wide parameter changes and it is often a mistake to believe that an "off the shelf" algorithm, even tested on the software market, will perform as well when included in a DSS.

The above discussion gives enough background to address the main theme of this section: what are the possible types of DSS? There are many possible dimensions of classification of DSS (Andriole, 1989); in the following list, we stress both applied and methodological aspects of such a classification. Thus, we can classify DSS according to:

1) *Application area* is perhaps the most important classification, because a modern specialized DSS should be user-oriented: its functions and detailed specification should be determined with the participation of future users, its user interface should rely on symbols and graphic representations typical for a given application area and thus well understood by the user, etc.

2) *Application type*, e.g., for strategic, tactical, and operational decisions. This classification is important methodologically; a DSS for strategic decisions should support learning about the problem and innovative ways of solving it, while the typical users are teams of analysts and policy advisers, whereas a DSS for operational decisions might concentrate on information processing and the optimization of typical solutions while the typical users are operational officers.

3) *Substantive model type*, e.g., expert systems with logical models and analytical systems with analytical (often, not quite precisely, called operations research type) models, with further subdivision - e.g., for linear, nonlinear, dynamic continuous models and discrete models of various classes. The way of representing uncertainty is another dimension of this classification; a prototype DSS, developed in order to test some aspects of the

developing methodology of decision support, might concentrate on a chosen type of the substantive model and of the uncertainty representation.

4) *Preferential model type*, or *rationality framework* together with selected algorithms of choice. This dimension is important from a decision-theoretical point of view, since practical decision behavior (see Rapoport, 1989) might suggest a broad rather than strict interpretation of the rationality framework and thus the use of incomplete but elastic preferential models, such as specifying only several objectives and applying multi-objective optimization not as an algorithm for choice but for supporting efficient option generation.

5) *Principles of interaction* with the user are a very important though often neglected dimension of DSS classification: it is very often crucial how the user can influence a decision suggested to him by the DSS and how a suggested decision is explained to him. For example, the reference point approach described later in more detail can be applied rather universally in order to stress the sovereignty of the user and to give him a full controllability of efficient option selection.

6) *The number of users and the type of their cooperation* is an extremely important dimension of a DSS classification for both practical and methodological purposes. A DSS might be designed to serve a *single user* or a *team of users* the members of which have different roles but the same interests and goals; quite different types of DSS would be designed to serve a *group of users* that might have different goals and would opt for different decisions but must achieve a joint decision (because only one decision is possible, such as in a committee deciding on a budget allocation); finally, yet different types of DSS are needed to support *bargaining and negotiations in a game-like situation*, when each user can not only opt for but also implement his own decisions.

We see that there can be many types of DSS; moreover, even the above classification is not exhaustive. A good DSS might be a mix of various types, (see e.g., Fedra et al. in this volume). We can expect in the future that a substantive model would have mixed logical and analytical aspects, that several types of preferential models or ways of representing uncertainty could be included, and that the user would choose between them.

3   Polish contributions to DSS methodology

The research on DSS in Poland has been quite intensive for over a decade, unlike the research being done in other countries, however, the research in Poland has been more methodological and theoretically oriented than applied. This is an anomaly, since the development of DSS in the world has been predominantly motivated by demand for applications; this anomaly resulted from the fact that until 1989, or even until now, the demand for industrial and business applications in Poland has been rather insignificant. On the other hand, the strong traditions of Polish mathematics, logic, and optimization techniques have provided a good background for this research, which resulted in original methodological reflection and considerable contributions to DSS methodology. The prevailing mood in the world, typical for this field since 1980, was skepticism in seeking new methodological approaches to DSS and very pragmatic and practical attitudes of applying any results already available from other studies (decision theory, AI, etc.,). This was seen not as a discouragement but as a challenge in Poland. As a result, the Polish contributions to DSS include various approaches to the representation of uncertainty, contributions to the development of multi-objective optimization techniques, the reference point methodology of interaction with the user, and advances in the methodology of multi-person DSS.

Strong Polish traditions in mathematical probability theory resulted in contributions to stochastic optimization; the most relevant results for the methodology of DSS are related to new approaches to multi-objective stochastic optimization (Ruszczynski, 1991). The multi-valued logic that was used by Zadeh (1978) in his fuzzy set theory was actually developed - though not used much - earlier in Poland by Lukasiewicz and others; following Zadeh, considerable results in fuzzy set theory were obtained by Kacprzyk, Slowinski, and others (Slowinski and Teghem, 1990). An original approach to representing uncertainty by rough sets was started by Pawlak (1990), and developed by others (Stefanowski in this volume) for various decision support applications.

The studies of multiple-criteria optimization also have a long tradition in Poland; many contributions to it were made starting in the 1960s. However, at a very abstract level, Rolewicz first noted in the mid-1970s that the sufficient conditions of multi-objective optimality are basically related to the selection of a proper type of *monotonicity of scalarizing functions*. Wierzbicki (1977) noted that the required monotonicity can be obtained by penalty function techniques and, moreover, that the necessary conditions of multi-objective optimality are basically related to a *property of conical separation of sets by scalarizing functions*.

While the considerable development of multi-objective optimization theory in the world since that time (Sawaragi et al., 1985; Yu, 1985; Steuer, 1986; Seo and Sakawa, 1988) has produced other, more detailed results, the principles of monotonicity and conical separation guided the development of a class of scalarizing functions that were sufficiently general and versatile to be used widely in multi-objective analysis of various classes of models for decision support. This class was later called *order-consistent achievement functions*. Functions of this class are monotonous in an appropriate sense with respect to the positive cone generating a partial order in the objective space, and some level sets of these functions separate (nonlinearly) the set of attainable outcomes in the objective space from the shifted positive cone that defines the multi-objective or Pareto optimality. Moreover, functions of this class are parameterized by a *controlling parameter vector called a reference point* that can be interpreted as a desirable point in the objective space but, in contrast to some older methods of multi-objective optimization also using such points, does not need to be restricted to the unattainable or to the attainable region in this space.

Combined with a study of interactive decision support methodology undertaken at IIASA in the early 1980s, this resulted in a development of the reference point method (Kallio et al., 1980; Wierzbicki, 1980) and a number of related prototype DSS. Applications in IIASA were related to several areas such as sectorial economic policies, future energy supply scenarios (Grauer et al., 1982), natural gas trade in Europe (Messner, 1985), as well as several other topics (Strubegger, 1985). While cooperating with IIASA, Polish researchers in this field concentrated on the development of several prototype DSSs using reference point (later called aspiration-based) methods. This development started with several variants of the DIDAS systems (Lewandowski et al., 1989; Kreglewski et al.,

1989), and included DINAS, HYBRID, and other systems (Ogryczak et al., Makowski et al., Bogucka et al. in this volume).

During the 1980s, methods similar or essentially equivalent to the reference point approach were internationally recognized as the most versatile tools for multi-objective model analysis (see Mikhalevich and Volkovich, 1982; Nakayama and Sawaragi, 1983; Nakayama in this volume; Steuer and Choo, 1983; Korhonen and Laakso, 1986; Seo and Sakawa, 1988). The reason for this development was that methods of this class preserve the advantages but overcome the drawbacks and can be treated as generalizations of several older methods of multi-objective analysis, such as compromise programming (Zeleny, 1973; Salukvadze, 1971, 1974), surrogate trade-off methods (Haimes and Hall, 1974), the maximal effectiveness principle (Khomenyuk, 1977), and especially goal programming (Charnes and Cooper, 1977; Ignizio, 1978). For the use in DSS, reference point methods have a specific desirable property: the *continuous controllability of the selection of efficient solutions by the user* (Wierzbicki, 1986).

In Poland, the development of prototype DSS has led to various related methodological and theoretical developments. Including the following:

- further refinement of robust nonlinear and nondifferentiable optimization algorithms (order-consistent scalarizing functions are usually nondifferentiable, Kiwiel and Stachurski, 1989; Altman in this volume);

- the extensions of reference point methods to discrete and mixed multi-objective programming problems (Ogryczak et al. 1989);

- further extensions to dynamic multi-objective programming problems - although one of the advantages of these methods is that they were applicable from the very beginning as tools of *multi-objective trajectory optimization* (Rogowski, 1989; Wierzbicki, 1991);

- some essential generalizations of the typical formulations of stochastic programming problems for the multiple criteria case with the help of the reference point approach and their use in DSS (Ruszczynski, 1991);

- the development of assessment methods of preferences of decision makers with the help of ordinal regression in the class of order-consistent achievement functions (Slowinski and Teghem, 1990);

- the development of multi-objective bargaining and negotiation methods based on the reference point approach (Bronisz et al., 1989; Krus et al. in this volume);

- some theoretical advancements of multicriteria game theory (Wierzbicki, 1990), combined with further advancements in the theory of proper efficiency (Kaliszewski in this volume).

Many lessons can be drawn from these and related developments. An important methodological lesson that was derived from various applications of the prototype DSS developed in Poland (Bogucka et al., this volume) is the importance of software tools supporting the specification, edition, and preliminary analysis of selected classes of models used in DSS. Even in the most standard class of linear programming models, off-the-shelf tools often do not meet the requirements of modern DSS and must be modified in order to support a versatile, multi-objective model analysis. In the prototype DSS system for nonlinear programming models DIDAS-N, we had to develop our own standards of nonlinear model specification as well as some special software for model edition, verification, automatic and symbolic computation of derivatives, etc. The typical standards of model specification that result either from model simulation approaches or from single-objective optimization become inadequate for the purpose of multi-objective model analysis. We shall address this question in more detail in the next section.

4 Multi-objective modeling and simulation

While single-objective mathematical optimization models are in a sense closed and distinct from simulation models that are typically used for analytical representations of knowledge in a given substantive field, multi-objective optimization models can be formulated as a natural, open extension of simulation models. If we admit that a decision maker in the real world can have multiple objectives and then we simulate a part of this world by a model, we can simply treat various quantities represented by variables of the model as possible objectives, while the final selection of the objectives will be made by the user - the analyst or the decision maker. The model might not be complete in the sense that it might not represent all the concerns of the user, but then either it must be reformulated even for simulation purposes or its incompleteness must be overtly admitted and accounted for in the analysis.

Thus, an analytical model of the substantive aspects of a decision situation typically contains:

- actions or decisions represented by *decision variables*;

- potential objectives represented by *outcome variables*;

- various *intermediate variables* (state variables, balance variables etc.) that are essential for a flexible model formulation;

- *parametric variables* or parameters that might remain constant during model simulations but are essential for model validation and alternative model variants;

- *constraining relations* (inequalities, equations etc.) that determine the set of admissible decisions and are usually divided into direct decision constraints that involve only decision variables and indirect constraints that involve also outcome and intermediate variables;

- *outcome relations* that determine how the outcome variables depend on the decision variables (often not directly, with the help of intermediate variables and equations such as state equations in dynamic models, often with the help of recursive or even implicit formulae);

- a *representation of model uncertainty* (in probabilistic, fuzzy set, set-valued, etc., terms); if such a representation is not explicit, we often call such a model "deterministic" and assume that it represents average situations.

While the typical models for single-objective optimization specify only one optimized outcome (*"the" objective function*) and treat all constraints inflexibly, the use of multi-objective modeling and optimization allows a flexible choice of objective variables between the outcome variables (if necessary, also between decision variables) and a much more elastic interpretation of constraints.

It is well known that (particularly indirect) constraints that are represented in single-objective optimization with a standard form, say, of an inequality, intend to model two quite different classes of phenomena of the real world. One of these classes contains balances that must be satisfied such as the balance of energy in a physical model, or domains of model validity such as the edges of a table for a model of the motion of a ball; these are so-called *hard constraints*. The other class contains balances that we would like to satisfy, such as the balance in a budget sheet. These constraints can be violated (at an appropriate cost) and are called *soft constraints*. Soft constraints can be modeled even in single-objective optimization by appropriate penalty terms in the objective function; but then the answer to the question what are their permissible violations calls for additional judgment. In multi-objective modeling and optimization, soft constraints are most naturally interpreted as additional objectives and their evaluation is thus included in the overall evaluation of a multi-objective solution.

Having formulated a multi-objective model, one has to estimate its parameters and validate it – that is, check whether the model represents adequately not only the formal, but also the intuitive side of expert knowledge in a given substantive field. While there are many methods of parameter estimation and formal model validation, depending on particular model type and described in a broad literature, the intuitive model validation usually relies on repetitive simulation: the model must be run many times by experts in the field of knowledge under changing assumptions about decisions (or their scenarios in case of dynamic models) or even parameters, and the outcomes obtained (or their trajectories in the dynamic case) must be compared against the formal knowledge and the intuition of the experts. It has often been stressed that most valuable are models that can produce also counter-intuitive results; but the experts must be able to internalize such results, that is, explain to themselves why these results are obtained and check with their intuition (also by additional research and experiments) whether these results can also occur in the real world; otherwise, counter-intuitive results are useless in learning.

The way that various constraints are treated during the simulation of a model is also essential for its validation. Typical approaches to simulation and existing simulation languages usually allow for only the inclusion of direct decision constraints that can be represented by admissible ranges of decision variables; they do not allow for an inclusion of indirect constraints nor for a distinction between hard and soft constraints. Moreover, expert users of simulation models are often interested in inverse simulation, in which desirable trajectories of model outcomes are specified by the user and decision variables should be chosen during the simulation to result in model outcomes close to the specified trajectories. Inverse simulation is particularly useful in scenario generation. Moreover, good simulation techniques should make it possible to perform sensitivity analysis of simulated solutions along with simulation runs. All of these issues - simulation under constraints, inverse simulation, scenario generation, and sensitivity analysis - can be included in sufficiently sophisticated methods of simulation that use optimization techniques and multi-objective approaches as tools of simulation support. Both IIASA and the Polish DSS community have contributed considerably to the development of such methods (Kallio et al., 1980; Grauer et al., 1982; Kurzhanski, 1986; Lewandowski et al., 1989; Makowski and Sosnowski in this volume).

We shall illustrate some of these issues on the example of possible standards of definition and methods of analysis of nonlinear dynamic multistage (time-discrete) models, for the sake of brevity without uncertainty representation, often used in various applications of systems analysis. Textbooks advise formulating them in the state *equation form*:

$$w[t+1] = f(w[t], x[t], t), \text{ with } t = 0, 1, ..., T \text{ and } w[0] \text{ - given} \tag{1}$$

where the square brackets stress the discrete nature of the variable t, usually interpreted as a discretized time, $w[t]$ is the *vector of state variables* (usually denoted by $x[t]$ in control theory, but we use $w[t]$ to stress the internal variable nature of the state), $x[t]$ is the *vector of decision variables* (control variables, usually denoted by $u[t]$ in control theory). Even if we use this form, there is a question of useful standards for the definition of the vector function f - should it contain only standard nonlinear functions say, as admitted by the languages PASCAL or FORTRAN, or should it also admit logical expressions (which makes the standard much more versatile, but increases the difficulties of analytical support).

Although it is essential that the user understands the importance and the properties of the concept of the state of a dynamic system, the form (1) *is the wrong standard of definition of nonlinear (or even linear) multistage dynamic models for a versatile multi-objective simulation* - a good analyst while modeling his substantive expertise does not think in such terms. He thinks rather in terms of various outcomes or *intermediate variables* $y_i[t]$ that depend on selected *actions or decision variables* $x_i[t]$ at a given stage t; moreover, he usually defines his outcome variables *recursively*: the next ones depend on the previously defined ones, not only directly on decision variables. He also specifies the dependence on *parametric variables* $z[t]$ that might be constant (or a specified function of time t) during a simulation, but might vary in sensitivity analysis and in consecutive simulation runs. He must not forget that he is building a dynamic model, hence he should keep in mind his state variables $w[t]$ – but usually included as part of his outcome variables, $w[t] = \{y_i[t]\}_{i \in I}$ – and note that the outcome variables might also depend on the state in the previous time instant. Thus, the *right standard* for this type of models is the following recursive definition:

$$y_1[t] = h_1(w[t-1], x[t], z[t], t),$$

$$y_2[t] = h_2(w[t-1], y_1[t], x[t], z[t], t),$$

....

$$y_n[t] = h_n(w[t-1], y_1[t], ..., y_{n-1}[t], z[t], t) \tag{2}$$

when the functions h_i, $i = 1, ..., n$, are consecutively specified by the modeler. The form (2) might be called an *explicit structural form* of the model; its structure, i.e. the way the functions h_i are specified, is a very important element of the substantive knowledge encoded by the modeler. But what if the modeler changes the order of definitions of these functions and, through an oversight or purposely, introduces a loop in the dependence between outcome variables y_i such that the model cannot be computed explicitly? Good simulation support software should automatically check such cases and ask the modeler, whether he purposely wants to introduce a model of a more difficult class – in the *implicit structural form* – that will be commented upon later.

After specifying the structural form of the model for one instant of time t, the modeler might check the dynamic structure of his model in the *structural form of state equations*, defined jointly by (2) and by the selection of state variables:

$$w[t] = \{y[t]\}_{i \in I}; \quad w[0] \text{ - given} \tag{3}$$

which the software for simulation support might check for inconsistencies, or even transform by symbolic computation to form (1) and display for the modeler. In a good software package for simulation support, more complicated time structures should also be admitted and supported; later we shall give later an example, but omit the details for the sake of brevity.

The modeler should not be obliged to repeat the definition of the model for all $t = 1, ..., T$; rather he should be supported by being asked for editorial changes of the model for consecutive time instants.

All constraints that the modeler wants to specify can be expressed as lower or upper bounds either on decision variables $x_i[t]$ (direct constraints) or on outcome variables $y_i[t]$ (indirect constraints; the modeler might include some outcome variables for the sole purpose of specifying constraints). As mentioned before, good simulation software should handle also simulation with constraints, while the modeler must be aware that *indirect constraints require actually the use of optimization tools during simulation.*

Simulation with constraints is actually equivalent to implicit structural model definition; in both cases, at each instant of time t the simulation software must employ additional iterative algorithms to specify such changes of decision and outcome variables that the indirect or implicit constraints are satisfied with a given accuracy. The *modeler should be aware not only of his expert interpretation of such constraints, but also of the fact that they actually introduce a second, faster time scale in the model.*

Consider, for example, an economic model where an implicit structural model definition expresses the assumption of market equilibration. A good simulation algorithm should handle this and define the equilibrium variables with the help of either an optimization or a fixed-point type algorithm; but this not only takes computer time, it somehow represents the actual market equilibration mechanism that has its own dynamics on a (hopefully) faster time scale that could be represented by an additional time index τ. Thus, the modeler should ask himself: am I right in the assumption that the time scale for τ is really much faster than that for t? The answer might often be negative, as it is actually shown by the transforming economies of Eastern Europe, and an implicit structural model definition should be avoided in such cases. We can then either include equilibration mechanisms directly in the state equations or explicitly build a model with two different time scales (a faster one for commodity markets with international commerce, a slower one for capital markets and investments).

Good simulation software must support not only simulation's with constraints, but also sensitivity analysis of the model. For the standard form of the state equation (1) it is well known that the gradients of various variables in a dynamic model can be usefully computed (Wierzbicki, 1984), if we simulate also the adjoint or co-state equation in the reverse direction of time:

$$p[t-1] = \frac{\partial f}{\partial w}(w[t], x[t], t)p[t] \tag{4}$$

while the conditions for $p[T]$ and the use of the adjoint vector $p[t]$ to compute various gradients depend on a more specific model definition. A typical expert in a substantive area might be a good modeler, but he usually will not be bothered by the intricacies of dynamic sensitivity analysis; even if he is willing to learn about this field, the specification, by hand, of many necessary derivatives is always a source of errors. Moreover, textbooks tell us how to perform sensitivity analysis for the standard textbook form of state equations; it is much more difficult for the structural form.

Yet today good simulation software could perform all these tasks for the modeler: determine all necessary derivatives by symbolic manipulation, take into account the dynamic structure of the model, and display to the user not only the analytic forms of partial derivatives but also sensitivity coefficients requested by him and computed during a simulation run. Ready, off-the-shelf symbolic manipulation software does not help much here - it does not account for complicated structures of dynamic models and must be incorporated into the simulation software. Until now, no commercial simulation software exists that meets all of these requirements; but some of them are already satisfied, e.g., in the prototype system DIDAS-N (Kreglewski et al., 1989), although the system supports rather static nonlinear models while fully dynamic simulation is not supported yet.

If the simulation software meets such high requirements, it is relatively easy to supplement it with tools for multi-objective model analysis. Sensitivity analysis and optimization modules are needed anyway; a multi-objective model analysis module for a good simulation system should make it possible for the user to run also an inverse or multi-objective dynamic simulation. In such a mode, the user should be able to select some *outcome trajectories* q as elements of objective space (similarly as he has chosen the state variables, though from a possible different index set $J \neq I$),

$$q = \{q[1], q[2], ...q[T]\}; \quad q[t] = \{y_i[t]\}_{i \in I}, \tag{5}$$

and then to specify what to do with these trajectories during the simulation - to maximize or minimize their components (for all time instants - though more flexible specifications of selected time instants should be also possible) or to *stabilize* some or all their components along a *reference trajectory* $\bar{q}$ defined by the user. The software system would then define the deviations:

$$\Delta q_i[t] = q_i[t] - \bar{q}_i[t] \text{ for maximized trajectories,}$$

$$\Delta q_i[t] = \bar{q}_i[t] - q_i[t] \text{ for minimized trajectories,}$$

$$\Delta q_i[t] = |\, q_i[t] - \bar{q}_i[t]\, | \quad \text{for stabilized trajectories} \tag{6}$$

use them in a standard way to define an order-consistent achievement scalarizing function (Kreglewski et al., 1989; Wierzbicki, 1991) and support multi-objective simulation by maximizing this function. The user might additionally specify which decision variables should be treated as parametric ones (determined only by the user) and which might be changed in optimization (possibly from initial values determined by the user as in ordinary simulation).

In such a simulation, either a nondifferentiable optimization algorithm must be used or a differentiable approximation of the nondifferentiable achievement scalarizing function utilized; there are positive experiences with both approaches for models of a modest scale. Such multi-objective simulation of a dynamic model produces a trajectory that is in a sense closest (or better, if this is possible for admissible trajectories in the case of maximization or minimization) to the reference trajectory. Several runs of multi-objective simulation help the user to explore the properties of the model much faster than ordinary simulation.

5 Conclusions

The theoretical and methodological developments, as well as the experiences from applications, gathered by the Polish DSS community working in close cooperation with IIASA and other international networks of researchers, not only result in reflections and refinements of DSS methodology, but also show the importance and directions for further work. One such direction is the further development of much more advanced, modular software systems that will considerably help modelers to encode their expertise in the form of specialized substantive analytical models. Such models constitute an important dimension of knowledge representation for decision support.

References

Andriole, S.J. (1989). *Handbook of Decision Support Systems*. TAB Professional and Reference Books, Blue Ridge Summit, PA.

Bronisz, P., L. Krus, and A. P. Wierzbicki (1989). Towards Interactive Solutions in a Bargaining Problem. In A. Lewandowski and A.P. Wierzbicki, eds., *Aspiration Based Decision Support Systems*. Lecture Notes in Economics and Mathematical Systems Vol. 331, pp. 251-268. Springer-Verlag, Berlin - Heidelberg.

Charnes, A., and W. W. Cooper (1977). Goal programming and multiple objective optimization. Journal for Operational Research, Vol. 1, pp. 39-54.

Cooke, S., and N. Slack (1984). *Making Management Decisions*. Prentice-Hall, Englewood Cliffs.

Dreyfus, H., and S. Dreyfus (1986). *Mind over Machine: The Role of Human Expertise and Intuition in the Era of Computers*. Free Press, New York, NY.

Grauer, M., A. Lewandowski, and L. Schrattenholzer (1982). Use of the reference level approach for the generation of efficient energy supply strategies. WP-82-19. IIASA, Laxenburg.

Haimes, Y. Y., and W. A. Hall (1974). Multiobjectives in water resource systems analysis: the surrogate trade-off method. *Water Resource Research*, Vol. 10, pp. 615-624.

Ignizio, J. P. (1978). Goal programming: a tool for multiobjective analysis. *Journal for Operational Research*, Vol. 29, pp. 1109-1119.

Kacprzyk, J., and M. Fredrizzi (eds.) (1988). *Combining Fuzzy Imprecision with Probabilistic Uncertainty in Decision Making.* Lecture Notes in Economics and Mathematical Systems Vol. 310. Springer-Verlag, Berlin - Heidelberg.

Kallio, M., A. Lewandowski, and W. Orchard-Hays (1980). An implementation of the reference point approach for multi-objective optimization. WP-80-35. IIASA, Laxenburg.

Keeney, R., and H. Raiffa (1976). *Decisions with Multiple Objectives: Preferences and Value Trade-Offs.* John Wiley, New York, NY.

Khomenyuk, V. V. (1977). *Optimal Control Systems* (in Russian). Nauka, Moscow.

Kiwiel, K. C., and A. Stachurski (1989). Issues of Effectiveness Arising in the Design of a System of Nondifferentiable Optimization Algorithms. In A. Lewandowski and A.P. Wierzbicki (eds.). *Aspiration Based Decision Support Systems.* Lecture Notes in Economics and Mathematical Systems Vol. 331, pp. 180-192. Springer-Verlag, Berlin - Heidelberg.

Korhonen, P., and J. Laakso (1986). Solving a generalized goal programming problem using visual interactive approaches. *European Journal of Operational Research 26*, pp. 355-363.

Kreglewski, T., J. Paczynski, J. Granat, and A. P. Wierzbicki (1989). IAC-DIDAS-N: A Dynamic Interactive Decision Analysis and Support System for Multicriteria Analysis of Nonlinear Models. In A. Lewandowski and A.P. Wierzbicki (eds.). *Aspiration Based Decision Support Systems.* Lecture Notes in Economics and Mathematical Systems Vol. 331. pp. 378-381, Springer-Verlag, Berlin - Heidelberg.

Kurzhanski, A. B. (1986). Inverse problems in multiobjective stochastic optimization. In Y. Sawaragi, K. Inue and H. Nakayama (eds.). *Towards Interactive and Intelligent Decision Support Systems.* Springer-Verlag, Berlin - Heidelberg.

Lewandowski, A., T. Kreglewski, T. Rogowski, and A. P. Wierzbicki (1989). Decision Support Systems of DIDAS Family. In A. Lewandowski and A. P. Wierzbicki, (eds.). *Aspiration Based Decision Support Systems.* Lecture Notes in Economics and Mathematical Systems Vol. 331, pp. 21-47. Springer-Verlag, Berlin - Heidelberg.

Messner, S. (1985). Natural Gas Trade in Europe and Interactive Decision Analysis. In G. Fandel et al., (eds.). *Large Scale Modeling and Interactive Decision Analysis*. Springer-Verlag, Berlin - Heidelberg.

Mikhalevich, V. S., and V. L. Volkovich (1982). *A Computational Method for Analysis and Design of Complex Systems* (in Russian). Nauka, Moscow.

Naisbitt, J. (1982).*Megatrends - Ten New Directions Transforming Our Lives*. Warner Books, New York, NY.

Nakayama, H., and Y. Sawaragi (1983). Satisficing Trade-off Method for Multi-Objective Programming. In M. Grauer and A.P. Wierzbicki (eds.). *Interactive Decision Analysis*. Lecture Notes in Economics and Mathematical Systems, Springer-Verlag, Berlin - Heidelberg.

Ogryczak, W., K. Studzinski, and K. Zorychta (1989). Solving Multiobjective Distribution-Location Problems with the DINAS System. In A. Lewandowski and A. P. Wierzbicki (eds.), *Aspiration Based Decision Support Systems*. Lecture Notes in Economics and Mathematical Systems, Vol. 331, pp. 230-250. Springer-Verlag, Berlin - Heidelberg.

Pawlak, Z. (1991). *Rough Sets: Some Aspects of Reasoning about Knowledge*. Kluwer Academic Publishers, Dordrecht (forthcoming).

Rapoport, A. (1989). *Decision Theory and Decision Behavior*. Kluwer Academic Publishers, Dordrecht.

Rogowski, T. (1989). Dynamic Aspects of Multiobjective Optimization in Decision Support Systems. In A. Lewandowski and A. P. Wierzbicki (eds.), *Aspiration Based Decision Support Systems*. Lecture Notes in Economics and Mathematical Systems Vol. 331, pp. 92-105. Springer-Verlag, Berlin - Heidelberg.

Ruszczynski, A. (1991). Reference Trajectories in Multistage Stochastic Optimization Problems. IIASA Workshop on User-Oriented Methodology and Techniques of Decision Analysis and Support, Serock near Warsaw, September.

Salukvadze, M. E. (1971). On the optimization of vector functionals, 1. Programming optimal trajectories, 2. Analytic construction of optimal controllers. *Automation and Remote Control*. Vol. 31, No. 7, 8.

Salukvadze, M. E. (1974). On the existence of solutions in problems of optimization under vector-valued criteria. *JOTA*, Vol. 13, No. 2, pp. 203- 217.

Sawaragi, Y., H. Nakayama, and T. Tanino (1985). *Theory of Multiobjective Optimization*. Academic Press, Orlando Fl.

Seo, F., and M. Sakawa (1988). *Multiple Criteria Decision Analysis in Regional Planning: Concepts, Methods and Applications*. D. Reidel Publishing Company, Dordrecht.

Simon, H. A. (1957). *Models of Man.* Macmillan, New York, NY.

Slowinski, R., and J. Teghem (eds.) (1990). *Stochastic versus Fuzzy Approaches to Multiobjective Mathematical Programming under Uncertainty.* Kluwer Academic Publishers, Dordrecht.

Steuer, R. E. (1986). *Multiple Criteria Optimization: Theory, Computation and Application.* John Wiley, New York, NY.

Steuer, R. E., and E. V. Choo (1983). An interactive weighted Chebyshev procedure for multiple objective programming. *Mathematical Programming.* Vol. 26, pp. 326-344.

Strubegger, M. (1985). An approach for integrated energy - economy decision analysis: the case of Austria. In G. Fandel et al. (eds.). *Large Scale Modeling and Interactive Decision Analysis.* Lecture Notes in Economics and Mathematical Systems, Vol. 273. Springer-Verlag, Berlin - Heidelberg.

Wierzbicki, A. P. (1977). Basic properties of scalarizing functionals for multiobjective optimization. Mathematische Operationsforschung and Statistik, s. *Optimization,* Vol. 8, pp. 55-60.

Wierzbicki, A. P. (1980). The use of reference objectives in multi-objective optimization. In G. Fandel, T. Gal (eds.). *Multiple Criteria Decision Making: Theory and Applications.* Lecture Notes in Economics and Mathematical Systems, Vol. 177, pp. 468-486. Springer-Verlag, Berlin - Heidelberg.

Wierzbicki, A. P. (1982). A mathematical basis for satisficing decision making. *Mathematical Modeling.* Vol.3, pp. 391-405.

Wierzbicki, A. P. (1984). *Models and Sensitivity of Control Systems.* Elsevier, Amsterdam.

Wierzbicki, A. P. (1986). On the completeness and constructiveness of parametric characterizations to vector optimization problems. *OR-Spektrum.* Vol. 8, pp. 73-87.

Wierzbicki, A. P. (1989). Multiobjective decision support for simulated gaming. In A.G. Lockett and G. Islei (eds.). *Improving Decision Making in Organizations.* Proceedings, Manchester. Lecture Notes in Economics and Mathematical Systems, Vol. 335, Springer-Verlag, Berlin - Heidelberg.

Wierzbicki, A. P. (1990). Multiple criteria solutions in noncooperative game theory. Part III: Theoretical foundations. Kyoto Institute of Economic Research, Discussion Paper No. 288.

Wierzbicki, A. P. (1991). Dynamic Aspects of Multi-Objective Optimization. In A. Lewandowski, and V. Volkovich (eds.). *Multiobjective Problems of Mathematical Programming.* Proceedings, Yalta 1988. Lecture Notes in Economics and Mathematical Systems Vol. 351, pp. 154-174. Springer-Verlag, Berlin - Heidelberg.

Yu, P. L. (1985). *Multiple-Criteria Decision Making - Concepts, Techniques and Extensions*. Plenum Press, New York and London.

Zadeh, L. A. (1978). Fuzzy sets as a basis for a theory of possibility. *Fuzzy Sets and Systems* Vol. 1, pp. 3-28.

Zeleny, M. (1973). Compromise programming. In J. L. Cochrane and M. Zeleny (eds.). *Multiple Criteria Decision Making*. University of South Carolina Press, Columbia, SC.

An Application of the Analytic Centers to a Generic Nondifferentiable Minimization Problem

Anna Altman

Systems Research Institute
Polish Academy of Sciences, Poland

Abstract

An application of the concept of analytic centers to generic nondifferentiable minimization problem is shown. The proposed method is based on a cutting planes technique defining a sequence of linear programming problems. Every such sub-problem is solved using a projective method for computing an analytic center of a polytope. It is equivalent to minimizing Karmarkar's potential. Supporting hyperplanes are generated in approximated analytic centers. The algorithm is compared with the older projective one. Numerical results are given.

1   Introduction

This paper deals the computer implementation of the method for nondifferantiable convex minimization (NDCM) proposed by (Goffin, J-L et al., 1991). It is an application of a variant of an interior-point algorithm for linear programming (LP) to a cutting planes method for the minimization of a nondifferentiable convex function defined by the supporting hyperplanes to its epigraph.

The polynomial interior-point methods can be classified into two broad categories. The primary concept for the projective algorithm of Karmarkar and its numerous variants and extensions is potential function. These methods are based on an efficient minimization procedure, which guarantees a linear decrease of this potential. Other methods, called path-following ones, use the idea of the analytic center to guarantee a linear decrease in the duality gap. Both categories are strictly connected.

Notions of the potentials and centers are strongly related. Given a LP problem in the form of the maximization of an objective function under a set of inequality constraints and given a known lower bound for the optimal value, one naturally associates the polytope, which is defined by the inequality constraints of the problem and the additional constraint that the objective value achieves a value larger than the lower bound. The anlytic center of this polytope minimizes the product of the slacks of the constraints, including the one associated with the objective function. By letting the lower bound increase to the optimal value of the linear program one obtains a trajectory of centers. If a similar approcach is taken on the dual, the center for the dual can be defined. Hence centers come by dual pairs. Any algorithm that minimizes Karmarkar's potential for the primal problem, computes also the analytic center of the dual polytope and v.v.

Cutting planes are generated at the analytic centers of special polytopes. These polytopes are determined by the outer approximation of the epigraph associated with the current LP relaxation and limited above by the best observed objective value. The adventage of the computing supporting hyperplanes at analytic centers stems from the fact that analytic centers summarize all the information accumulated in the past about the problem.

The analytic centers are especially useful in NDCM. Due to this approach the number of supporting hyperplanes can be reduced compared with the old Karmarkar-type algorithm described in (Altman, A., 1990). In the old projective algorithm for NDCM the upper bound is modified after every computed projection and , if necessary, the new supporting hyperplane is added. In the new one these two operations are made after every computation of the analytic center, which is equivalent to several computed projections.

The paper is organized as follows: In section 2 the analytic centers are defined. In section 3 NDCM problem and its approximation are described. In section 4 the algorithm for computing analytic centers and in section 5 the algorithm for NDCM are shown. In section 6 two projective methods for NDCM are compared. In section 7 results of some classical test problems are given. These results are compared with old ones from (Altman, A., 1990). Conclusions are in the last section.

2   Analytic centers

Consider the pair of LP problems

$$\min\{c^T x : Ax = b, x \geq 0\} \tag{1}$$

and its dual

$$\max\{b^T u : A^T u + s = c, s \geq 0\}, \tag{2}$$

where $s = c - A^T u$ is the vector of dual slack variables. Let up be a strict upper bound for the minimal value of (1) and low be a strict lower bound for the maximal value of the (2). Let us define two polytopes

$$F(up) = \{x : c^T x \leq up, Ax = b, x \geq 0\}$$

and

$$H(low) = \{u : b^T u \geq low, A^T u \leq c\}.$$

The analytic center of $F(up)$ is defined as the unique maximizer of $\prod_{j=0}^{n} x_j$ subject to $x_0 = up - c^T x, Ax = b$ and $x_j \geq 0, j = 0, \ldots, n$. Similarly the analytic center of $H(low)$ is defined as the unique maximixer of $\prod_{j=0}^{n} s_j$ subject to $s_0 = b^T u - low, A^T u + s = c$.

The analytic center depends on the analytic definition of the polytope. For two geometrically identical polytopes with different analytical definition analytic centers can be different. If some constraints in the definition of the polytope are duplicated the polytope will not be changed, but its analytic center moves further from the duplicated hyperplane.

3 The description of the problem

We consider the generic nondifferentiable convex optimization problem

$$\min_x \{f(x) : l \leq x \leq h\}, \tag{3}$$

where $f : \mathcal{R}^n \mapsto \mathcal{R}$ is convex function, $x, l, h \in \mathcal{R}^n$.

Every convex function can be defined as an envelope of its supporting hyperplanes with normal vectors g_i, i.e.

$$f(x) = \max_{i \in I} \{d_i - \sum_{j=1}^{n} g_{ij} x_j\} \tag{4}$$

and set I can be infinite.

Using definition (4) we can formulate the problem (3) in equivalent form

$$\min_x \max_{i \in I} \{d_i - \sum_{j=1}^{n} g_{ij} x_j : l \leq x \leq h\}. \tag{5}$$

We can approximate the problem (5) by a family of subproblems with finite subsets K of the set I,

$$\min_x \{f_K(x) : l \leq x \leq h\},$$

where F_K is CPL (convex piecewise linear) approximation of f, i.e.

$$f_K(x) = \max_{i \in K} \{d_i - \sum_{j=1}^{n} g_{ij} x_j\}.$$

Every such subproblem can be formulated as a linear programming problem in primal form

$$\min x_1 \tag{6}$$

$$\text{s.t. } \overline{G}_K x \geq d_K, l \leq \overline{x} \leq h,$$

$$\overline{G}_K = \{1_m, G_K\} \in \mathcal{R}^{m*(n+1)},$$

$$x = (x_1, \overline{x}) \in \mathcal{R}^{n+1}, l, h \in \mathcal{R}^n, d_K \in \mathcal{R}^m,$$

or in the dual one

$$\max \tilde{c}_K^T y \tag{7}$$

$$\text{s. t. } \tilde{G}_K y = 0,$$

where

$$\tilde{c}_K^T = (l^T, -h^T, d_K^T) \in \mathcal{R}^p, p = m + 2 * n,$$

$$\tilde{G}_K = \begin{bmatrix} & g^T & \\ I_n & -I_n & G_K^T \end{bmatrix} \in \mathcal{R}^{n*p},$$

$$g^T = (0_n^T, 0_n^T, 1_m^T) \in \mathcal{R}^p,$$

$$y^T = (v^T, w^T, u^T) \in \mathcal{R}^p.$$

To the problem (7) we add a homogenizing variable $s_1 \geq 0$ with the normality condition $s_1 = 1$. Knowing the upper bound up for the optimal value of (7) we formulate the linear feasibility problem parametrized with up

$$c_K^T s = 0, \tag{8}$$

$$A_K s = 0,$$

$$s_1 = 1,$$

where

$$s^T = (s_1, y^T) \in \mathcal{R}^{p+1},$$

$$c_K^T = (-up, \tilde{c}_K^T) = (-up, l^T, -h^T, d_K^T) \in \mathcal{R}^{p+1},$$

$$A_K = \begin{pmatrix} -1 & \tilde{G}_K \\ 0_n & \end{pmatrix} = \begin{pmatrix} -1 & 0_n^T & 0_n^T & 1_m^T \\ 0_n & I_n & -I_n & G_K^T \end{pmatrix} \in \mathcal{R}^{(n+1)*(p+1)},$$

and define the potential function

$$\phi_K(s) = (p+1)\ln(-c^T s) - \sum_{j=1}^{p+1} \ln s_j. \tag{9}$$

We also define the polytope

$$F(up) = \{x \in \mathcal{R}^{n+1} : A_K^T x \geq c_K\} = \{(x_1, \bar{x}) : l \leq \bar{x} \leq h, G_K \bar{x} \geq d_K, x_1 \leq up\}, \tag{10}$$

where up is upper bound for objective value for (6).

The analytic center of $F(up)$ is a point belonging to $ri(F(up))$ and maximizing $\prod_{i=1}^{p+1} z_i$.

$$\max_x \prod_{i=1}^{p+1} z_i,$$

$$\text{where } z_i = (A_K^T x - c)_i, \text{ for } i = 1, \ldots, p+1. \tag{11}$$

The first order optimality conditions for the problem (11) are:

$$A_K \lambda = 0, \tag{12}$$

$$Z^{-1} 1 - \lambda = 0,$$

$$A_K^T x - z = c_K,$$

where λ is taken as Lagrange's multiplier associated with the constraint $A_K^T x - z = c_K$ and Z is diagonal matrix with entries from vector z.

Computing the analytic center of $F(up)$ is related to minimizing Karmarkar's potential given by the formula (9). Minimizing Karmarkar's potential is aproximated by three first elements of the Taylor series. Our problem will have a form

$$\min_q \|q - 1_{p+1}\|_2 \tag{13}$$

$$\text{s. t. } c_K^T S q = 0,$$

$$A_K S q = 0,$$

$$s_1 = 0,$$

where S is diagonal matrix with entries from vector s.

Geometrically it means searching the projection of vector 1_{p+1} onto null space of matrix $\begin{bmatrix} c_K^T S \\ A_K S \end{bmatrix}$. Using Vial's formula (Vial, J-P., 1989) we can compute q using projections onto $ker(A_K S)$, then

$$q = 1_{p+1} - \frac{c_K^T}{\|\gamma\|_2^2} \gamma, \tag{14}$$

where γ is projection of vector Sc_K onto $ker(A_K S)$.

Finding γ is equivalent to the linear least square problem of searching

$$\min_x \| Sc_K - SA_K^T x \|_2.$$

Then $SA_K^T x_{min}$ is the projection of Sc_K onto $im(SA_K^T)$ and $\gamma = Sc_K - SA_K^T x_{min}$. A QR decomposition for solving this problem was used. A fully description of the can be found in (Altman, A., 1990). Subroutines from the LINPACK library (Dongarra, J. R., et al., 1978) were exploited.

4 Computing the analytic center

We will compute the analytic center for the polytope $F(up)$ defined in (10). Let $\bar{x}$ satisfy the box constraints for (3) and up be a value of the function in the point $\bar{x}$. Then the pair $(up, \bar{x})$ is feasible to the primal problem (6) for every $K \subset I$. The value up is a valid upper bound for dual restricted problem (7), also for every $K \subset I$. The interior feasible point s is constructed from any interior, but not necessarily, feasible point. We used the idea from (Goffin, J-L. and Vial J-P., 1990).

The input data for the projective algorithm for analytic centers are

- ε - tolerance for objective,

- $\theta \in (0, 1)$ - threshold number for q,

- up - upper bound for (7).

At first we compute vector q using (14) and its Euclidean norm. While $\|q\|_2 > \theta$ or $-c^T s (= up - \tilde{c}^T y) > \varepsilon$ we do, so called, inner loop of the algorithm.

If $\|q\|_2 \geq 1$ set $\bar{\alpha} := 1$, otherwise $\bar{\alpha} := arg\min_{\alpha > 0} \{ \phi_K(s + \alpha Sq, up) : s + \alpha Sq > 0 \}$. The new s will be generated in two steps. In the first one $\bar{s} := s + \alpha Sq$, in the next one $s := \bar{s}/s_1$. The choice of $\alpha = 1$ when $\|q\|_2 < 1$ is dictated by the fact that this condition on q defines a domain of quadratic convergence of the algorithm with unit step length.

5 The basic steps of the algorithm for NCDM

As a result of the projective algorithm for analytic centers we have the point s in dual space and the analytic center

$$x_{AC} = (A_K S^2 A_K^T)^{-1} A_K S^2 c_K. \tag{15}$$

Since f_K is a relaxation of f, one always has $f_K \leq f$. If $f(x_{AC}) > f_K(x_{AC})$, then a new supporting hyperplane for $epi(f)$ must be generated. We construct it in the point x_{AC}, computing a subgradient of f there. A new index will be added to the set K. All matrices, vectors and functions with subscript K will be changed. We will also update upper bound up, $up := \min\{f(x_{AC}), up\}$.

The crucial problem in this procedure is obtaining an interior feasible point for (8) after updating set K. The method inspired by Mitchell (Mitchell, J. E., 1988) described in (Goffin J-L. and Vial J-P., 1990) was used. Let A_K denote the constraints matrix for old set K. The new matrix is obtained by adding a column $\begin{bmatrix} 1 \\ a \end{bmatrix}$ to the old one. Let δ be a vector in $\mathcal{R}^{p+1}$ satisfing $A_K S \delta = - \begin{bmatrix} 1 \\ a \end{bmatrix}$. The point $\begin{bmatrix} s + \alpha S \delta \\ \alpha \end{bmatrix}$ for all $\alpha > 0$ will be feasible for (8) with new K. We choose vector δ to minimize its norm and parameter α to minimize Karmarkar's potential (9). In this procedure a linear least square problem must be solved. We use here, as in computing q in (13), a QR decomposition.

Our algorithm terminates when

- maximum of supporting hyperplanes is exceeded,

- stopping criterion $-c_K^T s \leq 10^{-6} \max\{1, |up|\}$ is satisfied.

6 Comparing two projective methods for NDCM

We will compare the current algorithm, called the analytic center algorithm AC, with the projective algorithm PA described in (Altman, A., 1990). A new normalizing variable $s_1 = 1$ in AC was introduced and therefore the dimension of a dual feasible region increases.

In PA after every computation of q the value of the function is checked, the upper bound up is modified and, if necessary, a new supporting hyperplane is added. The number of supporting hyperplanes grows up very quickly and with every new supporting hyperplane the dimensions of constraints matrix grow up too. The upper bound up is modified more often in PA, but changes are smaller. The PA is more time- and memory-consuming.

The Goffin and Vial method (Goffin, J-L and Vial, J-P., 1990) for generating a new interior feasible point was used in PA. In AC Mithell's method (Mitchell, J. E., 1988) was used. The Mitchell method is more time consuming but gives much better points, i. e. values of Karmarkar's potential in these points are smaller.

7 Results

Some classical NDCM problems were solved using two projective methods PA and AC. All computer programs were written in FORTRAN 77 and implemented on IBM PC/AT and SUN SPARCserver 470. The subroutines from (Kiwiel, K. and Stachurski, A., 1989) for computing values and subgradients minimized functions were used.

In both methods we give the number of generated supporting hyperplanes. In PA the number of iterations means the number of computed projections. In AC the number of inner loops means the number of loops for computing analytic centers by algorithm described in section 4. We do not quote times of computing. The computations were made on the two different computers and times are not comparable. If somebody is interested in times of computation we can provide them.

Example 1. Shor's MinMax problem (Shor, N. Z., 1985), p. 138

$$f(x) = \max_{1,\dots,10}\{b_i \sum_{j=1}^{5}(x_j - a_{ij})^2\},$$

$n = 5, f_{min} = 22.60016, x_j^0 = 0, j = 1,\dots,4, x_5^0 = 1, l = -10^{-2} * \mathbf{1}, h = -2 * \mathbf{1}.$
For PA
number of generated supporting hyperplanes: 55,
number of iterations: 129.
For AC
number of generated supporting hyperplanes: 36,
number of inner loops: 30.

Example 2. Lemarechal's MinMax problem MAXQUAD (Lemarechal, C., 1978), p. 151.

$$f(x) = \max_{1,\dots,5}\{u^T A^i u - x^T b^i)\}$$

$n = 10, f_{min} = -.841408, x^0 = \mathbf{1}, l = -1 * \mathbf{1}, h = 3 * \mathbf{1}.$
For AC
number of generated supporting hyperplanes: 94,
number of inner loops: 42.

Example 3. The polyhedral problem TR48 (Kiwiel, K., 1987).

$$f(x) = \sum_{j=1}^{n} d_j \max_{1,\dots,n}\{x_i - a_{ij}\} - \sum_{i=1}^{n} s_i x_i,$$

$n = 48, f_{min} = -638565, x^0 = 0, l = -10^4 * \mathbf{1}, h = 10^4 * \mathbf{1}.$
For AC
number of generated supporting hyperplanes: 246,
number of inner loops: 55.

Example 4. L^1 approximation L1APR (Goffin, J-L. et al., 1991).

$$f(u, v, w) = \sum_{k=1}^{p} \sum_{l=1}^{q} |a_{kl} - u_k - v_l + w|,$$

where the data a_{kl} are randomly generated according to $N(\mu_{kl}, 1)$ distribution with $\mu_{kl} = k + l - \frac{p+1}{2} - \frac{q+1}{2}$, $p = 15, q = 4, n = 20, x^0 = 0, f_{min} = 33.1152, x^0 = 0, l = -100 * 1, h = 100 * 1$.

For PA $(\varepsilon = 10^{-3})$
number of generated supporting hyperplanes: 93,
number of iterations: 247.

For AC
number of generated supporting hyperplanes: 76,
number of inner loops: 58.

Example 5. Nonlinear multicommodity flow problem (Goffin, J-L., et al., 1991).

$$\min \sum_a \frac{y_a}{c_a - y_a}$$

$$s.\ t. Ex^c = f^c, x^c \geq 0, c \in C$$

$$\sum_{c \in C} x^c \leq y, y \geq 0,$$

where C is the set of commodities, E is the node-arc incidence matrix, f^c are the requirements of commodity c, and x_a^c is the flow of commodity c on arc $a, n = 22, f_{min} = 103.41202, x^0 = 1, l = 0, h = 10^4 * 1$.

For AC
number of generated supporting hyperplanes: 76,
number of inner loops: 58.

Example 6. Polyhedral problem GO0 – a simplicial pyramid (Goffin, J-L., et al., 1991).

$$f(x) = n \max_{1,\ldots,n} \{x_i - \sum_{i=1}^{n} x_i,$$

$n = 48, f_{min} = 0, x_j^0 = j - \frac{n+1}{2}, j = 1,\ldots,n, l = -10^4 * 1, h = 10^4 * 1$.
For PA $(\varepsilon = 10^{-3})$
number of generated supporting hyperplanes: 48,
number of iterations: 136.
For AC
number of generated supporting hyperplanes: 48,
number of inner loops: 96.

8 Conclusions

The numerical results show that the described method is attractive for NDCM. The AC is better than the PA method. It can be also improved in some ways. The weighted analytic centers can be introduced. The weights are interpreted as the number of duplicated constraints in the polytope definition. The analytic centers can be used to construct inner and outer ellipsoids for the polytopes. We will try to use these ellipsoids to eliminate redundant supporting hyperplanes.

References

Altman Anna (1990). A Computer Implementation of the Projective Algorithm for Convex Piecewise Linear and Generic Convex Functions Minimization. (in Polish) Working Paper of Systems Research Institute, ZTSW-6-18/90, 19 pp.

Dongarra, J., Bunch, J. R., Moller, C. B. and G. W. Stewart (1978). LINPACK Users Guide, SIAM Publications, Philadelphia.

Goffin Jean-Louis, Haurie, A. and Jean-Philippe Vial (1991). Decomposition and Non-differentiable Optimization with the Projective Algorithm. McGill University, Research Working Paper No. 91-01-17.

Goffin Jean-Louis and Jean-Philippe Vial (1990). Cutting Planes and Column Generation Techniques With the Projective Algorithm. Journal of Optimization Theory and Applications, vol. 65, No. 3, pp. 409-429.

Kiwiel Krzysztof (1987). Proximity control in bundle methods for convex nondifferentiable optimization. Manuscript, Systems Research Institute Polish Academy of Sciences.

Kiwiel Krzysztof and Andrzej Stachurski (1989). Issues of effectiveness arising in the design of a system of nondifferentiable optimization algorithms. In Lewandowski, A. and A. P. Wierzbicki (eds): Aspiration Based Decision Support System. Springer-Verlag, Berlin.

Mitchell, J. E. (1988). Karmarkar's Alforithm and Combinatorial Optimization Problems, PhD Thesis, Cornell University.

Lemarechal Claude (1978). Non-smooth Optimization. Pergamon Press, Oxford.

Shor, N. Z. (1985). Minimization Methods for Nondifferentiable Functions. Springer-Verlag, Berlin.

Vial Jean-Philippe (1989). A unified approach to projective algorithms for linear programming. Optimization - Fifth French-German Conference Castel Novel 1988, Lecture Notes in Mathematics 1405, pp. 191-220, Springer Verlag.

Classification Support Based on the Rough Sets Theory

Jerzy Stefanowski

Institute of Computing Science
Technical University of Poznań, Poland

Abstract

Problems of knowledge analysis for decision systems by means of the rough sets theory are considered in this paper. Knowledge coming from experience concerns classification of data and is represented in a form of an information system. Application of the rough sets theory to the analysis of information systems enables a reduction of superfluous information and the derivation of a decision algorithm. These results are used to support a classification of new facts. An idea of using "the nearest" rules to support this classification is presented.

1 Introduction

The paper is devoted to the problems of knowledge analysis for decision support systems. We consider a kind of human knowledge concerning classification of data obtained from observation, measurements etc. which can be represented in a form called an information system. The information system stores data about objects (examples, cases, observations etc.) described by multivalued attributes (features, tests, characteristics, variables etc.). The set of objects is classified into a disjoint family of classes. This classification results from the expert's knowledge in considered field.

The formal definition of the information system can be found in (Pawlak, Z., 1982). In this paper we use an interpretation of the information system as a finite table, columns of which are labelled by attributes, rows are labelled by objects and each row in the table represents information about an object. The set of objects and the set of attributes are finite. Attributes can be divided into condition ones which describe the state of objects and decision ones which establish a classification of objects.

Let us note that the domains of condition attributes are finite and of rather low cardinality. In this paper, we assume that attributes can have nominal or ordinal character. For ordinal attributes a ranking of their values can be established and for nominal ones one can only determine if there is a difference between values.

A simple example of information system is presented in Figure 1. It consists of 14 objects described by 6 condition attributes ($a1, ..., a6$) and classified into two classes $(1, 2)$ by a decision attribute.

no.	a1	a2	a3	a4	a5	a6	d
x1	2	3	2	1	1	1	1
x2	2	3	3	1	1	1	1
x3	1	2	3	2	1	2	1
x4	1	1	3	1	2	2	1
x5	2	1	2	1	1	2	1
x6	2	2	3	1	1	1	1
x7	1	1	3	2	2	1	1
x8	2	2	1	2	1	2	2
x9	1	1	1	2	2	1	2
x10	2	2	2	1	1	1	2
x11	1	1	2	1	1	2	2
x12	2	1	2	2	1	2	2
x13	2	1	3	1	2	2	2
x14	2	2	3	1	1	1	2

Figure 1: An example of an information system

Two stages are the most interesting in the analysis of knowledge contained in' a information system. The first one refers to:

- a reduction of all superfluous attributes and objects in the information system,

- an identification of relationships between the description of the object and its assignment to the certain class as well as the presentation of these relationships e.g. in a form of decision rules.

The second one refers to decision support in classifying of new objects (unseen in the expert's experience i.e. in the information system) on the basis of conclusions derived from existing expert's experience (i.e. results of solving of the first stage).

The rough sets theory, proposed by Pawlak (Pawlak, Z., 1982), has been proved to be an efficient tool for the first stage of the analysis. In this paper we present a new idea of using the results of knowledge analysis obtained by the rough sets theory in decision support for classifying of new objects.

2   The rough sets approach

We do not present basic concepts of the rough sets theory. Exhaustive information can be found in (Pawlak, Z. 1982; Pawlak, Z. 1991; or Fibak, J. et al., 1986; Słowiński, K., et al. 1988; Krusińska, E. et al., 1992; Słowiński, R. and Stefanowski, J., 1989).

The rough sets approach makes it possible to solve such major problems in the analysis of knowledge represented in information systems as: the evaluation of importance of attributes in relationships between the description of the object and its assignment to a certain class; reduction of all redundant objects and attributes so as to get a minimal

#1	if			(a2=3)					then	(d=1)		
#2	if	(a1=1)	and	(a2=2)					then	(d=1)		
#3	if	(a1=1)	and	(a2=1)	and	(a3=3)			then	(d=1)		
#4	if	(a1=2)	and	(a2=1)	and	(a3=2)	and	(a4=1)	then	(d=1)		
#5	if					(a3=1)			then	(d=2)		
#6	if			(a2=2)	and	(a3=2)			then	(d=2)		
#7	if	(a1=1)	and	(a2=1)	and	(a3=2)			then	(d=2)		
#8	if	(a1=2)	and	(a2=1)	and	(a3=3)			then	(d=2)		
#9	if	(a1=2)	and	(a2=1)	and	(a3=2)	and	(a4=2)	then	(d=2)		
#10	if	(a1=2)	and	(a2=1)	and	(a3=2)	and	(a4=1)	then	(d=1)	or	(d=2)

Figure 2: The decision algorithm

subset ensuring the same quality of classification as the set of all attributes. Moreover, the reduced information system can be identified with a decision table which shows relationships between the minimal subset of condition attributes and a particular class. Then, the decision algorithm can be derived from the decision table. It consists of decision rules which are logical statements. The decision rules can be deterministic, if they uniquely imply decisions or non–deterministic if they imply few possible decisions. Non–deterministic rules result from imprecisions in available descriptions of objects. The procedure for the derivation of the decision algorithm are presented in (Fibak, J. et al., 1986; Pawlak, Z., 1991)

For example, let us consider the information system presented in Figure 1. There is one minimal subset $\{a1, a2, a3, a4\}$. So, the information system is reduced using these attributes. Then, the decision algorithm can be derived from it. It is presented in Figure 2.

In recent years, the rough sets approach was applied to analysis of a large variety of information systems, in particular medical ones (e.g. Fibak, J. et al., 1986; Słowiński, K. et al., 1988; Słowiński, K. and Słowiński, R., 1990; or in technical diagnostics Nowicki, R. et al. 1990; Nowicki, R. et al., 1992). An exhaustive list of references is given in (Pawlak, Z., 1991). The approach was implemented as a very efficient program RoughDAS (Gruszecki, G. et al., 1990).

3 Classification of new objects

The results obtained using the rough sets theory, in particular the reduction of attributes and the decision algorithm have a great practical importance. For instance, the reduction may cause in medicine an elimination of superfluous clinical tests harmful for patients (cf. Fibak, J. et al, 1986; Słowiński, K., 1988) or it may decrease the cost of diagnosis in technical diagnostics. The decision algorithm shows all important relationships using the minimum number of decision rules and/or the minimum number of attributes appearing in all decision rules. Hence, the decision algorithm is more readable for the user than the initial information system.

Let us notice, however, that these results on the representation of important characteristic facts and relationships refer to existing experience (i.e. the analysed information system) and do not refer directly to new facts. In this first stage, the analysis is done under the Closed World Assumption (cf. Pawlak, Z., 1991).

On the other hand, these results represent knowledge gained by a specialist on all cases from his experience and in fact it is interesting and desirable for him to use this knowledge for supporting decisions concerning a classification of new objects (cf. Słowiński, K., 1991).

By new objects we understand objects unseen in the expert's experience (i.e. information system) which are described by values of attributes only (all or from the reduced set). The assignment of these objects to any class is unknown. The aim of the specialist or a decision maker is to predict this assignment on the basis of his knowledge coming from previous experience.

A concept of such decision support is presented in the next parts of the paper. We assume that experience represented in the information system is representative for the purposes of such classification support.

Let us also assume that the decision maker in the first stage of analysis of the information system has found minimal subsets of attributes and derived the decision algorithm. It should be noted, however, that this stage may be not simple and with unique results. It is possible to obtain several minimal subsets of attributes (each equally good) and/or several possible decision algorithms. There are no strict rules how to choose the proper one. Hence, this choice is done according to expert's opinions and his preferences (cf. Słowiński, K. et al., 1988; Nowicki, R. et al., 1990; Nowicki, R. et al., 1992).

The application of the obtained decision algorithm seems to be most natural in classification support of new objects. Knowing the description of the objects, by values of attributes, the decision maker is trying to find in the decision algorithm a decision rule matching the description of the new object, i.e. the rule in which conditions are fulfilled by the description of the new object. Two results of such matching are possible :

- the new object confirms present knowledge (i.e. its description of the new object matches a rule in the information system),

- the new object is completely new (i.e. its description does not match any rule in the information system).

In the first case, moreover, the new object may match a deterministic rule which determines uniquely a class or a non–deterministic one which assigns the object to several possible classes.

The second case seems to be unsolved yet. However, taking into account the results of previous applications of the rough sets theory, it can be noticed that it is very often possible to find rules which differ from the description of the new object on the value of one attribute.

Rules which are similar to the description of the new object in the sense of a chosen certain distance will be called "the nearest rules".

Hence, in the second case we suggest to use heuristic approach that makes it possible to find the nearest rules to the new object. The set of several nearest rules should be presented to the decision maker. Rules can be also described by the values of the chosen distance to the new object, information about the "strength" of the rule, etc. The decision maker classifies the new object using these parameters and his preferences.

Before presenting the procedure for finding the nearest rules, let us consider the decision algorithm again.

The known procedures for derivation of decision algorithms (cf. Fibak, J. et al., 1989; Pawlak, Z., 1991) are very efficient in the analysis of the information systems under the Closed World Assumption but the obtained algorithms may not work properly during the classification of new objects. In the case of a classification of new objects we should use the Open World Assumption, i.e. the experience represented in the information system may be not complete. Some possible descriptions of objects may be not included in this experience and new objects having such descriptions may appear during classification. The obtained algorithms may not be useful for the classification because of apparent inconsistencies. For example, (in the decision algorithm from Figure 2.) consider the two following rules :

    #2 if (a1=1) and (a2=2) then (d=1)
    #6 if (a2=2) and (a3=2) then (d=2)

Note that if we assume the experience represented in the information system to be complete, both rules work correctly. Suppose, however, that the experience is not complete, and consider a new object that may have the description $(a1 = 1)$ and $(a2 = 2)$ and $(a3 = 2)$. This object can not be classified uniqely. This inconsistence results from the fact that the procedures for the derivation of the decision algorithm work under the Closed World Assumption, hence they try to build the shortest decision rules taking into account the characteristic uniqueness of values of certain attributes, pairs of attributes etc. For instance, the rule #2 is build using conditions $(a1 - 1)$ and $(a2 - 2)$ because only objects from class 1 have this combination of values $a1$ and $a2$ in this information system. Hence, the conditions for $a3$ and $a4$ are not added. This observation leads to a simple solution of deleting apparent inconsistencies. The idea consists in augmenting the apparently inconsistent decision rule, by additional conditions for the attributes occurring in an other rule. One should choose to extend such a rule from a given pair which belongs also to other inconsistent pair.

In the presented example we decided to extend the rule #2 and we add the attribute $a3$. As a result, the following rules were obtained:

    #2' if (a1=1) and (a2=2) and (a3=3) then (d=1)
    #6   if (a2=2) and (a3=2) then (d=2)

Let us also note, that decision rules in a decision algorithm can be build on various numbers of objects from the information system. One rule may refer to a single object, others to many objects. Hence, the first rule is "weaker" others are "stronger", i.e. more reliable for the decision maker. We propose to extend information about each rule by introducing the coeffcient called the "strength" of the rule. The "strength" expresses the number of objects in the information systems which confirm the given rule. The use of such a coefficient may help in the analysis of the non–deterministic rules. For such rules, coeffcients are defined for each possible class. If any class significantly dominates others, one can assume that the object should be classify to this class according to this rule.

4 Finding the nearest rules

As it was indicated in Section 3, the nearest rules are rules which are similar to the description of the new object in the sense of a chosen distance. In this section, we propose a distance measure.

Let A denote all considered attributes $(a_1, a_2, ..., a_n)$ describing objects. Let $A = O \cup U$, where O denotes ordinal attributes and N nominal attributes.

Assume that the a given new object x is described by values of attributes $(x_1, x_2, ..., x_n)$. We are going to calculate its distance to any rule y described by values of attributes $(y_1, y_2, ..., y_m)$. Note that the decision rule may be shorter than the description of the new object i.e. some attributes from the description may not appear in the conditions of the decision rule. For these attributes we assume that there is no difference between the new object x and the rule y. For other attributes the distance D is defined as:

$$D^p = \frac{1}{m}(\sum_{i \in N} k_i d_i^p + \sum_{j \in O} k_j d_j^p)$$

where:

$p = 1, 2, ...$ – natural number to be chosen by the analyst,

m – the number of attributes,

d_i – the normalized distance for nominal attributes defined as $d_i = 0$ if $\mid x_i - y_i \mid = 0$ and $d_i = 1$ if $\mid x_i - y_i \mid > 0$;

d_j – the normalized distance for ordinal attributes defined as $d_j = \mid x_j - y_j \mid / (v_{jmax} - v_{jmin})$; where v_{jmax} denotes the maximal value of the attribute a_j and v_{jmin} denotes the minimal value of the attribute a_j, whereas d_j belongs to $[0, 1]$;

k_i and k_j are weights of attributes and can be determined on the basis of significance of attributes for the classification as in (Pawlak, Z., 1991) (i.e. the significance of the attribute a_i is expressed by the difference between the quality of classification for all considered attributes and the quality of classification for all attributes excluding the attribute a_i).

The squared values of distances are used to show that a greater difference on one (ordinal) attribute can be more important than smaller differences on two (ordinal) attributes.

The procedure for finding the nearest rules is now simple. For each rule the distance D is calculated and rules are ranked according to it if the distances are not greater than a certain threshold τ (which is used to distinquish really nearest rules from any near rules).

Finally, the k nearest rules are presented to the decision maker. They are described by the values of distance measure, information about their strength (sometimes one rule can be confirmed by one object only and other rules by many objects). Other parameters can be used additionally, e.g. the number of differences between the new object and the rule. The decision maker uses this information to classify the new object.

5 A computational experiment

In order to check whether the idea of finding the nearest rules works properly, we decided to perform a simulation of classification support. A real data set was used. It represented data about 1443 objects described by 10 condition attributes and classified into 3 classes. The data set was analysed by the rough sets approach. The minimal subset of attributes consists of seven first ones. Then, for the aim of experiment we decided to divide the data set into two subsets in a random way. The first subset (922 objects) gave a base to derive the decision algorithm consisting of 347 rules (deterministic and non–deterministic). The apparent inconsistencies were deleted from it. The second subset (491 objects) was the testing one. We tried to classify objects from it using the obtained decision algorithm. Having no real decision maker we used the following procedure. If we could not find the proper rule, we considered k nearest rules to the given object and summed up supports for each class expressed by the coeficeints of strength of these rules. Then, we chose the class with the greatest supporting number. If there were two or more classes with similar number of supporting objects, we treated this case as a non–deterministic rule. On the other hand if we could not find the nearest rule because of the threshold value τ, we assumed that we can not classify the given object.

Results of the classification of the testing subset without trying to find the k–nearest rules (i.e. if no rule matches the description of the new object, the object is not classified) are as follows:

number of objects:

classified correctly	362	(73.7 %)
classified incorrectly	21	(4.3 %)
not classified	28	(5.7 %)
non–deterministic case	80	(16.3 %)

Results of classification with decision support by finding the k–nearest rules are as follows:

number of objects:

classified correctly	393	(80.1 %)
classified incorrectly	40	(8.1 %)
not classified	0	(0 %)
non–deterministic case	58	(11.8 %)

where $k = 5$ and $\tau = 0.2$ were used.

The experiment shows that the proposed idea can be used for decision support in a classification of new objects.

References

Gruszecki, G., Słowiński, R. and Stefanowski, J. (1990). RoughDAS – Data and Knowledge Analysis Software based on the Rough Sets Theory. User's manual. APRO SA, Warsaw, 49 pp.

Fibak, J., Pawlak, Z., Słowiński, K. and Słowiński, R. (1986). Rough sets based decision algorithm for treatment of duodenal ulcer by HSV, Bull. Polish Acad. Sci., Biol. Ser., vol. 34 (10-12), 227.

Krusińska, E., Słowiński, R. and Stefanowski, J. (1992). Discriminant versus rough sets approach to vague data analysis. Applied Stochastic Models and Data Analysis, vol. 8, (to appear).

Nowicki R., Słowiński, R. and Stefanowski, J. (1990). Possibilities of an application of the rough sets theory to technical diagnostics. In: Materialy IX Sympozjum Techniki Wibracyjnej i Wibroakustyki, Krakow 12-14.12.1990, AGH Press, Krakow, 149-152.

Nowicki, R., Słowiński, R. and Stefanowski, J. (1992). Rough sets analysis of diagnostic capacity of vibroacoustic symptoms. Computers and Mathematics with Applications (to appear).

Pawlak, Z. (1982). Rough Sets. International Journal of Information and Computer Sciences, vol. 1(5), 341-356.

Pawlak, Z. (1991). Rough Sets. Some aspects of reasoning about knowledge. Kluwer Academic Publishers, Dordrecht.

Słowiński, K. (1991). Wykorzystanie teorii zbiorow przyblizoanych do analizy leczenia wrzodu dwunastnicy wysoce wybiorcza wagotomia i ostrego zapalenia trzustki plukaniem otrzewnej. Habilitation Thesis Medical Academy of Poznan (in Polish).

Słowiński, K. and Słowiński, R. (1990). Sensitivity analysis of rough classification. International Journal of Man–Machine Studies, vol. 32, 693-705.

Słowiński, K., Słowiński, R. and Stefanowski, J. (1988). Rough sets approach to analysis of data from pertioneal lavage in acute pancreatitis. Medical Informatics, vol. 13 1(3), 143-159.

Słowiński, R. and Stefanowski, J. (1989). Rough classification in incomplete information systems. Mathematical and Computing Modelling, vol. 12 (10-11), 1347-1357.

An Interactive Program for
Defining Two–Dimensional Irregular Figures for
Decision Support Cutting System

J. Błażewicz, M. Drozdowski, A. Piechowiak, R. Walkowiak
Institute of Computing Science
Technical University of Poznań, Poland

Abstract

An approach and a program for interactively defining two-dimensional irregular objects is presented in this paper. Firstly, various schemes for representation of geometrical entities in computer system are described. Usefulness of existing models for our application is discussed and features of a chosen boundary representation are presented in detail . Then, a description of an interactive method for defining two-dimensional figures is given. Finally, the implementation of this method is presented.

1 Introduction

A two–dimensional cutting problem (see Błażewicz, J. et al., 1989) consists in cutting a set of two-dimensional objects from a sheet of rectangular material in order to minimize a waste. The problem arises in various production processes, hence it is interesting not only from theoretical but also from the practical point of view. Recently an automatic system for two-dimensional irregular cutting problem has been developed (Błażewicz, J. et al., 1990).

A module for the description of input data,which enables defining, in a user friendly way, all geometrical structures used by the system, is a very important part of the decision support system.A data file containing description of a set of objects to be allocated within a sheet of material, and some additional information specific for automatic cutting methods is a final result of processing. Such a program module has been designed and developed. In this paper we discuss theoretical aspects of graphic data representation in connection with the problem. Then, a description of a proposed two-stage method for graphical data input is given.

The organization of the paper is as follows. Section 2 contains a survey of existing representation schemes used in geometric modelling. Then, the chosen boundary representation of two-dimensional figures is describes in detail. In Section 3 the basis of the proposed interactive method for defining two-dimensional elements is presented. In Section 4 an implementation of the method is described in a form of the guide for users of the program.

2 Representations of objects in geometric modelling

At present, six main techniques for object representation in geometric modelling can be distinguished (see Baer, A. et al., 1979; Brenker, F. and Ecker, K., 1987; Bin Ho, 1986; Requicha, A., 1980). We will describe them briefly below.

2.1 Primitive Instancing Schemes

Primitive instancing approach is based on the concept of families of objects. Each family of geometrical objects consists of elements called primitive instances. Objects in the family have different values of the parameters describing their shape. The meaning and the number of parameters characteristic for the family is fixed by the definition of the family.

This scheme is unique, concise, easy to use and promote standardization. On the other hand, it has at least two disadvantages. There is no way for combining families or their elements to create new, more complex ones. Moreover, programs written for transforming objects represented in this way contain big amount of representation-oriented knowledge. Uniform treatment of elements from different families is impossible.

2.2 Spatial Occupancy Enumeration

The idea of a list of cells occupied by the object is the main feature of this representation scheme. The cells, are cubes or squares of a fixed size (depending on the dimension of objects) and lie in a fixed grid. Each cell may be represented by the coordinates of a single point characteristic for it.

This scheme is unique, but its reasonable application is found only for some cases. Its request for a large amount of memory is the main drawback. This representation fails in cases when objects should be defined precisely.

2.3 Cell Decomposition

In this scheme objects are represented by their decomposition into cells (less complex objects) e.g. polyhedron into tetrahedra, and defining each cell in the decomposition. This approach is more general then spatial occupancy enumeration as cells need not to be cubical or lie in a fixed grid.

Though cell decompositions is (particularly for non rectilinear objects) neither concise nor easy to create, it is used in 3-D finite element method and in some problems of object's property (i.e. connected object, "holes" in object, or objects intersections) determination.

2.4 Constructive Solid Geometry

In constructive solid geometry scheme geometrical objects are represented by the use of three different types of objects: a collection of primitives, a set of transformations and Boolean operations. Simple geometrical objects such as cube, pyramid, cylinder, cone and sphere can be used as primitives. Special features of the primitives are adjusted by transformations enabling translation, rotation and scaling. The construction of the object is represented by a graph called "CSG" tree. It is a binary tree with primitives as leaves and subobjects as nodes. Boolean operations, such as union, difference and intersection are used to label nodes. A subobject is the result of the node operation executed on subobjects or primitives taken from the lower level of the tree.

This scheme is a very simple way for defining objects, but final form of the representation is not a convenient source of graphical data to use in a graphic presentation and interaction-type processes.

2.5 Sweep representation

In this representation one can describe objects which are the result of translation or rotation of two-dimensional shape along or around a straight line. The 3-D object representation is hence reduced to description of the 2-D object and the line. This scheme is concise, but its use is limited because of the lack of algorithms that enable processing such a representation.

2.6 Boundary Representation

This most often used scheme is based on a description of faces bounding the object. Several kinds of regular surfaces such as planar, cylindrical, conical and spherical ones can be used as the face of the object. Each surface is defined by its type and a set of directed edges bounding it. An edge is represented by its type (taken from a group of analytically given curves) and two vertices - ending points which are defined by its coordinates. In such a way one gets three level structure of the description: face, edge and vertex.

In boundary representation data describing objects needed for computational and graphic display purposes are ready to use and hence this form of object representation is widely spread in computer graphic.

2.7 Boundary Representation for the DSS

Most of the above representation schemes have only theoretical meaning and have never been widely applied. The reason for using the boundary representation scheme in great number of geometrical computational systems is obvious. This approach is more convenient for people who design such systems and use them. Programmers are looking for schemes which are first of all concise and easy to implement, and which enable uniform treatment of various classes of objects. Moreover , it is essential that the task is realized on the data supplied in a given form. Hence, a question arises whether to adapt existing algorithms or to design new ones. Furthermore, program should allow for a convenient way of defining input data. The boundary representation is natural for all users familiar with a draught. If in the main program, a representation other then boundary one is used, special computations for changing data form would be required.

In the DSS for two-dimensional cutting problem the two-dimensional objects, whose edges are either sections of a line or sections of a circle, are taken into consideration.

They must belong to the "one-piece" object class and thay must have no "holes" inside. Detection of overlapping figures is the main computational problem in the system. It is solved by checking whether lines composing boundaries of different objects have common points. Hence, any part of the object boundary should be accessible for the program. This information is directly available in boundary representation.

Owing to the above mentioned advantages the boundary representation was chosen for the system.

3 An Interactive Method for Defining Two-Dimensional Object

3.1 Data Representation

Two-dimensional geometrical objects are described by two level edge-vertex structure. This is because these objects lie on a plane. The edge of an object may consist of two types of elements. One of them is a section of a line. It is uniquely defined by its two ending vertices. Second type of object boundary element - section of a circle is also defined by two ending vertices, but in this case an additional data are needed i.e. coordinates of the center point and information which section of a circle is to be used.

3.2 Data Structures

Similarly to the geometrical object representation, also program data structures are influenced by the application. The data structure should reflect the structure of object representation, allow for interactions in the process of object defining and for modifications of the object at any stage of this process. Graphically oriented data structures must first of all be dynamic. The size of structures created interactively is not known a priori. For data structures designing both the way and the frequency of data usage is important.

Pointer structures such as list, hierarchical or association structures belong to the most frequently used in graphical systems.

The method of defining figures consists of two stages. At the first stage basic geometric entities such as points, lines and circles are defined. Then, using these elements, objects represented by their boundaries are created. Hence, data structures for points, lines, circles and final objects are needed.

For all types of basic geometrical entities separate bidirectional lists are applied. Each element of the list is defined by the coordinates of characteristic points. A line is defined by two points lying on it. A circle is defined by its center and one point on its arc.

A structure of defining objects is hierarchical. It consists of one additional bidirectional list. Each record of the list defines one particular element of the object's boundary. There are three types of object's boundary elements: section of a line , section of a circle and a circle. Each element of the list must be capable to define any of them. Each record consists of the following fields:

- a type of an element

- an index of a starting point (0 for a circle)

- an index of a center of a circle (0 for a line)

- a parameter defining if the element causes a concavity of the object.

    - 0 for a section of a line

    - a radius for a circle

    - a radius for a section of a circle if it causes element to be convex

    - negative radius for a section of a circle if it causes element to be concave

The concavity at this point means that a line section joining starting and ending points of this part of the boundary lies outside the object.

The index of a point is equal to its number on the list of points, hence, each point is defined only once. The list of points is the lower level of the hierarchical structure of data that defines an object.

For all the objects considered we assume that:

- every ending point of any element of object boundary is also a beginnig of the next boundary element,

- object's boundary elements are to be given in a clockwise order,

- if the object's boundary consists of a circle it is its only element.

These assumptions allow for a unique definition of objects by the boundary representation scheme.

3.3   Method Description

In the method, basic geometric entities such as points, lines and arcs are meant as elements. As we mentioned in Section 3.1. two-dimensional irregular objects are defined by their boundaries. The boundary of an object is composed of the elements. A point is a basic entity to define element of a higher order such as lines, arcs and circles.

In order to define an object we use a two-stage method. At the first stage, one has to define all boundary elements - especially characteristic points i.e vertices of object's boundary. The process of defining a boundary is based on the fact that every characteristic point is a point of an intersection or a point of osculating of boundary elements. Thus, every characteristic point is determined by previously defined elements (lines, arcs, circles) and relations between them (intersection, tangency). An appropriate procedure transforms this information into the coordinates of the characteristic point in question.

At the second stage, boundary itself is defined on the skeleton of the points. This consists in choosing characteristic points of object's boundary and determining a type of their link, chosen from previously fixed basic geometric entities (such as circles, arcs, lines).

4 Implementation of the method

4.1 Introductory remarks

The program was designed to define two-dimensional figures. It produces a data file containing description of a set of objects to be allocated within a sheet of material. This file supplies data to the DSS for two-dimensional cutting. The program enables in particular:

- storing of the object's structure at any stage of the defining process,

- updating, deleting, and changing elements of the object's structure.

This program can be run on any IBM PC and all compatible computers for which a graphic driver for Borland's Turbo Pascal version 5 exists. This program uses mouse or keyboard and handles about 50 errors and exceptions.

4.2 Program options

Every program option accessible at a current execution level is displayed in the menu form in the left top screen corner. One can choose an option with mouse or keyboard " ↑", " ↓", "Enter" keys. To abandon and return to the preceding option one should press "Esc" on keyboard or appropriate mouse button.

In order to move cursor when choosing graphic entities one should use mouse or keyboard " ↑", " ↓", " ←", " →" keys, to confirm a choice "Enter" key or a corresponding mouse button.

All basic graphic entities have their markers (points). A point can be chosen unambiguously with one marker. A line or a circle need two markers. Thus, it may be necessary (when there is a common marker for two elements of the same type) to tag two markers to choose a line or a circle. All options where graphic element is to be chosen have a built in "zoom" mechanism. If two markers (points) are too close to distinguish one from the other, the enlarged surrounding is displayed on the screen. Then, the necessary point should be chosen once again. A speed of cursor movements in graphic mode can be changed with "Ins" key.

We are going to describe now all program options. This is summarized in the Appendix A in the tree form.

At the beginning of program execution the following menu appears on a screen:
NEW
LOAD
SAVE
SKELETON
SHAPE
OUT
QUIT

OPTION: \ NEW This option is designed to set dimensions of working area for object's definition. Maximal x and y dimensions of the object should be written into a window that appears in the screen. Then a rectangle defined in the above way appears. This option has to be chosen if an object is defined from the very beginning.

OPTION: \ LOAD This option enables loading previously defined elements to edit them. The file name (without extension) should be written in a window that appears in the center of the screen. If the name is unambiguous, then the file is loaded. Otherwise (i.e. "A:" or "*.*" are entered), all file names matching specification are displayed. Then, one has to choose a file to be loaded in. When no drive is specified, the current drive is chosen.

OPTION: \ SAVE This option allows for saving edition results at any stage of defining an object. The file name is to be written in the center of the screen. It must be unambiguous and must not have extension. One should be aware that any file with the same name will be overwritten.

At this moment files with extensions .SZK, .SEL, .SPT are created depending on the stage of object definition. File .SZK is always created and contains an information about elements of the object. This information is defined at the first stage of defining an object (see option \ SKELETON). File .SEL is created if a definition of the object reached a second stage (i.e. contour setting in option \ SHAPE). File .SPT is created only if the whole object contour (boundary) has been defined. This file contains coordinates of all characteristic points of object's boundary.

OPTION: \ SKELETON This option is used at the first stage of object's definition. It allows for a construction of a boundary elements . If this option is chosen, then a submenu with the following items appear
DRAW
DELETE
MEASURE

    OPTION: \ SKELETON \ DRAW This option is designed to introduce basic geometric entities. Following submenu appears
    POINT
    LINE
    CIRCLE

        OPTION: \ SKELETON \ DRAW \ POINT This option sets a point. It has a following submenu
        RELATIVE
        LINE
        INTERSECTION
        TANGENCY

OPTION: \ SKELETON \ DRAW \ POINT \ RELATIVE This option introduces a point in the easiest way, i.e. by coordinates relative to a certain base point. The base point is chosen by moving marker "+" appearing in the working area with cursor keys or with a mouse. A confirmation of its position is by "Enter" key or a corresponding mouse key. If two points are so close to each other that thay are difficult to distinguish, the program displays enlarged surroundings and then the point in question is to be marked once again. If the basic point is chosen then relative coordinates of a new point should be given.
At the start of the definition process four points in the corners of the working area are predefined.

OPTION: \ SKELETON \ DRAW \ POINT \ LINE This option allows for definition of points on the line under the condition that an other point has been defined on this line. One has to choose the line (if a point belongs to more than one line then two points marking the line must be given), then a point on the line. After that program asks for a distance from that point. If the distance is positive then the point being defined is above or to the right of the base point. If the distance is negative then the point in question is below or to the left of the base point.

OPTION: \ SKELETON \ DRAW \ POINT \ INTERSECTION This option deals with points defined by intersection of other boundary elements. It has the following submenu
LINE-LINE
LINE-CIRCLE
CIRCLE-CIRCLE

OPTION: \ SKELETON \ DRAW \ POINT \ INTERSECTION \ LINE-LINE In this option a point is defined as an intersection of two lines. One has to choose two lines that intersect one another.

OPTION: \ SKELETON \ DRAW \ POINT \ INTERSECTION \ LINE-CIRCLE In this option a point is defined by an intersection of a line and a circle. One is prompted to tag a line and a circle. If the circle choice is ambiguous, then one has to mark an other point - a pointon on an arc of the circle. Both intersection points are displayed (if there are any), otherwise an error message is issued.

OPTION: \ SKELETON \ DRAW \ POINT \ INTERSECTION \ CIRCLE-CIRCLE This option enables one to introduce a point as a point of two circles intersection. One has to mark both circles in question. If a tangency occurs error message is displayed.

OPTION: \ SKELETON \ DRAW \ POINT \ TANGENCY This option deals with points of osculation. There is the following

submenu in this option
LINE-CIRCLE
CIRCLE-CIRCLE

OPTION: \ SKELETON \ DRAW \ POINT \ TANGENCY \ LINE-CIRCLE In this option a point of tangency of a circle and a line is defined. One has to mark the line and the circle in question.

OPTION: \ SKELETON \ DRAW \ POINT \ TANGENCY \ CIRCLE-CIRCLE This option enables one to define a point as a point of circle-circle tangency. Circles in question have to be marked.

OPTION: \ SKELETON \ DRAW \ LINE This option deals with line definitions. After a choice of this option the following submenu is displayed
POINT-POINT
POINT-ANGLE
PARALLEL
TANGENT

OPTION: \ SKELETON \ DRAW \ LINE \ POINT-POINT A line is defined by two marked points.

OPTION: \ SKELETON \ DRAW \ LINE \ POINT-- ANGLE This option is designed to define a line intersecting an other line at a given point with a certain slope angle. One has to choose a line, an intersection point and an angle. The intersection angle should be positive and less than 180 degrees.

OPTION: \ SKELETON \ DRAW \ LINE \ PARALLEL Parallel lines are defined in this option. Previously defined line have to be chosen then the program asks for a distance to the new line. If the distance is positive, then the new line is above or to the right of the base line . If it is negative, then the new line is below or to the left of the given line.

OPTION: \ SKELETON \ DRAW \ LINE \ TANGENT This option enables one to define a line tangent to a circle. We have two suboptions here
POINT-CIRCLE
CIRCLE-CIRCLE

OPTION: \ SKELETON \ DRAW \ LINE \ TANGENT \ POINT-CIRCLE From a given point lines tangent to a certain circle are drawn. One has to mark the point and the circle. The point can not be inside the circle.

OPTION: \ SKELETON \ DRAW \ LINE \ TANGENT \ CIRCLE-CIRCLE For the two given circles all existing tangent lines are drawn. Circles can not include one another or overlap. One has to choose two circles in question.

OPTION: \ SKELETON \ DRAW \ CIRCLE This option defines circles. In order to define a circle one has to choose previously defined point as a center of a circle and enter a circle's radius.

OPTION: \ SKELETON \ DELETE This option is used to remove (delete) elements defined previously. We have two suboptions here
POINT
LINE
CIRCLE

OPTION: \ SKELETON \ DELETE \ POINT This option deletes points.

OPTION: \ SKELETON \ DELETE \ LINE This option deletes lines.

OPTION: \ SKELETON \ DELETE \ CIRCLE This option deletes circles.

OPTION: \ SKELETON \ MEASURE This option deals with measurements of defined elements. A following submenu is displayed
POINT-POINT
POINT-LINE
LINE-LINE-DIST
LINE-LINW-ANGLE

OPTION: \ SKELETON \ MEASURE \ POINT-POINT A distance between two points is determined. It is given in the form of a distance in X and Y coordinates and also as a ordinary euclidean distance.

OPTION: \ SKELETON \ MEASURE \ POINT-LINE Determines distance from a given point to a given line.

OPTION: \ SKELETON \ MEASURE \ LINE-LINE-DIST This enables one to measure a distance between parallel lines.

OPTION: \ SKELETON \ MEASURE \ LINE-LINE-ANGLE In this option a slope of the two intersecting lines is determined. This angle is positive and not greater than 90 degrees.

OPTION: \ SHAPE This option implements the second stage of object's defining, i.e. the creation of its boundary. If this option is chosen then the following submenu appears
CONTOUR

OPTION: \ SHAPE \ CONTOUR This option defines external boundary of the object. The following submenu appears in this option
DRAW
DELETE

OPTION: \ SHAPE \ CONTOUR \ DRAW This option appends elements to the object's boundary. Elements of the boundary are to be introduced in the clockwise order. User may choose one of the following options
LINE
ARC
CIRCLE

OPTION: \ SHAPE \ CONTOUR \ DRAW \ LINE In this option one can append a section of a line to the string defining object's boundary. If it is the first element of the boundary one has to mark starting and ending points, respectively. If it is appended to the string of other elements, then only ending point is needed.

OPTION: \ SHAPE \ CONTOUR \ DRAW \ ARC This option allows for a definition of a section of a circle - arc as an element of the boundary. One has to tag starting and ending points of the arc and to define the circle to which the arc belongs. Then the program asks wether this arc causes concavity of the object i.e. if a line joining starting and ending point's respectively is outside the object.

OPTION: \ SHAPE \ CONTOUR \ DRAW \ CIRCLE This option is used to define a contour (boundary) as a circle. One has to choose a circle center.

OPTION: \ SHAPE \ CONTOUR \ DELETE This options enables one to delete the last appended boundary element. Each time space bar is pressed the last element is deleted.

OPTION: \ OUT This option prepares a data file for the two-dimensional cutting DSS. If this option is chosen, the following submenu appears
PARAMETERS
ELEMENTS
EXECUTE

OPTION: \ OUT \ PARAMETERS In this option parameters for the two-dimensional cutting program are read in. In the center of the screen a window appears where parameters should be written in consecutive lines. If the erronous value occurs, the program does not let to change a line.

OPTION: \ OUT \ ELEMENTS This option deals with members of the set of objects. The following submenu appears
APPEND
CHANGE

OPTION: \ OUT \ ELEMENTS \ APPEND This option appends a new object to the set of objects. Program asks for the name of a file with the object's description. This name is read in and the file is loaded in the same way as in the \ LOAD option. Then the chosen object is displayed and the program asks for a number of such objects to be cut.

OPTION: \ OUT \ ELEMENTS \ CHANGE This option changes the number of previously chosen objects.If the new number is 0 then the object is deteted.

OPTION: \ OUT \ EXECUTE In this option a file for the two-dimensional cutting system is created and saved (Błażewicz, J. 1990).

OPTION: \ QUIT A choice of this option breaks an execution of the program.

5 Additional comments on the implementation

The program name is DRAFT.EXE and to run it one should write DRAFT and press "Enter" key. It is necessary to start program execution from the directory containing appropriate (for the graphic adapter) Borland's graphic driver (files with .BGI extension for Turbo Pascal ver.5). If a mouse is to be used an appropriate driver should be installed first.

In the second stage of object's definition (option \ SHAPE), all considered basic entities (points, circles etc.) should be already defined (in option \ SKELETON). In order to define an arc, a circle must be previously defined. To make a circular boundary its center must be known. If an element of a boundary is created in \ SHAPE option, one can not add new elements in the \ SKELETON option. If it is necessary to do so, all the elements of the boundary have to be deleted first.

X and y dimensions are within the range (0,99999.999] and an accuracy of computations is 0.0001.

There are two appendices enclosed. Appendix A summarizes the option's tree, and appendix B gives an example of object's definition.

References

Baer, A., Eastman C. and Henrion M. (1979). Geometric modelling: a survey, *CAD, Vol 11, No 5*,253-272.

Bin Ho (1986). Inputing constructive solid geometry representations directly from 2D orthographic engineering drawings, *CAD, Vol 18, No 3* , pp. 147-155.

Błażewicz J., Drozdowski M., Soniewicki B., Walkowiak R. (1989). Two-dimensional cutting problem basic complexity results and algorithms for irregular shapes, *Foundations of Control Engineering, Vol 14, No 4*, pp. 137-159.

Błażewicz J., Drozdowski M., Soniewicki B., Walkowiak R. (1990). Decision support system for cutting irregular shapes - implementation and experimental comparison, *Foundations of Computing and Decision Sciences, Vol 15, No 3-4* , pp. 121-129.

Brenker F., Ecker K. (1987). CHARM- Ein Ansatz für ein hierarchisch aufgebautes graphisches System, *Technical Report, Institut für Informatik, Technische Universität Clausthal 87/2*

Requicha A. A. G. (1980). Representations of rigid solids: theory, methods and systems, *Computing Surveys, Vol 12, No 4*, pp. 437-464

Appendix A

OPTIONS TREE.

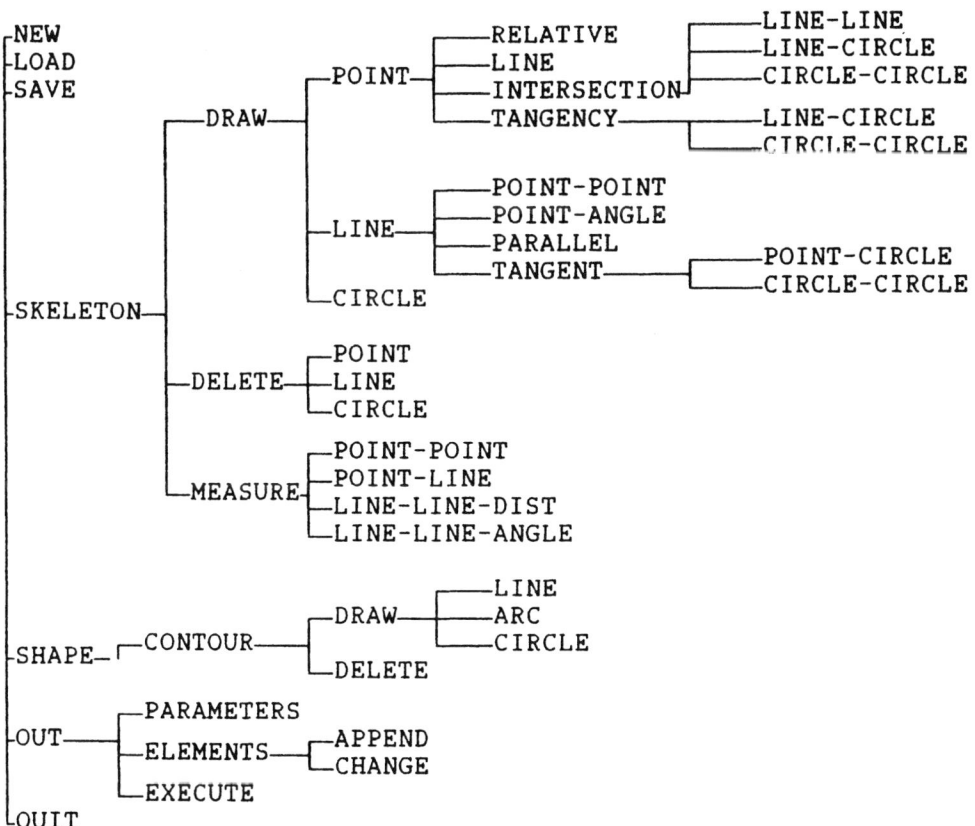

Appendix B

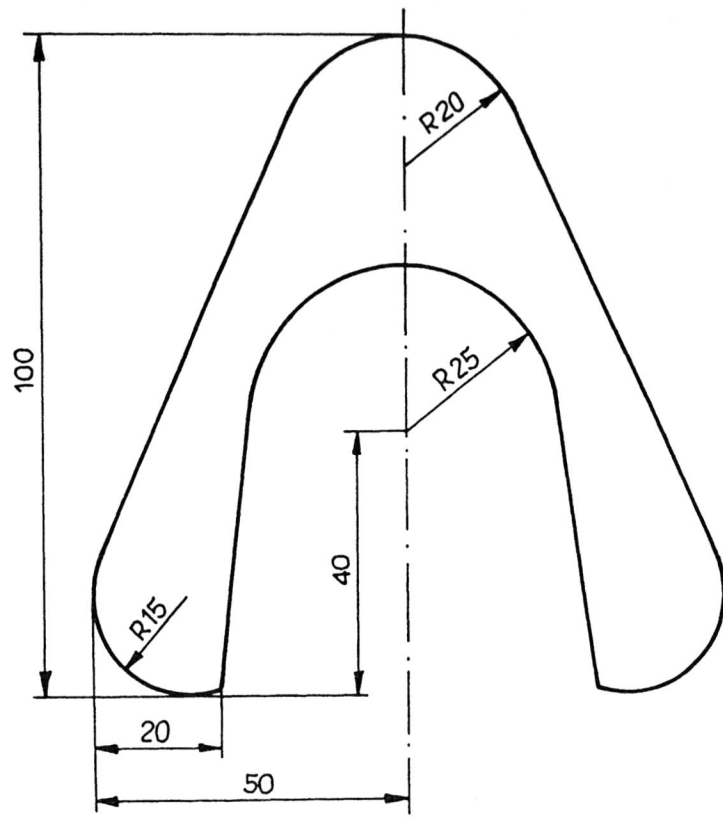

Figure 1: Figure to be defined.

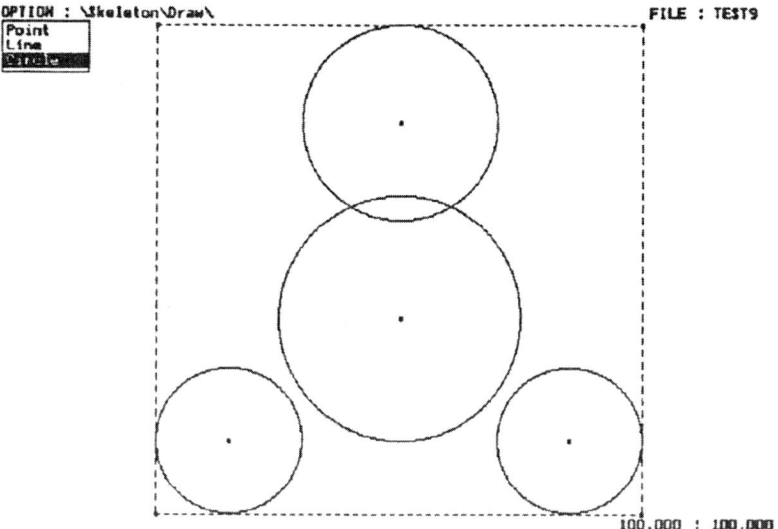

Figure 2a: First stage of figure defining – circles.

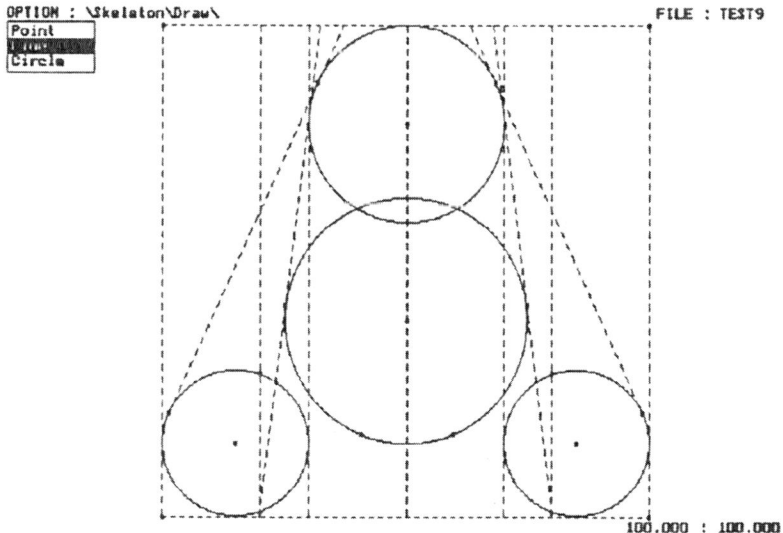

Figure 2b: First stage of figure defining – lines.

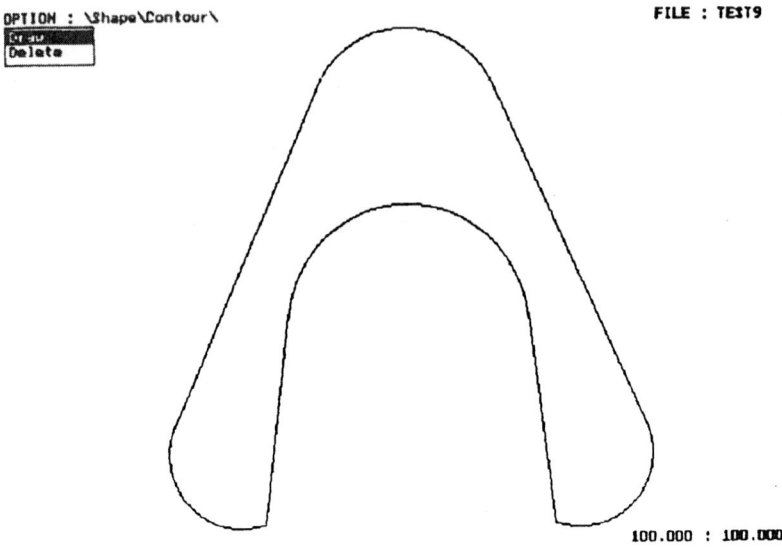

Figure 2c: Second stage of figure defining – defined object.

DINAS
Dynamic Interactive Network Analysis System
A Tutorial Example

Wlodzimierz Ogryczak, Krzysztof Studzinski, Krystian Zorychta

Institute of Informatics, Warsaw University, Poland

Abstract

This paper describes a tutorial example for the Dynamic Interactive Network Analysis System (DINAS) which enables the solution of various multi-objective transshipment problems with facility location using IBM-PC microcomputers. The system utilizes an extension of the classical reference point approach to handling multiple objectives. DINAS is prepared as a menu-driven and easy to use system armed with a special network editor which reduces to minimum effort associated with defining real-life problems. As the tutorial problem we use an artificial part of the real-life model connected with a sugar-beet transshipment system.

1 Introduction

DINAS (Dynamic Interactive Network Analysis System) is a scientific transferable software tool which enables the solution of various multi-objective transshipment problems with facility locations. For a given number of fixed facilities and customers and for a number of potential facilities to be optionally located, DINAS provides you with a distribution pattern of a homogeneous product under multi-criteria optimality requirements. While working in an interactive mode, you get optimal locations of the potential facilities and a system of optimal flows of the product between nodes of the transportation network.

With DINAS you can analyze and solve such problems as:

- the transportation problem with new supply and/or demand points location,

- the problem of warehouses location,

- the problem of stores location for the agricultural production,

- the problem of service centers location and districts reorganization,

and many other real-life distribution-location problems.

DINAS is implemented on IBM-PC XT/AT/386/486 as a menu-driven and easy to use system armed with a special network screen editor for a friendly data input and results examination. While working with DINAS you will get a permanent assistance by the help lines which will inform you about operations available at this moment. Moreover, at any moment you will have opportunity to get more general information from the help file. The DINAS menu and all commands of the system are briefly described in Appendix.

To illustrate the DINAS methodology a small testing example is presented. As the test problem an artificial part of the real-life model connected with a sugar-beet transshipment system is considered.

2   The problem statement and methodology

DINAS works with problems formulated as multi-objective transshipment problems with facility location. A network model of such a problem consists of nodes connected by a set of direct flow arcs. The set of nodes is partitioned into two subsets: the set of fixed nodes and the set of potential nodes. The fixed nodes represent "fixed points" of the transportation network, i.e., points which cannot be changed, whereas the potential nodes are introduced to represent possible locations of new points in the network.

Some groups of the potential nodes represent different versions of the same facility to be located (e.g., different sizes of a warehouse etc.). For this reason, potential nodes are organized in the so-called selections, i.e., sets of nodes with the multiple choice requirements. Each selection is defined by the list of included potential nodes as well as by a lower and upper number of nodes which have to be selected (located).

A homogeneous good is distributed along the arcs among the nodes. Each fixed node is characterized by two quantities: supply and demand on the good, but for the mathematical statement of the problem only the difference supply—demand (the so-called balance) is used. Each potential node is characterized by a capacity which bounds maximal good flow through the node. The capacities are also given for all the arcs but not for the fixed nodes.

A few linear objective functions are considered in the problem. The objective functions are introduced into the model by given coefficients associated with several arcs and potential nodes (the so-called cost coefficients, independently of their real character). The cost coefficient connected to an arc is treated as the unit cost of the flow along the arc. The cost coefficient connected to a potential node is considered as the fixed cost associated with locating of the node (e.g., an investment cost).

Summarizing, the following groups of input data define the transshipment problem under consideration:

- objectives,

- fixed nodes with their supply-demand balances,

- potential nodes with their capacities and (fixed) cost coefficients,

- selections with their lower and upper limits on number of active potential nodes,

- arcs with their capacities and cost coefficients.

In the DINAS system there are two restrictions on the network structure:

- there is no arc which directly connects two potential nodes;

- each potential node belongs to at most two selections.

The first restriction does not imply any loss of generality since each of two potential nodes can be separated by an artificial fixed node, if necessary. The second requirement is not very strong since in practical models usually there are no potential nodes belonging to more than two selections.

The problem is to determine the number and locations of active potential nodes and to find the good flows (along arcs) so as to satisfy the balance and capacity restrictions and, simultaneously, optimize the given objective functions. The mathematical model of the problem is described in details by Ogryczak et al. (1989).

The problem under consideration is a specialized form of Multiple Criteria Decision Making (MCDM). The basic concept of the multiple objective optimization was introduced by Pareto over 80 years ago. He developed the idea of the so-called efficient (or Pareto-optimal) solution, that is a solution which cannot be improved in any objective without some other objective being worsened. However the set of all the efficient solutions is, in practice, extremely huge. Therefore development of practical MCDM tools has begun from the 70's when the computer technique reached a sufficient level for an efficient implementation of various interactive (decision support) systems.

The interactive system does not solve the multi-objective problem. It rather makes the user selecting the best solution during interactive work with the system. According to some user's requirements, the system generates various efficient solutions which can be examined in details and compared to each other. The user works with the computer in an interactive way so that he can change his requirements during the sessions.

DINAS is such an interactive decision support system. The DINAS interactive procedure utilizes an extension of the reference point optimization introduced by Wierzbicki (1982). The basic concept of that approach is as follows:

- the user forms his requirements in terms of aspiration and reservation levels, i.e., he specifies acceptable and required values for given objectives;

- the user works with the computer in an interactive way so that he can change his aspiration and reservation levels during the sessions.

- after editing the aspiration and reservation levels, DINAS computes a new efficient solution while using an achievement scalarizing function as a criterion in single-objective optimization (see Ogryczak et al. 1989 for more details).

- each computed efficient solution is put into a special solution base and presented to the decision maker as the current solution to allow him to analyse performances of the current solution in comparison with the previous ones.

3 Tutorial example

As an illustration, the problem of depots location in a sugar-beet distribution system may be considered (Jasinska, Wojtych 1984). Consider an agricultural region containing a number of farms that produce sugar beet. Each farm is considered as a supply point and is characterized by its total supply during the sugar-beet harvesting period. The sugar-beet is delivered to a number of sugar-mills. Each sugar-mill has some limited total production capacity during the production season.

Climate conditions, poor storage facilities or an underdeveloped transportation network may cause losses of sugar-beet volume, losses of sugar content in the sugar-beet, or extremely high transportation costs. To avoid these difficulties, a part of the sugar-beet supply has to be delivered to special depots and stored there temporarily. The depots are considered as purchasing centers. Some of them are already in use and an amount of the sugar-beet is shipped through them. For further improvement of the sugar-beet transportation system, some additional depots have to be opened or some existing ones should be modernized. Each potential depot is characterized by the upper bound on its throughput as well as by the corresponding investment cost. The investment cost is treated as the fixed cost associated with locating (or modernizing) of the potential depot. Moreover, unit shipping costs are connected with all the delivery routes: from farms to sugar-mills. Each of the routes is also characterized by a capacity which bounds the maximal flow of the sugar-beet along the route.

The problem is to determine the number, location and sizes of the depots in use. Moreover, the corresponding sugar-beet flow from farms to sugar-mills directly or through depots has to be found so as to minimize the total transportation cost or/and the depots investment cost (provided that the total amount of the sugar-beet is delivered from farms to sugar-mills).

The problem represents a class of transshipment problems with facility location. It is a single-objective optimization problem if only one of the objective functions, the total transportation cost or the total investment cost, is minimized. However, it should be treated as a double-objective optimization problem if both the objectives are considered simultaneously. Single- or double-objective optimization can be insufficient in real-life circumstances and some additional objectives should then be taken into consideration. For instance, the total amount of the sugar-beet flow through depots is sometimes considered to be minimized as direct flow from farms to sugar-mills is technologically more efficient. As another objective maximization of the total amount of sugar-beet delivered by railway or maximization of the sugar production volume can be considered. The objectives are, in general, not comparable and thereby our problem should be considered as a multi-objective optimization problem.

Consider a small artificial part of the problem. Eight fixed nodes (seven farms: Udry, Dubiel, Smrock, Gondek, Pogaj, Runo, Cuple; one sugar-mill Klew) and three potential depots (Jurga4, Jurga7, Tyn) are considered. Supply amounts for farms and a demand on the sugar-beet for the sugar-mill are presented in Table 1.

fixed node	supply	demand	balance
Cuple	1.6	0	1.6
Dubiel	1.3	0	1.3
Gondek	1.4	0	1.4
Pogaj	1.6	0	1.6
Runo	1.2	0	1.2
Smrock	1.4	0	1.4
Udry	1.5	0	1.5
Klew	0	10	-10.0

Table 1: Fixed nodes

Note that the sum of the farms supplies is equal to the demand of the sugar-mill. There are considered three potential depots: Jurga4, Jurga7 and Tyn. In fact, the potential depots Jurga4 and Jurga7 represent two versions of the same depot which has to be located. These versions differ only in their capacities and hence they are considered as one selection Jurga with its lower and upper bounds equal to 0 and 1, respectively. The potential depots with their data are presented in Table 2.

potential node	selection	capacity	objective coefficients		
			INVEST	SHIP	RAIL
Jurga4	Jurga (0,1)	4	200	0	0
Jurga7	Jurga (0,1)	7	300	0	0
Tyn		4	220	0	0

Table 2: Potential nodes

A network scheme of this example is presented in Figure 1. The arcs in the network represent possible flows of the sugar-beet and are directed from farms to the sugar-mill or to depots, and from the depots to the sugar-mill. All the arcs with their capacities are listed in Table 3. The arcs connecting the nodes representing depots with the nodes representing farms or sugar-mill have essentially unlimited capacities. However, in practice, flows along these arcs are also bounded by capacities of the corresponding depots and we use them as arcs capacities.

arc		capacity	objective coefficients		
from	to		INVEST	SHIP	RAIL
Cuple	KLew	10	0	80	0
Dubiel	Klew	10	0	132	0
Gondek	Klew	10	0	113	0
Pogaj	Klew	10	0	150	0
Runo	Klew	10	0	167	0
Smrock	Klew	10	0	113	0
Udry	Klew	10	0	201	0
Tyn	Klew	4	0	70	23.3
Jurga4	Klew	4	0	67	22.3
Jurga7	Klew	7	0	67	13.2
Dubiel	Jurga4	4	0	60	0
Smrock	Jurga4	4	0	41	0
Udry	Jurga4	4	0	129	0
Dubiel	Jurga7	7	0	55	0
Gondek	Jurga7	7	0	35	0
Pogaj	Jurga7	7	0	78	0
Smrock	Jurga7	7	0	36	0
Udry	Jurga7	7	0	124	0
Cuple	Tyn	4	0	0	0
Pogaj	Tyn	4	0	70	0
Runo	Tyn	4	0	87	0

Table 3: Arcs

One must decide which potential depots have to be built so as to meet the total demand on the sugar-beet in the sugar-mill. The decision should be optimal with respect to some objectives. Three objectives are considered in the problem (see Table 4): minimization of the INVEST function, minimization of the SHIP function and maximization of the RAIL function.

objectives	max/min
INVEST	min
SHIP	min
RAIL	max

Table 4: Objectives

INVEST represents an investment cost associated with the location of the potential depots. SHIP represents total transportation cost in the network. The RAIL function is used by the railway administration to evaluate shipping of the sugar-beet from the depots to the sugar-mill (these routes are serviced by the railway). The objective coefficients corresponding to the potential nodes as well as to the arcs are presented in Table 2 and 3, respectively.

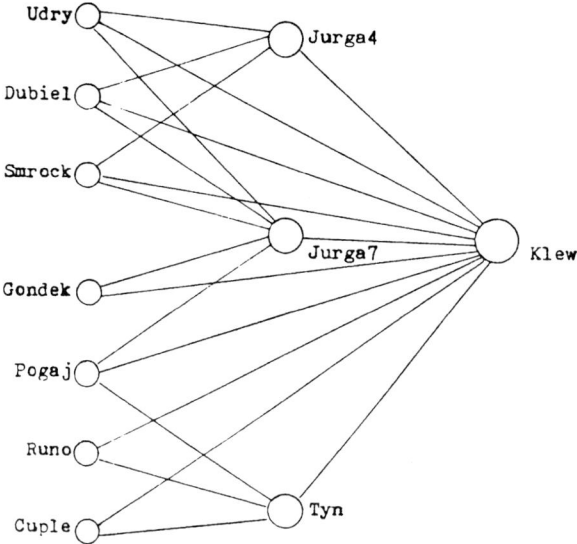

Figure 1

4   Multi-objective analysis

The multi-objective analysis is performed in two stages. In the first stage the decision maker is provided with some initial information which gives him an overview of the problem. In the second stage, an interactive selection of efficient solutions is performed.

First stage

Having edited and converted the input data of the problem as the first step of the multi-objective analysis one must perform the PAY-OFF command (see Appendix). It executes optimization of each objective function separately. In effect, we get the so-called pay-off matrix presented in Table 5. The pay-off matrix is a well-known device in multi-objective programming. It gives values of all the objective functions (columns) obtained while solving several single-objective problems (rows) and thereby it helps to understand the conflicts between different objectives.

optimized function	objective values		
	INVEST	SHIP	RAIL
INVEST	0	1357.9	0
SHIP	520	1258.5	172.4
RAIL	420	1297.9	182.4

Table 5: Pay-off matrix

Execution of these optimizations provides also us with two reference vectors: the utopia vector and the nadir vector (see Table 6). The utopia vector represents the best values of each objective considered separately. Usually, the utopia vector is not attainable, i.e., there are no feasible solutions with such objective values. The nadir vector expresses the worst values of each objective, noticed during optimization of another objectives. However, coefficients of the nadir vector cannot be considered as the worst values of the objectives over the whole efficient (Pareto optimal) set. We found out in further analysis that these estimations can be sometimes overstep.

reference vector	objective values		
	INVEST	SHIP	RAIL
utopia	0	1258.5	182.4
nadir	520	1357.9	0

Table 6: Initial references

While analyzing Tables 5 and 6 we find out that the objective values vary significantly depending on selected optimization. Moreover, we recognize a strong conflict between the investment cost and both the other objectives. The objective functions SHIP and RAIL take the worst values while minimizing the investment cost. Therefore there is a need to perform the interactive analysis in order to find some satisfactory compromise solution.

Second stage

Having computed both the utopia and nadir vectors we can start the interactive search for a satisfying efficient solution. As we have already mentioned DINAS utilizes aspiration and reservation levels to control the interactive analysis. More precisely, you specify acceptable values for several objectives as the aspiration levels and necessary values as the reservation levels, and then DINAS searches an efficient solution corresponding to these values. The succeeding modifications of the reference values are presented in Table 7.

stage	solution	objectives					
		INVEST		SHIP		RAIL	
		Aspir.	Reser.	Aspir.	Reser.	Aspir.	Reser.
1	Sol.1						
	Sol.2						
	Sol.3						
2	Sol.4	utopia	nadir	utopia	nadir	utopia	nadir
	Sol.5	utopia	250	utopia	nadir	utopia	nadir
	Sol.6	utopia	250	utopia	nadir	utopia	90

Table 7: Succeeding values of the aspiration and reservation levels

Due to the special regularization technique used while computation of the pay-off matrix each generated single-objective optimal solution is also an efficient solution to the multi-objective problem. DINAS stores these solutions in a special solution base as well as all the efficient solutions generated during a session. So, after the first stage we have

already available in the solution base three efficient solutions connected with several rows
of the pay-off matrix (see Table 8: Sol.1, Sol.2 and Sol.3).

objective	Sol.1	Sol.2	Sol.3	Sol.4	Sol.5	Sol.6
INVEST	0	520	420	300	200	220
SHIP	1357.9	1258.5	1297.9	1293.5	1337.9	1317.9
RAIL	0	172.4	182.4	92.4	89.2	93.2

Table 8: Objective values for the efficient solutions

Now, we can examine in details these solutions. You may notice that the first solution which minimizes the investment cost does not use any depot (see Table 9). On
the other hand, the second solution minimizing the shipping cost uses both the most
expensive depots (Jurga7, Tyn). So the investment cost for the solution is maximal (IN-
VEST=520). Maximization of the RAIL function (Sol.3) get also too high investment
cost (INVEST=420). Since these solutions are not acceptable we have compute a new
efficient solution.

depot	Sol.1	Sol.2	Sol.3	Sol.4	Sol.5	Sol.6
Jurga4	No	No	Yes	No	Yes	No
Jurga7	No	Yes	No	Yes	No	No
Tyn	No	Yes	Yes	No	No	Yes

Table 9: Potential depots activity for the efficient solutions

At the beginning of the interactive analysis the so called neutral solution is usually
computed. For this purpose you should accept the utopia vector as the aspiration levels
and the nadir vector as the reservation levels. In result the fourth solution is computed.
The investment cost (INVEST=300) for this solution is still too large for us. Suppose
that the necessary value for the INVEST function is 250. So we put 250 as the reservation
level corresponding to INVEST and compute the next solution (Sol.5). The solution is
based on the sole depot Jurga4 which has the smallest investment cost (INVEST=200).
Unfortunately, the shipping cost as well as the railway evaluation are not satisfying for
this solution (SHIP=1337.9, RAIL=89.2). To avoid too small value of the RAIL function
in the next solution we modify the reservation level for this objective putting 90 as its
new value. After repeating the computation we get the sixth efficient solution based on
the sole depot Tyn. In comparison with the previous solution the investment cost is
only 10% worse. But both the shipping cost as well as the rail evaluation are slightly
better (SHIP=1317.9, RAIL=93.2). The rail evaluation in Sol.6 is better even than for
the fourth solution (RAIL=92.4, INVEST=300). So, Sol.6 guarantees the investment cost
less than 250, relatively small shipping cost and middling rail evaluation. The solution
seems to be the best among the efficient solutions based on a sole depot. Adding a second
depot leads to objectionably high values of the investment cost (Sol.2 and Sol.3). Note
that all the feasible combinations of the depots were considered (see Table 9) and there
are no another crucial efficient solutions. So, the analysis is finished.

References

Grauer, M., Lewandowski, A., Wierzbicki, A. (1984): DIDASS — theory, implementation and experiences. In: Grauer, M., Wierzbicki, A. P. (eds), Interactive Decision Analysis. Springer, Berlin 1984.

Jasinska, E., Wojtych, E. (1984): Location of depots in a sugar-beet distribution system. *European Journal of Operational Research* **18**, pp. 396–402.

Ogryczak, W., Studzinski, K., Zorychta, K. (1989). A solver for the multi-objective transshipment problem with facility location. *European Journal of Operational Research* **43**, pp. 53–64.

Ogryczak, W., Studzinski, K., Zorychta, K. (1989). A generalized reference point approach to multi-objective transshipment problem with facility location. In: Lewandowski, A., Wierzbicki, A. P. (eds.), Aspiration Based Decision Support Systems. Springer, Berlin (331).

Ogryczak, W., Studzinski, K., Zorychta, K. (1991). DINAS — Dynamic Interactive Network Analysis System, v. 3.0. Collaborative Paper CP–91–012, IIASA, June 1991.

Wierzbicki, A. P. (1982). A mathematical basis for satisficing decision making. *Math. Modelling* **3**, pp. 391–405.

Appendix

The Interactive System

- MENU BRANCHES

  PROCESS   commands connected with processing of the multiobjective transship-ment–location problem and generation of several efficient solutions (PROBLEM, CONVERT, PAY–OFF, EFFICIENT, QUIT).

  SOLUTION   commands connected with the Current Solution (SUMMARY, BROWSE, SAVE, DELETE).

  ANALYSIS   commands connected with operations on the Solution Base which collects up to nine efficient solutions (COMPARE, PREVIOUS, NEXT, LAST, RESTORE).

- PROCESS COMMANDS

  PROBLEM   edit or input of a problem with the Network Editor. If the network file for the problem has been prepared before DINAS starts, PROCESS can be initialized directly using the CONVERT command.

  CONVERT   convert the network file with error checking.

  PAY–OFF   compute the pay–off matrix and the utopia and nadir vectors.

  EFFICIENT   compute an efficient solution depending on introduced aspiration and reservation levels:

       1.edit aspiration and reservation levels;

       2.DINAS computes a new efficient solution depending on the aspiration and reservation levels;

       3.the solution is put into the Solution Base and is presented as the Current Solution with special tables and bars.

  The Current Solution is presented similarly as while using the SUMMARY command.

  QUIT   leave the DINAS system.

- SOLUTION COMMANDS

  SUMMARY   present short characteristics of the Current Solution such as:

       table of selected locations,

       table of objective values,

       bars in the aspiration/reservation scale,

       bars in the utopia/nadir scale.

The tables characterize the current and previous efficient solutions. The bars show a percentage level of each objective value with respect to the corresponding scale.

Change the kind of characteristic using the <PGDN> or <PGUP> key.

BROWSE   examine or print the Current Solution with the Network Editor.

SAVE   save out the Current Solution on a separate file for using in next runs.

DELETE   delete the Current Solution from the Solution Base.

- ANALYSIS COMMANDS

COMPARE   perform comparison of all the efficient solutions from the Solution Base or of some subset of them. (A solution can be temporarily deleted from the comparison using the <DEL> key.)
The following characteristics are used for the comparison:

   table of selected locations,

   table of objective values,

   bars in the aspiration/reservation scale,

   bars in the utopia/nadir scale.

The tables and bars characterize all the efficient solutions selected from the Solution Base. The bars are given for each objective and show a percentage level of the objective value with respect to the corresponding scale. Use the <PGDN> or <PGUP> key to change the kind of the current characteristic. Scroll bars for different objectives using the <UP> or <DOWN> arrows.

PREVIOUS   take the previous efficient solution as the Current Solution.

NEXT   take the next efficient solution as the Current Solution.

LAST   take the last efficient solution as the Current Solution.

RESTORE   restore some efficient solution (saved earlier with the SAVE command) to the Solution Base.

The Network Editor

- MENU COMMANDS

LOAD   select a file to be used to edit.
If the file name is typed without any extension the extension .NET is automatically appended to the name.

SAVE   write the edited file out without creating a .BAK file.

MPS   generate the MPS file and write it without creating a .BAK file.

QUIT   leave the Network Editor and return to the Main Menu.

PRINT NETWORK   print the entire data file and the solution included in the file.

LIST NODES   display a list of nodes (may be empty at the start).
Type or select a node name to activate the edit mode. Then the CURRENT NODE window including the selected (or typed) node is activated.

NETWORK   display a scheme of the transportation network. Select a node to activate the edit mode. Then the CURRENT NODE window is activated with the selected node as the current node.

SELECTIONS   list the nodes belonging to several selections. Use <ENTER> to display the BOUNDS window corresponding to the pointed out selection.

OBJECTIVES   define, modify or examine objectives.

• WINDOWS

CURRENT NODE   the node selected (or typed) with the LIST NODES or NETWORK command becomes the current node. Then it can be defined and its data can be edited or examined.

NODE FROM (list of predecessors)   list names of nodes and the corresponding arcs which precede the current node. Select a node or type a node name.

NODE FROM (a selected predecessor) edit or examine data of the selected (or typed) node. If the <ESC> key is used after the edit is finished, the question ARC? appears. Press <Y> if an inspection of the corresponding arc is needed, or <N> otherwise.

NODE TO (list of successors)   list names of nodes and the corresponding arcs which succeed the current node. Select a node or type a node name.

NODE TO (a selected successor)   edit or examine data of the selected (or typed) node. If the <ESC> key is used after the edit is finished, the question ARC? appears. Press <Y> if an inspection of the corresponding arc is needed, or <N> otherwise.

BOUNDS   edit or examine lower and upper bounds on a number of potential nodes which can be used in the given selection.

• ALT COMMANDS

ALT C change the actual node from/to into the current node.

ALT D delete an objective, selection, node or arc using the OBJECTIVES, SELEC-
TIONS, CURRENT NODE or NODE FROM/TO command, respectively.

ALT L execute the LIST NODES command.

ALT N execute the NETWORK command.

ALT O execute the OBJECTIVES command.

ALT S execute the SELECTIONS command.

- CTRL COMMANDS

CTRL LEFT ARROW   move from CURRENT NODE to NODE FROM.

CTRL RIGHT ARROW move from CURRENT NODE to NODE TO.

CTRL RIGHT ARROW display the ARC window for a node selected from the
NODE FROM/TO list, and reversely.

- USING ANOTHER KEYS

ESC   leave the current command/window, e.g., from NODE FROM/TO to
CURRENT NODE, from CURRENT NODE to the Editor Menu, and
out of HELP.

ENTER   command execution or node selection.

ARROWS move the pointer or scroll data.

HYBRID: Multicriteria Linear Programming System for Computers under DOS and Unix

Marek Makowski

IIASA, Laxenburg, Austria.[*]

Janusz S. Sosnowski

Systems Research Institute, Polish Academy of Sciences

ul. Newelska 6, 01–447 Warsaw, Poland.

Abstract

HYBRID is a mathematical programming package which includes all the functions necessary for the solution of single and multicriteria LP problems. HYBRID uses a non-simplex method which combines the proximal multiplier method and an active set algorithm and has been prepared in two versions: one for UNIX (currently implemented for Sun Sparc running under SunOS 4.1) and one for a PC compatible with the IBM PC.

1 Introduction

HYBRID can serve as a tool which helps in choosing a decision in a complex situation in which many options may and should be examined. Such problems occur in many situations, such as problems of economic planning and analysis, technological or engineering design problems, and problems of environmental control. We assume that such problems can be defined as multicriteria linear programming problems. The method adopted in HYBRID for multicriteria optimization is the interactive reference point approach (cf Wierzbicki, 1980). In this method a sequence of Pareto-optimal solutions is computed by minimization of a piecewise linear achievement functions subject to linear constraints. These functions are defined for each reference point[1]. The problem of minimization a piecewise linear function can be converted to a linear programming (LP) problem.

For solving a linear programming problem HYBRID uses a non-simplex algorithm – a particular implementation of the proximal multiplier method. This method differs from

---

[*]On leave from the Systems Research Institute of the Polish Academy of Sciences, Warsaw.

[1]A reference point is composed of aspiration values for each criterion. These values are set by a user.

the ordinary multiplier method by adding a quadratic form to the augmented Lagrangian function and making the function strongly convex with respect to primal variables. The minimized function is strongly convex and provides an unique minimizer.

In the proximal multiplier method, the LP problem is solved by minimizing a sequence of piecewise quadratic, strongly convex functions subject to simple constraints (lower and upper bounds). This minimization is achieved by using a method which combines the conjugate or preconditioned conjugate gradient method and an active constraints set strategy. As an option for solving the problem of minimizing piecewise quadratic function a numerically stable method which combines the proximal multiplier method and an active set algorithm with QR factorization or Cholesky factorization is implemented.

In these methods we solve a sequence of minimization quadratic functions without constraints. After a finite number of steps, a set of indices of constraints, which are active in the solution, is found.

Our algorithm differs from the active set algorithm described by Fletcher (1981) and by Gill, Murray, and Wright (1981) because in addition to upper and lower bounds, we also take the piecewise quadratic form of the minimized function into account.

In the case of QR factorization, the Q–R transformation is carried out by using the Givens rotations. In our implementation we do not store the orthogonal matrix Q.

The working sets are changed during steps of the active set algorithm. Consequently, we update R factor whenever a index is added to or deleted from the working set. Computing a new factorization ab initio would be too expensive, so we adopted effective methods for updating the R factor. The details of the algorithms are described in Sosnowski (1990b).

2 Multicriteria linear programming

A linear multicriteria programming problem can be formulated as follows

$$\min q \tag{1}$$

$$q = Px \tag{2}$$

$$Ax = b \tag{3}$$

$$d \leq x \leq g \tag{4}$$

where P is a given $k \times n$ matrix of objective functions coefficients; and $x, d, g \in R^n$, $b \in R^m$, and A is an $m \times n$ matrix.

2.1 Finding Pareto-optimal solutions

A Pareto-optimal solution can be found by the minimization of the achievement scalarizing function in the form

$$\max_{i=1,\ldots,k} \left(w_i(q_i - \bar{q}_i) \right) + \varepsilon \sum_{i=1}^{k} w_i(q_i - \bar{q}_i) \tag{5}$$

where:

k is the number of criteria,

q_i is the i-th criterion,

$\bar{q}_i$ is the aspiration level for the i-th criterion,

w_i is a weight associated with the i-th criterion, and

ε is a given non-negative parameter.

It should be noted that if specified aspiration level $\bar{q}$ is the ideal point, then the Pareto-optimal point is the nearest, in the sense of an augmented Tchebycheff weighted norm (Steuer, 1982), to the aspiration level. If the aspiration level is attainable, then the Pareto-optimal point is uniformly better than $\bar{q}$. Properties of the Pareto-optimal point depend on the localization of the reference point (aspiration level) and on weights associated with criteria.

The scalarizing function (5) is nondifferentiable piecewise linear but the problem of minimization (5) subject to (2)–(4) can be transformed to a linear programming problem. We introduce an additional variable x_0 and auxiliary constraints:

$$- x_0 + w_i q_i \leq w_i \bar{q}_i \quad i = 1, \ldots, k. \tag{6}$$

The objective function for the auxiliary problem takes the form

$$\min x_0 + \varepsilon \sum_{i=1}^{k} w_i(q_i - \bar{q}_i). \tag{7}$$

The problem of minimization (7) subject to (2)–(4) and (6) can be treated as a special case of the following general linear programming problem:

$$\min cx \tag{8}$$

$$A_1 x = b_1 \tag{9}$$

$$A_2 x \leq b_2 \tag{10}$$

$$l \leq x \leq u \tag{11}$$

where $x = (x_0, x_1, \ldots, x_n)$ and A_1, A_2, b_1, b_2, c are respectively defined. Some variables are unbounded (e.g., x_0); this means that some of the bounds (l_i or u_i) are assumed to be equal to minus, or plus infinity.

3   Interactive reference point optimization

The method adopted in HYBRID for multicriteria optimization is the reference point approach (cf Wierzbicki, 1980). This approach may be summarized in the following stages:

1. The user of the model specifies a number of criteria (objectives).

2. The user specifies an aspiration level for each criterion. An aspiration level is also called a reference point.

3. The problem is transformed by HYBRID into an auxiliary parametric LP problem and then solved. Its solution gives a Pareto-optimal point. If the specified aspiration level is not attainable, then the Pareto-optimal point is the nearest (in the sense of a Chebyshev weighted norm) to the aspiration level. If the aspiration level is attainable, then the Pareto-optimal point is uniformly better. Properties of the Pareto-optimal point depend on the localization of the reference point (aspiration level) and on weights associated with criteria.

4. The user explores the various Pareto-optimal points by changing either the aspiration level and/or weights attached to criteria and/or other parameters related to the definition of the multicriteria problem.

5. The procedure described in points 3 and 4 is repeated until a satisfactory solution is found.

4 Solution technique of LP problem

For solving LP problems HYBRID uses a non-simplex method – proximal multiplier method.

The proximal multiplier method for the convex problems was introduced in (Rockafellar, 1976). The method differs from ordinary multiplier method (Bertsekas, 1982) due to the addition of a quadratic form to the augmented Lagrangian function, making the function strongly convex with respect to primal variables.

4.1 Proximal multiplier method

We introduce the following notation: $\|x\|$ denotes L_2–norm of x and $(u)_+$ denotes the vector composed of components $\max(0, u_i)$.

The following scheme describes a modification of the proximal multiplier method for the linear programming problem (8)–(11).

ALGORITHM. *Select initial vector of multipliers $y^0 = (y_1^0, y_2^0)$ (e.g., $y^0 = 0$) and $\rho^k, \gamma^k \in R$, $\rho^k, \gamma^k > 0$. Then for $k = 0, 1, \ldots$ determine successive x^{k+1}, $y^{k+1} = (y_1^{k+1}, y_2^{k+1})$ where*

$$x^{k+1} = \arg\min_{l \le x \le u} L(x, y^k) + \frac{1}{2\rho^k\gamma^k}\|x - x^k\|^2 \tag{12}$$

and

$$y_1^{k+1} = y_1^k + \rho^k(A_1 x^{k+1} - b_1) \tag{13}$$

$$y_2^{k+1} = (y_2^k + \rho^k(A_2 x^{k+1} - b_2))_+ \tag{14}$$

where

$$L(x, y^k) = (1/\rho^k)cx + \frac{1}{2}\|((1/\rho^k)y_1^k + A_1 x - b_1)\|^2 + \frac{1}{2}\|((1/\rho^k)y_2^k + A_2 x - b_2)_+\|^2 \tag{15}$$

until a stopping criterion is satisfied.

The minimized function (12) with $\gamma^k > 0$ is strongly convex and provides unique minimizer. Strong convexity is always further required during minimization (see Section 5). In the proximal multiplier method we lose the finite convergence property, but the sequence $\{(x^k, y^k)\}$ has better properties, and convergence to the Kuhn–Tucker point can be obtained under weak assumptions.

The crucial point of the method is the problem of minimization of piecewise quadratic functions. The ordinary or preconditioned conjugate gradient algorithms can be used for finding the minimum (Makowski and Sosnowski, 1989). Theoretically, those algorithms guarantee that the exact minimizer will be found after finite number of iterations. However, during computations the rounding errors often cause numerical problems.

To overcome these difficulties, a numerically stable method for piecewise quadratic strongly convex function was proposed in Sosnowski (1990a). This method combines an active set algorithm with QR factorization. As an option the Cholesky factorization is suggested in the new implementation of the HYBRID package.

5 Minimization piecewise quadratic functions with lower and upper bounds

In this section we discuss methods for finding the minimizer of (12).

Let us consider the following problem of minimizing piecewise quadratic function subject to both lower and upper bounds:

$$\min f^k(x) \tag{16}$$

$$f^k(x) = c^k x + \frac{1}{2}\|A_1^k x - b_1^k\|^2 + \frac{1}{2}\|(A_2 x - b_2^k)_+\|^2 \tag{17}$$

$$l \leq x \leq u \tag{18}$$

where: $A_1^k \in R^{r \times n}, A_2 \in R^{s \times n}, b_1^k \in R^k, b_2^k \in R^s, c^k \in R^n$, and $l, u \in R^n$ are given lower and upper bounds

The above formulation generalized the problems of minimization the piecewise quadratic function in the proximal multiplier method. Note that if we introduce the following notation

$$A_1^k = \begin{pmatrix} A_1 \\ (1/\sqrt{\rho^k \gamma^k})\, I \end{pmatrix}, \quad b_1^k = \begin{pmatrix} b_1 - (1/\rho^k) y_1^k \\ (1/\sqrt{\rho^k \gamma^k})\, x^k \end{pmatrix} \tag{19}$$

$$b_2^k = b_2 - (1/\rho^k) y_2^k, \quad c^k = (1/\rho^k) c \tag{20}$$

then the minimizer of (12) is equal to the minimizer of (16)–(18).

For the sake of simplicity, we drop k index in the formulation (16)–(18) and we will describe methods for solving the following problem:

$$\min f(x) \tag{21}$$

$$f(x) - cx + (1/2)\|A_1 x - b_1\|^2 + (1/2)\|(A_2 x - b_2)_+\|^2 \tag{22}$$

$$l \leq x \leq u. \tag{23}$$

Additionally for $l = 1, 2$ the following notation will be used:

$a_i^{(l)}$ denotes the i–th rows of matrix A_l,

$b_i^{(l)}$ denotes the i–th components of the vector b_l,

x_j denotes j–th component of x, and

A_l^T denotes the transposition of matrix A_l.

The minimized function (22) is convex, piecewise quadratic, and twice differentiable beyond the set:

$$\{x \in R^n : a_i^{(2)}x - b_i^{(2)} = 0, \quad i = 1, \cdots, s\}. \tag{24}$$

5.1   Active set algorithm

We define two types of working sets. At the given iteration of the active set algorithm, I will be a working set of the function f. That set defines a quadratic function as follows:

$$f_I(x) = cx + (1/2)\|A_1 x - b_1\|^2 + (1/2)\sum_{i \in I}(a_i^{(2)}x - b_i^{(2)})^2. \tag{25}$$

The second working set defines those variables which are fixed at bounds:

$$J = \{j : x_j = l_j \text{ or } x_j = u_j\}. \tag{26}$$

Additionally, the complements of the working sets will be defined as follows

$$\bar{I} = \{1, 2, \ldots, s\} \setminus I. \tag{27}$$

$$\bar{J} = \{1, 2, \ldots, n\} \setminus J. \tag{28}$$

Using the notation defined above for given working sets $\bar{I}$ and $\bar{J}$, the following minimization subproblem can be formulated

$$\min f_I(x) \tag{29}$$

$$x_j = \bar{x}_j \quad j \in J \tag{30}$$

where

$$\bar{x}_j = \begin{cases} l_j & \text{if fixed lower bound} \\ u_j & \text{if fixed upper bound.} \end{cases} \tag{31}$$

The active set algorithm, in the form described below, solves a sequence of the subproblems. For given working sets I and J, we minimize the quadratic function f_I in respect to variables x_j whose indices j belong to set $\bar{J}$. These variables will be free. The variables whose indices belong to set J are fixed on their bounds. This is an unconstrained quadratic subproblem. Its solution defines a search direction. The step length is determined to provide feasibility. The piecewise quadratic form of the function f is also taken into account while the step length is computed.

Let A_2^I and b_2^I be a submatrix and a subvector composed of rows and coordinates corresponding to indices $i \in I$:

$$A^I = \begin{pmatrix} A_1 \\ A_2^I \end{pmatrix} \quad b^I = \begin{pmatrix} b_1 \\ b_2^I \end{pmatrix}. \tag{32}$$

Using the above notation, problem (29)–(31) can be rewritten as follows:

$$\min f_I(x) \tag{33}$$

$$f_I(x) = cx + (1/2)\|A^I x - b^I\|^2 \tag{34}$$

$$x_j = \bar{x}_j \quad j \in J. \tag{35}$$

We divide the vector x into two vectors corresponding to the working set J and its complement:

x_J – vector of the free variables

x_J – vector of the fixed variables.

We have

$$x = (x_J \quad , \quad x_J). \tag{36}$$

Then we divide the matrix A^I into two submatrices whose rows correspond to the fixed and free variables respectively:

$$A^I = (A_J^I \quad A_J^I). \tag{37}$$

So we have:

$$f_I(x) = c^J x^J + c^J x^J + (1/2)\|A_J^I x^J + A_J^I x^J - b^I\|^2. \tag{38}$$

Let us consider the problem of finding free variables x^J as a result of minimization (38) without constraints. Because the matrix A_J^I has full column rank, the problem of minimizing function (38) has a unique solution. The minimum of the function (38) can be obtained by solving the following system of equations:

$$(A_J^I)^T A_J^I x^J = (A_J^I)^T (b^I - A_J^I x^J) - c^J. \tag{39}$$

The classical approach to solving this problem is via the system of normal equations

$$\bar{B} x^J = \bar{b} \tag{40}$$

where $\bar{B}$ is the symmetric positive definite matrix in the form:

$$\bar{B} = (A_J^I)^T A_J^I \tag{41}$$

and

$$\bar{b} = (A_J^I)^T (b^I - A_J^I x^J) - c^J. \tag{42}$$

Equation (40) can be solved via the conjugate gradient algorithm or by the preconditioned conjugate gradient algorithm. Those methods can be especially useful for large and sparse problems, but unfortunately the algorithms converge slowly when the problem is ill-conditioned.

5.1.1 Cholesky factorization

Another approach to solving the normal equation is based on the factorization of the matrix $\bar{B}$ using Cholesky method:

$$\bar{B} = R^T R \tag{43}$$

where R is upper triangular, and then x^J is computed by solving the two triangular systems

$$R^T y = \bar{b} \tag{44}$$

$$Rx^J = y. \tag{45}$$

5.1.2 QR decomposition

To simplify the discussion we write (38) in the following form:

$$f_I(x) = c^J x^J + (1/2)\|A_J^I x^J - h_J^I\|^2 + g_J^I \tag{46}$$

where

$$h_J^I = b^I - A_J^I x^J \tag{47}$$

$$g_J^I = c^J x^J \tag{48}$$

In the orthogonal factorization approach a matrix Q is used to reduce A_J^I to the form

$$Q^T A_J^I = \begin{pmatrix} R_J^I \\ 0 \end{pmatrix} \quad Q^T h_J^I = \begin{pmatrix} p_1 \\ p_2^I \end{pmatrix} \tag{49}$$

where R_J^I is the upper triangular. We have

$$f_I(x) = c^J x^J + (1/2)\|R_J^I x^J - p_1\|^2 + (1/2)\|p_2\|^2 + g_J^I. \tag{50}$$

The application of the orthogonal matrix Q does not change L_2–norm and an advantage of such a transformation is that we do not need to save the matrix Q. It can be discarded after it has been applied to the vector h_J^I. Moreover, the matrix R_J^I is the same as the Cholesky factor of $\bar{B}$ (41) apart from the possible sign differences in some rows.

The above QR transformation can be carried out by using the Givens rotations (see Golub and Van Loan, 1983). In our implementation we do not store the orthogonal matrix Q and the obtained matrix R_J^I is used for solution (44)–(45), where the vector $\bar{b}$ is given by (42).

5.1.3 Update of R factor

In the active set algorithm, the working sets are changed during sequential steps. Changes of working sets result in changes of the matrix A_J^I but only one row or one column can be added to or removed from that matrix at a time. This means that the matrix A_J^I which defined the Hessian of minimizing function (46) is changed. Consequently, we should update the R_J^I factor whenever a index is added to or deleted from the working set. Computing a new factorization ab initio would be too expensive. The modification of the R_J^I factor is described in (Sosnowski, 1990a).

5.2 Description of the software package and data structure

The package is composed of modules that provide a reasonably high level of flexibility and efficiency. This is crucial for the rational use of computer resources and for planned extensions of the package and the possible modification of the algorithm.

HYBRID is oriented toward an interactive mode of operation in which a sequence of problems is to be solved under varying conditions (e.g., different objective functions, reference points, values of constraints or bounds). Criteria for multiobjective problems may be easily defined and updated with the help of the package.

The method chosen of allocating storage in the memory takes maximal advantage of the available computer memory and of the features of typical real-world problems. In general, the matrix of constraints is large and sparse; therefore, a sparse-matrix technique is applied to store the data that define the problem to be solved. The memory management is handled in a flexible way. HYBRID is coded in C and is composed of four mutually linked modules:

1. A preprocessor processes the initial formulation of the problem. In the standard distribution of the package, a version of the preprocessors that handles input file in the MPS format is provided. However, it is recommended that for real-life application this preprocess be replaced by a specialized problem generator (cf, e.g., Makowski and Sosnowski, 1991).

2. A preprocessor generates multicriteria task. It also transforms a multicriteria problem into a parametric single-criteria optimization problem. The second preprocessor allows for the analysis of a solution and for the interactive change of various parameters that may correspond to choice of some option, change of parameters in definition of multicriteria problem. This module also optionally scales the LP problem.

3. The optimization module is called solver.

4. Driver, eases the usage of all modules. The PC version of driver provides context sensitive help which helps an inexperienced user in efficient usage of the package.

In addition to this, HYBRID offers many options useful for diagnostic and verification of a problem being solved. The data format for the input of data follows the MPS standard adopted by most commercial mathematical programming systems.

HYBRID has been made operational in two versions: one for UNIX currently implemented for Sun Sparc running under SunOS 4.1 and one for a PC compatible with IBM PC.

6 Test problems

In this section, we report the computational results of running the HYBRID package on a set of single objective linear programming test problems. All but one of the test problems are from computer networks (NETLIB system); one (Rains-Op) is an environment model developed at IIASA.

Name	Rows	Columns	Time(min.)
Adlitte	57	97	0.07
Bandm	306	472	5.08
Beaconfd	174	256	0.19
Capri	272	353	2.44
E226	224	282	1.43
Israel	175	142	0.29
Share1B	118	225	0.77
Share2B	97	79	0.11
Rains-Op	711	108	2.90
Scagr25	472	500	12.74
Scagr7	130	140	0.23
Scorpion	389	358	4.90
Scsd1	78	760	1.10
Scsd6	148	1350	6.03
Sctap1	311	480	5.07

All test runs were carried out on the Sun Sparc 1+ Workstation under SunOS 4.1. The GNU C compiler (ver. 1.40) has been used and the default parameters (which include the Cholesky partition option) have been selected.

7 Final remarks

HYBRID is oriented toward an interactive mode of operation in which a sequence of problems is to be solved under varying conditions (e.g., different objective functions, reference points, values of constraints or bounds). Criteria for multiobjective problems may be easily defined and updated with the help of the package.

The simple constraints (lower and upper bounds) for variables are not violated during optimization and the resulting sequence of multipliers is feasible for the dual problem. Constraints other then those defined as simple constraints may be violated, however; therefore the algorithm can be started from any point that satisfies the simple constraints.

HYBRID has been designed and implemented for applications to real life optimization problems. More attention has therefore been paid to the preprocessing of data, generation of sequence of related problems, diagnostics and robustness of the software. This, to some extent, increases the overall execution time needed for processing data and solving an LP problem. The algorithm outlined in this paper is also capable of solving badly conditioned problems (e.g., some of those reported here as test problems). Different algorithms have been implemented in the solver (cf Makowski, Sosnowski, 1989). The authors will continue to work on improving the solver performance while keeping its robustness.

References

Bertsekas, D.P. (1982). *Constrained Optimization and Lagrange Multiplier Methods.* Academic Press, New York.

Flecher, R. (1981). *Practical Methods of Optimization.* Vol. II: *Constrained optimization.* John Wiley, New York.

Gill, P.E., W. Murray, and M.H. Wright (1981) *Practical Optimization.* Academic Press. London New York.

Golub, G.H., and C.F. Van Loan (1983). *Matrix Computations.* Johns Hopkins University Press, Baltimore, Maryland.

Lewandowski, A., and A.P. Wierzbicki (eds.) (1989). *Aspiration Based Decision.* Lecture Notes in Economics and Mathematical System, Vol. 331. Springer–Verlag, Berlin.

Makowski, M., and J.S. Sosnowski (1989). *Mathematical Programming Package* HYBRID. In: A. Lewandowski, and A.P. Wierzbicki, (eds.) *Aspiration Based Decision.* Lecture Notes in Economics and Mathematical System, Vol. 331. Springer–Verlag, Berlin.

Makowski, M., and J.S. Sosnowski (1991). HYBRID-FMS: An Element of DSS for Designing Flexible Manufacturing Systems, In: P. Korhonen, A. Lewandowski, and J. Wallenius (eds.) *Multiple Criteria Decision Support.* Lecture Notes in Economics and Mathematical Systems, Vol. 356. Springer-Verlag, Berlin.

Rockafellar, R.T. (1976). Augmented Lagrangians and Applications of the Proximal Point Algorithm in Convex Programming. *Mathematics of Operations Research,* No 1, pp. 97–116.

Sosnowski, J.S. (1990a). A stable method for solving multicriteria linear programming problems in a decision support system. In: M. Fedrizzi and J. Kacprzyk (eds.) *Systems Analysis and Decision Support in Economics and Technology.* Proceedings of 8th Italian–Polish Symposium. Omnitech Press, Warsaw.

Sosnowski, J.S (1990b). A method for minimization of piecewise quadratic functions with lower and upper bounds. CP-90-003. IIASA, Laxenburg.

Steuer, R.E. (1982). On sampling the efficient set using weighted Tchebycheff metrics. In: M. Grauer, A. Lewandowski, and A.P. Wierzbicki (eds.) *Multiobjective and Stochastic Optimization.* Collaborative Proceedings CP–82–S12. IIASA, Laxenburg.

Wierzbicki, A. (1980). A mathematical basis for satisficing decision making. WP-80-90. IIASA, Laxenburg.

Intelligent Software Tools
for Building Interactive Systems

Rumena Kaltinska, Zvetan Petrov

Institute of Mathematics, Bulgarian Academy of Sciences
"Acad. G.Bonchev" Str, bl. 8 , Sofia 1113, Bulgaria

1 Introduction

The paper considers an approach to the automatic building of *Interactive Systems* for solving mathematical problems (Sections 2 and 3). It discusses the intelligent RZtools package (Sections 4–9) realizing this approach on IBM PC/XT/AT and compatibles. By means of RZtools an interactive user friendly INTERFACE shelling LIBRARIES of *Turbo Pascal* and *MS/Fortran* programs representing numerical methods can be automatically created and the corresponding *Interactive System* formed. The INTERFACE built by means of RZtools is oriented properly towards both the classes of problems and numerical methods included in certain *Interactive System*. It is menu-driven and provides an easy-to-use Borland-like environment.

An important feature of the *Interactive Systems* built by RZtools is their unified human-machine interaction. They actually form a family of *Interactive Systems*. The users of such *Systems* are also facilitated. Once accustomed to a particular *System* they need minimal efforts to start working with another *System* from the family.

1.1 Motivation

An *Interactive System* intended to solve certain classes of mathematical problems is supposed to consist of the following components:

– LIBRARY (LIBRARIES) of programs representing numerical methods for solving the classes included;

– interactive INTERFACE, which carries out the dialogue with the user and provides him with tools to define the problems and to supervise the computational process.

There are many programs written in high-level program languages (*Fortran, Pascal,...*) representing numerical methods for solving different classes of mathematical problems. Many of them are published or even distributed in source code and can be easily adapted and used on different kinds of computers.

The case with the INTERFACE appears to be totally different. To create a really user friendly INTERFACE is a rather complicated problem, depending on the kind of computer and requiring different kinds of skills in using Operating Systems, programming languages, etc.

The well known *Interactive Systems* usually have a static INTERFACE specially oriented towards the classes of problems included. Some of them are open for the inclusion of new classes, but only if the INTERFACE proves to be suitable. Otherwise all the work of projecting and programming a new INTERFACE has to be done anew. On the other hand the specialists developing, programming and testing numerical methods are not allowed to include their method-programs in such *Systems*. Therefore they are forced to spend a lot of time to supply their programs with a self-made INTERFACE. Such an INTERFACE is usually a very simple one, which does not satisfy thoroughly the needs and which delays the experiments and researches. Moreover such a program cannot usually be used by other people.

Hence the great necessity of software tools that ease and automate the process of creating interactive INTERFACES. RZtools is an intelligent package intended to satisfy such a need.

1.2   Methodology and implementation

The approach for the automatic building of *Interactive Systems* considered here is based on both their unified structure (see Section 2) and some principles of building (see Section 3). According to this approach both the LIBRARIES and the INTERFACE are autonomous modules, the source language of which is allowed to be different. A LIBRARY is defined as a combination of programs representing numerical methods and no restrictions are imposed on these programs. The INTERFACE is composed of basic interactive modules, that are shared by all *Interactive Systems* and are easily oriented towards different classes of problems and numerical methods. They can be created beforehand and used as 'bricks' to assemble the INTERFACES needed for different *Interactive Systems*. This approach also enables to the automation of the whole building process.

The RZtools package (see Section 4) is based on the approach considered above. It consists of both basic interactive INTERFACE-modules and an interactive *INSTALLATION SYSTEM*. The latter is intended to assemble automatically the INTERFACE needed from the INTERFACE-modules and form the corresponding *Interactive System*, ensuring proper INTERFACE-LIBRARIES 'tuning'. It is supposed that these LIBRARIES exist and contain *MS/Fortran* or (and) *Turbo Pascal* programs.

Some features of the *Interactive Systems* built by means of RZtools are briefly given in Section 4.2. These features are a result of the possibilities of the RZtools INTERFACE-modules, the more important of which are discussed in Sections 5 and 6. The modules represent *Turbo Pascal 5.0* procedures, which could also be used independently and could be incorporated into different user programs. The *INSTALLATION SYSTEM* is presented in Section 7. Further explanations of how to form a LIBRARY can be found in Section 8. Some RZtools applications and possible users are given in Section 9.

2   Structure of interactive systems

The automatic building of *Interactive Systems* is based on their unified module structure. Such a structure is shown in Figure 1 (see also Kaltinska, 1989). For each class of problems the structure is one and the same.

The program *MAIN* is intended only to offer a menu mentioning the classes of problems included in the *System* and to call the module *CLASS_MONITOR*. The last one offers the menu with the basic activities: problem description and realization of the computational process. Depending on the choice *CLASS_MONITOR*, invokes the corresponding program tools, orienting them in advance to the particular class. 'Problem Description' in Figure 1

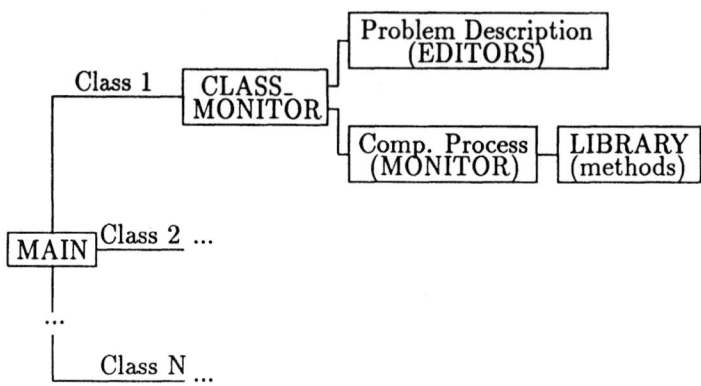

Figure 1. Structure of an *Interactive System*

stands for the program tools for input, editing and saving of problem descriptions. Usually such tools are some kind of editors. 'Comp. Process' stands for the module *MONITOR* which visualizes and controls the computational process and invokes the corresponding LIBRARY. Each LIBRARY contains method-programs for solving the corresponding class of problems.

All program tools mentioned up to here with exception of the LIBRARIES form the INTERFACE of an *Interactive System*.

3 Principles of building

The structure given in Section 2 and Figure 1 allows the following principles to be used:

– The whole INTERFACE as well as each LIBRARY of an *Interactive System* to be formed as an *.EXE* file.

– No restrictions on the method-programs in the LIBRARIES to be imposed in order that any method-program can be easily included in a LIBRARY, provided the source language is the same.

– All information needed to orient the INTERFACE's modules *EDITORS* and *MONI-TOR* to the classes of problems as well as the numerical methods included in an *Interactive System*, to be written properly in a *System*'s configuration file. *CLASS_MONITOR* reads the information for a certain class from the configuration file and adjusts the *EDITORS* and *MONITOR* conformity with it.

Some corollaries of these principles are:

1) The source languages of the different LIBRARIES and the INTERFACE in an *Interactive System* are not obligatorily the same.

2) The creation of the INTERFACE and the LIBRARIES of an *Interactive System* are independent activities.

3) The whole process of creating INTERFACES of *Interactive Systems* can be automated. For this goal two kinds of programs written in some programming language have to be prepared. First, a set of suitable interactive INTERFACE-modules (*CLASS_MONITOR*, some *EDITORS*, *MONITOR*, etc.), that are easily oriented towards different classes of problems and methods. These modules are further used to assemble INTERFACES needed for particular applications. Secondly, an interactive Installation module has to be prepared with the intension for automatically creating such an INTER-FACE, i.e. to obtain all information needed for adjusting the INTERFACE-modules and to create the corresponding *System*'s configuration file and *MAIN* program (see Figure 1).

4) Provided LIBRARIES of programs for solving different classes of problems are available, then various *Interactive Systems* can be created depending on the combination of classes included. That is a very flexible possibility to shape easily *Interactive Systems* containing classes of problems pertaining to specific applications.

5) Each *Interactive System* is open for inclusion/exclusion of classes of problems and methods. To include a class means both to form its LIBRARY as an *.EXE* file and to join the corresponding information to the configuration file of the *System* (to exclude a class means only to delete this information from the file). To include (exclude) a method means to put (delete) both the information for the method into (from) the configuration file and the corresponding method-program into (from) the LIBRARY.

4   The RZtools package

The RZtools package is a realization of the approach given in Sections 2 and 3 on IBM PC/XT/AT and compatibles. MS DOS version 3.10 and higher is required.

Each *Interactive System* built by means of RZtools consists of the components IN-TERFACE and LIBRARIES (see Figure 1). The INTERFACE is assembled from the *Turbo Pascal 5.0* INTERFACE-modules of the RZtools package (see Section 4.1) and formed as an *.EXE* file. The INTERFACE is to be automatically produced by means of the RZtools *INSTALLATION SYSTEM* (see Section 4.1). Each LIBRARY is a combination of programs representing numerical methods for solving a certain class (classes) of problems. No restrictions are imposed on these method-programs. It is only supposed that they are *Turbo Pascal* procedures or *MS/Fortran* subroutines and that the source language is the same for all programs in a LIBRARY. Each LIBRARY is also formed as an *.EXE* file (see Section 8).

No principal objections exist against other source languages for the method-programs in the LIBRARIES (see Section 4.1). The RZtools package is also open to the addition of extra INTERFACE-modules.

4.1 RZtools modules

RZtools contains modules separated in three groups:

   – A set of basic interactive INTERFACE-modules representing *Turbo Pascal 5.0* procedures and forming the *RZtools.TPU* unit. Each of them can be called from any *Turbo Pascal* program and oriented properly by means of their arguments. Therefore they could also be used independently and be incorporated into different user programs. Some of these INTERFACE-procedures represent (see Figure 1) the modules *CLASS_MONITOR*, *EDITORS* (see Section 5) and *MONITOR* (see Section 6). Other ones are auxiliary procedures intended to support the man-machine interaction: to create pull-down (nested) menus, to provide hot-key context-sensitive help, etc.

   – The modules *RZtoolsS.TPU* and *RZtoolsS.LIB* intended to pass information between the INTERFACE and the LIBRARIES written in *Turbo Pascal* and *MS/Fortran* respectively (see Section 8.1). Note that other source languages to be admissible for the LIBRARIES only requires the creation of similar modules, which transform information between the INTERFACE and LIBRARIES written in those languages.

   – The interactive *INSTALLATION SYSTEM RZinst.EXE*. It is supposed that *Turbo Pascal* and/or *MS/Fortran* LIBRARIES (LIBRARY) containing method-programs for solving some classes of problems exist. During an interactive session *RZinst* prompts for information for the classes of problems and the methods included in the LIBRARIES, creates a configuration file corresponding with this information, assembles the INTERFACE needed from the INTERFACE-procedures and forms the *Interactive System* (see Section 7).

4.2 Interactive systems built by the RZtools package

The INTERFACE of such *Interactive Systems* is menu-driven. All customary means to generate, edit, save, list or print problems are available (see Section 5). Problem functions are to be defined by the user by means of their analytical expressions. Symbolic differentiation of any order is supported (see Section 5). The computational process can be optionally visualized and controlled by the user (see Section 6). Numerical results can be optionally saved and are easily retrieved for further processing. Hot-key context-sensitive help (key *F1*) about the key definitions, options, classes of problems included, methods, etc. is also provided.

   Note that all flexible possibilities 1), 2),4), and 5) discussed in Section 3 are also available. According to 2) it is possible to create and run an *Interactive System* before some (all) of its LIBRARIES have been completed. A class of problems for which the LIBRARY is missing might be chosen in order to create and save files with data. Even if an attempt is made to solve a problem from a class for which the LIBRARY does not exist, only a message 'Solving Program not found' is displayed.

5 Problem description tools

Two interactive *EDITORS*, namely *TABLE EDITOR* and *FUNCTION EDITOR*, are included in *RZtools.TPU* (see Section 4.1). They are intended to support the input, editing and saving of mathematical problems (see also Kaltinska and Petrov, 1989). The *EDITORS* exist also as standalone *.EXE* files, which can be run directly from DOS (see Section 5.2).

5.1 Purpose

The spread-sheet *TABLE EDITOR* manipulates (see Section 5.3) data represented in the form of two-dimensional tables. It is useful for classes of problems that can be described by means of vectors and matrices. Such classes for instance are Linear, Quadratic and Discrete Programming problems, some Transportation problems, etc. The matrices and vectors describing the problems of a given class can be arranged as a sequence of tables, each one with its own structure and characteristics. Real, integer and long integer (four bytes) numbers are admissible.

The *FUNCTION EDITOR* is useful for problems described by means of functions, the analytical expressions of which depend on the particular problem. Such classes for example are Nonlinear Programming, Functions' Approximation, Optimal Control, etc. The *FUNCTION EDITOR* is intended to:

– Manipulate (see Section 5.3) arbitrary ASCII files and in particular functions' analytical formulas depending on variables and parameters.

– Perform automatic differentiation (Griewank, 1989) and thus find optionally functions' partial derivatives of any order. The automatic differentiation procedure is a kind of symbolic differentiation, which creates a special computational graph for efficient evaluation of the functions and their partial derivatives. The *FUNCTION EDITOR* builds this graph and writes it in an internal language understandable for the special interpreter-program serving the method-programs (see Section 8.1).

– Create optionally *Turbo Pascal* or *MS/Fortran* programs for evaluating the functions and their partial derivatives. These programs actually represent the computational graph written in the corresponding language.

For instance a set of functions $f_1(X, P), f_2(X, P), \ldots, f_s(X, P)$ depending on variables $X = (x_1, x_2, \ldots, x_n)$ and parameters $P = (p_1, p_2, \ldots, p_m)$ can be manipulated during an interactive session with the *FUNCTION EDITOR*. The letters x and p (small or capital) followed by figures are the compulsory names of the variables and parameters. The computational graph for evaluating the functions and their partial derivatives up to some order r can be optionally found (see *MAKE* option, Section 5.3) and written either in the internal language or as *Turbo Pascal* or *MS/Fortran* programs Fk , $k = 1, 2, \ldots, s$:

(i)   *Function Fk (j:integer; var X,P,D1,D2,...,Dr) : real*
(ii)   *Function Fk (j,X,P,D1,D2,...,Dr)*

Input parameters of the functions Fk are the order j $(0 \leq j \leq r)$, X and P. When Fk is invoked, Fk is passed the value of the function $f_k(X, P)$, and, if $j > 0$ its partial derivatives up to the j-th order are evaluated and their values are filled in the arrays $D1, D2, \ldots, Dj$ respectively.

The two *EDITOR*-procedures are adjusted to their mode of work, data attributes and the way information is to be displayed either directly through their arguments or by the *INSTALLATION SYSTEM* (see Section 7.1).

5.2   Stand-alone versions

The stand-alone versions of the *TABLE EDITOR* and *FUNCTION EDITOR* procedures are formed as *ETA.EXE* and *EFU.EXE* respectively. They can be run from DOS by ordinary users in order to obtain data files and files containing programs (i) or (ii) for miscellaneous applications. Programs (i) will be produced when the editor *EFU* is invoked by typing *EFU* at the DOS prompt. To obtain *MS/Fortran* programs (ii) an additional *F* qualifier has to be specified, i.e. *EFU F* to be typed. Note that *EFU* can be used also as an ordinary text editor (see Section 5.3), so that the files containing (i) or (ii) programs can be loaded and observed also by means of *EFU*.

ETA and *EFU* are adjusted through a configuration file named *CONFIG.ETA* and *CONFIG.EFU* respectively. They can be created/edited by the *INSTALLATION SYSTEM* (see Section 7.1). If such files are missing in the current directory, *ETA* and *EFU* use built-in default settings.

5.3   Possibilities

The *EDITORS* are menu-driven and provide a familiar and easy-to-use Borland-like environment. The main menu bar consists of the *FILE, EDIT, HELP* and *QUIT* options. The *MAKE* option is added in the case of the *FUNCTION EDITOR*. To choose an option from a menu means both to position a highlighted box over it using the arrow-keys and to press <*ENTER*>. Each menu (except the main menu) can be left when <*ESC*> is hit. Any of the pull-down menus appears and remains on the screen only when the corresponding main menu option is chosen.

The *FILE* option pull-down menu permits to create new data (*NEW* option), load old data from a file (*LOAD* option), save data in the current data file (*SAVE* option, key F2) or write them in another one (*WRITE TO* option). MS DOS 'wildcards' ('*' and '?') in file names are also allowed. Data files are in ASCII format.

The *EDIT* option provides all customary means to enter and edit data, i.e. tables or text/functions respectively, and to visualize them by means of the keypad (Up/Down/Left/Right/PgUp/PgDn, etc.) as well as to use the whole screen (zoom). The block commands allow to mark data blocks and have extremely convenient functions to print/copy/move/delete data blocks and write/read them to/from a file, thus permitting to create new data on the base of old data. The *TABLE EDITOR* also allows to: display the full mantissa of a real number; find a column/row by its name; insert/delete rows and columns; fill in a row, column or data block with an entered number; generate data, i.e. fill in the tables with uniformly distributed pseudorandom numbers; etc. The *FUNCTION EDITOR* provides an 'insert' and 'overwrite' mode as well as procedures for finding a text and replacing a text with another one.

The *EDITORS* provide a proper error handling and hot-key context-sensitive help (key F1) about the key definitions and options, as well as about the functions syntax in the case of the *FUNCTION EDITOR*. Syntax checking of the entered functions is also provided. The MAKE pull-down menu of the *FUNCTION EDITOR* main menu offers various choices to find optionally the computational graph for the evaluation of the functions and their partial derivatives up to a chosen order and write the corresponding programs either in the internal language or in the form of (i) or (ii) programs. In the last two cases they are automatically saved in a file with the same name as the name of the file containing the functions but extended with *PAS* or *FOR* respectively.

6 The computational process tool

The interactive procedure *MONITOR* from *RZtools.TPU* (see Section 4.1) is developed to visualize and control the computational process of the current problem, which has to be solved.

MONITOR forms four windows on the screen with lists of:
– the available numerical methods
– the input parameters (tolerances, steps, etc.) required by the current method as well as their current settings;
– the output parameters (elapsed time, iterations, etc.) of the current method and their current values, if they are evaluated;
– some 'result-variables', i.e. objects characterizing the solution, and their current values, if evaluated.

Through the *VIEW/EDIT* pull-down menu options the user is able to enter any of these windows and move a highlighted box through the list using the keypad (Up/Down/ PgUp/PgDn/Home/End) in order to visualize the list as well as to print it (Ctrl P). Some other possibilities depending on the window entered are:

– Choose/change the method. As a result the lists of its input and output parameters are automatically displayed in the corresponding windows.

– Change the current values of the input parameters. *MONITOR* keeps the information about their default settings and ranges. When a method is chosen its input parameters are displayed and their default settings are given. A new value is accepted only if it belongs to the parameter's range.

– Obtain some handy intermediate and final information. When a method is working, the values of some/all output parameters and result-variables could be continuously updated on the screen.

– Obtain hot-key context-sensitive help (key *F1*) about any element in the lists. The text of this help information is taken from the ASCII file, the name of which is an argument of procedure *MONITOR*. This file has to be prepared by the authors of the corresponding LIBRARY of methods and could contain more or less information about each of the methods (authors, algorithms,...), their input parameters (meaning, default settings, ranges), output parameters, etc. If such a file is missing, only a message 'No Help' appears.

Note that *MONITOR* also supports hot-key context-sensitive help of its own (key *F1*) about its options and key definitions.

MONITOR permits computation to be interrupted by the user (pressing <ESC>) in order to change the method and/or some input parameters' values and afterwards the computation to be continued.

When the current method stops or its work is interrupted by the user, *MONITOR* displays the current values of all output parameters and result-variables as well as a message about the kind of result reached by the method. The rest of the results, which are vectors or matrices, can be optionally displayed in the form of tables. All possibilities to visualize and print such a table, load and save it, as well as to change its elements are provided by the *TABLE EDITOR* (see Section 5.3).

MONITOR is oriented properly, i.e. with respect to LIBRARY methods' and parameters' names, default settings, etc., through its arguments directly or through the *INSTALLATION SYSTEM* (see Section 7.2).

7   The installation system

The *INSTALLATION SYSTEM RZinst* of the RZtools package (see Section 4.1) is an interactive system intended to orient properly the *TABLE EDITOR* and *FUNCTION EDITOR* procedures (see Section 7.1) and *MONITOR* (see Section 7.2) as well as to create automatically *Interactive Systems* (see Section 7.3). The installation allows the problem description, computational process and the output results to be interpreted in the terms of a certain class of problems and the specific task. *RZinst* is also a menu-driven system. It supports a proper error handling and hot-key context-sensitive help (key *F1*).

7.1   EDITORS installation

Two options of *RZinst* main menu activate the installation of the *TABLE* and *FUNCTION EDITOR* respectively. The user is shown a list of the *EDITOR*'s characteristics and the current values of the corresponding arguments of the *EDITOR*'s procedure. In the beginning the characteristics are given their default settings, which provide the corresponding *EDITOR* with its full possibilities. The user is prompted to adjust the characteristics in a desired way. For instance: some *EDITOR*'s options may be hidden or excluded; the sizes of the tables as well as the number of the variables and/or parameters in the functions may be fixed or defined by given expressions; the default names of the rows and columns as well as of the functions may be defined; the window where the information is displayed and edited may be moved and resized; etc. In the case of the *TABLE EDITOR*, the structure of the tables and the type of the numbers within them can also be adjusted. A demonstration mode in order to show the *EDITOR* screen layout, colour settings, etc. is also provided.

The characteristics adjusted can be optionally printed and saved in a configuration file (in ASCII format), which could be loaded when necessary. If such a configuration file is named *CONFIG.ETA* or *CONFIG.EFU* respectively, it can be used to adjust the corresponding stand-alone version *ETA* or *EFU* (see Section 5.2)

7.2   Class installation

This installation permits to create a configuration file containing the whole information needed to orient the INTERFACE both to a certain class of problems and to the methods included in the corresponding LIBRARY (see Figure 1). Using the menu options the user is prompted to:

– Adjust *MONITOR* characteristics. For instance the whole information needed for the lists, which *MONITOR* supports on the screen (see Section 6), etc. A demonstration mode in order to show the adjusted *MONITOR* screen is also provided.

– Enter a common information for the class: its name, *EDITORS* needed, the name of the LIBRARY.EXE file and the file containing help information, etc.

– Save the information in a configuration file (in ASCII format). If it is indicated that some *EDITORS* are needed, their characteristics are automatically joined to the other information in the configuration file. Therefore it is obligatory that the *EDITORS* are installed first.

Any configuration file obtained through such a kind of installation can be loaded. If it contains characteristics of some *EDITORS*, the values of these characteristics are also loaded and can be shown when the corresponding *EDITOR* installation option is chosen (see Section 7.1).

Additional inclusion (exclusion) of methods in the LIBRARY of a class, which is already installed (i.e. its configuration file exists), only requires the information for the methods (names, parameters, ranges, etc.) to be put into (deleted from) its configuration file. That could be done again through *RZinst* installation or by any text editor.

7.3   Interactive system installation

By means of this installation an *Interactive System* for solving some classes of problems can be automatically built. It is supposed that the configuration files for these classes already exist (see Section 7.2). The user is prompted to:

– enter the names of the classes' configuration files

– enter the name of the *Interactive System* (under which the *System* is to be run from DOS) and the name of its configuration file;

– design the title page of the *System*.

Using this information and the classes' configuration files, *RZinst* creates both the *MAIN* program (see Figure 1) written in *Turbo Pascal 5.0* and the configuration file of the *Interactive System*. Thus the INTERFACE of the *Interactive System* is prepared. It is carried out entirely by the *MAIN* program and the *RZtools.TPU* procedures (see Section 4.1) used.

The *MAIN* program is saved automatically under the entered *Interactive System*'s name and it has to be compiled by means of the *TPC* compiler. The INTERFACE obtained is an *.EXE* file and can be run directly from DOS. The *Interactive System* is completed when the LIBRARIES for the classes included are also formed as *.EXE* files (see Section 8).

8   To form a LIBRARY

A LIBRARY is supposed to contain *MS/Fortran* subroutines or *Turbo Pascal* procedures representing numerical methods and to be formed as an *.EXE* file. No restrictions are set on these method-programs.

To form a LIBRARY as an *.EXE* file means both to write the LIBRARY's Main program and to add some instructions within the method-programs, if necessary. Special utilities *RZtoolsS.TPU* and *RZtoolsS.LIB* (see Section 8.1) are included in the RZtools package in order to assist the specialists in these activities. They are intended for LIBRARIES written in *Turbo Pascal* and *MS/Fortran* respectively.

8.1 INTERFACE-LIBRARY communication utilities

One group of programs (formed as functions or procedures/subroutines) in *RZtoolsS.TPU* and *RZtoolsS.LIB* is intended to be used by the Main program of a LIBRARY in order to transfer some information between the INTERFACE and the LIBRARY. For instance the program *GETCOMMUNICATE* passes information from the INTERFACE about the numbers of the chosen class and method, dimensions of the current problem, etc. Similar programs to obtain data tables, current values of the method's input parameters, unknowns' initial values, etc., are also available. The same holds for the analogous programs concerning method's output parameters, results, messages etc. to be passed back to the INTERFACE.

The second group of programs is intended to serve the method-programs:

– Programs (procedures/subroutines) permitting the values of some result-variables and output parameters to be continuously updated on the *MONITOR* screen (see Section 6) as well as some messages to be displayed when necessary. It is a matter of choice which of the values are to be 'refreshed' continuously.

– When a problem is described by means of the *FUNCTION EDITOR*, a computational graph for any of the functions entered is created (see Section 5.1). The program *INTERPRETER* is intended to evaluate the functions and optionally their partial derivatives, using the corresponding computational graphs.

8.2 Writing a main program

The Main program of a LIBRARY is intended both to support the transfer of information between the INTERFACE and the LIBRARY and enable the LIBRARY to be compiled as an *.EXE* file.

The transfer is carried out by means of the *RZtoolsS.TPU* or *RZtoolsS.LIB* utility (see Section 8.1) depending on the source code of the LIBRARY. The corresponding utility has to be mentioned in the source text or in the LINK command line respectively. The User Guide of the RZtools package contains a thorough description of the *RZtoolsS.TPU* and *RZtoolsS.LIB* programs, i.e. their arguments and activities, as well as a dummy of a Main program in *Turbo Pascal* and *MS/Fortran* together with some examples.

It is clear that inclusion (exclusion) of a program representing some numerical method in (from) a LIBRARY means to include (exclude) both the instructions for passing the corresponding information needed from/to the INTERFACE (see Section 8.1) and the call of the method-program.

9 Possible users. Applications.

The RZtools package is intended to facilitate:

– Scientists, specialists and students, that create and/or test numerical methods. RZtools permits them to concentrate on the significant part of their work, since RZtools will automatically provide their method-programs with an interactive user friendly INTERFACE, i.e. input, output, datasaving, computational process control, etc., thus obtaining completely stand-alone software, which can be used further in the scientific practice and education.

– Specialists developing *Interactive Systems* based on LIBRARIES of numerical methods in different mathematical fields. They could use the RZtools package to create automatically the INTERFACE of these *Systems* or include some RZtools INTERFACE-procedures in INTERFACES of their own.

Because of the unified human-machine interaction of the *Interactive Systems* built by means of the RZtools package, they actually form a family of *Interactive Systems*. The ordinary users of such *Systems* are also helped once accustomed to a particular *System*, they need minimal efforts to start working with another *System* from the family.

The RZtools package is already used by scientists, post graduate students and students from both the Institute of Mathematics of the Bulgarian Academy of Sciences and Sofia University 'Kl. Ohridski'. The following two families are developed respectively:

– A family of *Interactive Optimization Systems*. It contains the systems *LIFQU* (Linear, Fractional and Degenerated Quadratic Programming problems), *TRANS* (Linear and Fractional Transportation problems), *QUADRO* (Quadratic Programming problems), *SQUARE* (Least Squares Optimization and Approximation, see Petrov and Kaltinska, 1990), etc.

– A family of *Interactive Systems* representing Students' Diploma Works *OPTI1* (some particular classes of nonlinear optimization problems), *DIFINC* (Differential Inclusion Problems), *FALO* (Facility Location Problems), *FRED* (Fredholm integral equation), APPRO (Functions' Uniform Approximation), etc.

The classes of problems mentioned above are combined in the corresponding *Interactive Systems* depending on the authors of the method-programs in the LIBRARIES. But according to the principles and structure of the *Interactive Systems* underlying the RZtools package (see Sections 2 and 3), any other combination of the classes can be chosen and the corresponding *Interactive System* is automatically formed by means of the *INSTALLATION SYSTEM* (see Section 7.3). For instance a common *Interactive Optimization System* uniting all the classes from the first family is also available.

References

Griewank, A. (1989). On automatic differentiation. M. Iri and K. Tanabe (eds), *Mathematical Programming*, KTK Scientific Publishers, Tokyo, pp. 83-107.

Kaltinska, R. (1989). A structure of the interactive optimization systems. A. Lewandowski and I. Stanchev (eds.) Methodology and Software for Interactive Decision Support, Lecture Notes in Economics and Math. Systems, vol. 337, Springer-Verlag, Berlin.

Kaltinska, R. and Petrov, Z. (1989). Problem Formulation Software Tools for Interactive Optimization Systems. Proc. 18th Spring Conf. Un. Bulg. Math., Sofia, pp.334-338.

Petrov, Z. and Kaltinska, R. (1990). SQUARE – An Interactive System for Least-Squares Optimization. 19th Spring Conf. Un. Bulg. Math., Sofia, pp. 392-397.

Software Tools for Multi–Criteria Programming

Vassil Vassilev, Atanas T. Atanassov, Vassil Sgurev,
Milosh Kichovich, Anton Deianov, Leonid Kirilov
Institute of Informatics
Bulgarian Academy of Sciences
Sofia, Bulgaria

1   Introduction

In the last few decades researchers' efforts have concentrated particularly on the theory
and methodology of Multiple Objective Programming. A considerable amount of theoretical properties and extensions of traditional mathematical programming to the case
of multiple objective optimization, methodological approaches, methods, algorithms and
procedures have been developed (Benayoun et al., 1971; Fishburn et al., 1990; Korhonen
and Laakso, 1986; Lewandowski and Wierzbicki, 1989; Nakayama and Sawaragi, 1984;
Sawaragi et al., 1985; Shin and Ravindran, 1991; Steuer, 1986; Wierzbicki, 1982; Zionts,
1988). Quite a lot real life Multiple Objective Linear Programming (MOLP) problems
have been solved by specific implementations of the developed methodology. Generally,
many multiple objective interdisciplinary problems relevant in practice have multiple nonlinear objectives and nonlinear constraints. Within the past two decades research interest
grows to involve Multiple Objective Nonlinear Programming (MONP). The progress has
been mainly theoretically and methodologically oriented. Only few real-life applications
were reported in literature (Nakayama and Furukawa, 1985; Nakayama and Sawaragi,
1984; Roy and Wallenius, 1992). Still, effective computer codes for MONP models are
insufficiently available. Even if the MONP problem is well-structured as to settle itself
to algorithmic procedures, there are inherent restraints to the practical achievement of
optimal solutions. MONP problems need algorithms that are known to require exponentially much computer time - such problems are said to be NP-hard. The very high
computational expense of attaining an optimal solution would be sufficient to discourage
the end-user or the Decision Maker (DM) from searching a solution that is guaranteed to
be optimal. Therefore, the objective is to obtain a good or satisfactory solution rather
than an optimal solution. An accelerated advance in software tool development will stimulate increased application of these methods. The authors' point of view on the activity
for the development of MOP software follows the approach of designing and developing
general purpose software that implements well-known methods and algorithms, as well as

improved and new techniques that can be easily incorporated and that can be used as efficient solvers of specific real-life MOP problems.

In this paper we consider a number of general requirements in the development of software tools for multiple objective programming. The described software tools represent two independent program packages, named MOLP-16 and MONP-16, designed for solving multicriteria linear and nonlinear problems respectively. The considerations are given, obtained from theoretical analysis and experimental research, forincorporating in the packages structure the chosen interactive procedures and single criterion optimization methods. The input/output data structures of the two packages are considered. The ideas for future development of the software tools are pointed out.

2 General requirements in the development of the software tools for multicriteria programming

In this section we present a list of desirable features for multicriteria software tools. The list of features implies our view on multicriteria software, based on analysis of existing Decision Support Systems and several years of experience in the development and implementation of mathematical programming packages. Suggestions and observations from users are taken into account. Good multicriteria tools must be easy to use; functionally complete, modular and augmentable; portable and robust.

2.1 Multilevel user interface

The applicability of a multicriteria software tool depends mostly on the characteristics of the user interface. Its final implementation must be oriented towards end-users of different qualifications with respect to mathematical programming, concrete real-life areas including research experimentations, software engineering and education. A hierarchical or multilevel structure of the user interface is essential for users concerning the ease of multicriteria software tool operation. The first or top-level is occupied by a menu-driven control or other supervisor or management program. It provides the user with the opportunity for an easy selecting of the separate program modules, full help information for the whole multicriteria software tool and usually access to the commands of the operating system. This level may include one or several menus as well as the opportunity for implementation of a specific command language. The second level of the user interface considers dedicated program modules called editing programs. These programs allow the input data to be entered and to be displayed on the screen; multiple editing sessions; easy access to separate arrays or expressions and their elements; inserting or deleting new arrays or expressions and their parts. The basic program and problem parameters are set up. These two levels are to a certain extend subsidiary concerning multicriteria problem solving. The most essential level of the user interface is the interaction of the end-user with the multiple objective optimization program. Usually the screen is divided into several parts or fields designed for displaying results and auxiliary information, selection of criterion and necessary functions for solving and investigating the multiobjective problem, messages output, entering alphanumeric and digital information required for solving the problem. Visual and alphanumeric data, scaled properly, should be displayed simultaneously.

2.2 Functional completeness, modularity and augmentability

These three requirements are closely related to one another. Because of the mutually exclusive demands, full functional completeness is very hard to be reached. On one hand a set of working interactive procedures, methods and algorithms, and efficient single optimization methods should be included into the software. Switching of the different procedures and methods is desirable. On the other hand in many practical problems rapid convergence, quick response, and easy and simple use are of great importance. One way of solving the problem is the implementation of the consolidated approach to multiple criteria optimization (Steuer and Whisman, 1986). We follow the principle of developing different program modules ready to be integrated on user's request in the software tools, having in mind the unification of the specific input and output structures. So the software augmentability is achieved too. As a base one of the most efficient approaches for solving real-life problems - grounded on the reference point/aspiration level concept, is implemented. Concerning the single objective optimization programs at least one or two methods may be incorporated in the software package and selected by the user independently. As well as the multicriteria interactive procedures, they may be replaced by more efficient ones or new methods may be appended. Approximate algorithms may be included for quick obtaining of solutions, near to the optimal ones for problems of high complexity.

2.3 Portability and robustness

Still, the personal computers IBM PC/AT, PS/2 and their compatibles are the most common microcomputers used for solving multicriteria optimization problems in real-life applications. We suppose that, although they are not so fast as it was expected, the 32-bit , Unix-based workstations will become more attractive for multicriteria optimization with respect to their performance, price and availability. The shortcomings and memory restrictions of the DOS environment will force the replacement of the personal computers by the more powerful microcomputers. The presented software tools are intended for IBM personal computers and they run under control of MS-DOS 3.30 and up. One of the main goals observed in the design of the software is the possibility of comparatively easy portability to microcomputers of different types. For this reason some of the optimization modules are written in ANSI FORTRAN, others, the control and editing programs - in ANSI C. A special library of routines (written in ANSI C) has been developed to accelerate the efficient transfer of the optimization programs to computers of different types. The following functions are implemented by these routines: dynamic allocation of main memory; cursor control; use of windows; keyboard processing; data file management. It is clear that this library is dependent on the computer type and the operating system. Although the user interface is written in ANSI C, actually it is hardly possible to make it portable.

Robustness mainly, but not only, of the optimization programs is the other important goal that should be aimed at. A quick and smooth response is needed to user's inaccuracy and numerical problems. Appropriate messages should be generated by the program modules and operating system.

3   MOLP-16

The MOLP-16 package is designed for solving the basic class of multiobjective linear programming problems. It consists of the following program modules:

- a control program;

- editing programs;

- optimization programs - multiobjective interactive procedures and single objective optimization subroutines;

- utilities.

The control program is menu-oriented. It selects the required mode of operation and the corresponding functional modules of the product. MOLP-16 runs in four modes of operation: input data editing; problem solving; Help; DOS commands execution. The user's interface with the control program is implemented by three menus. The main menu selects the mode of operation. The other two menus select editing and optimization procedures. On terminating operations, each activated program (optimization, editing, or utilities) returns control to the control program which displays the main menu to the user again in order to select a mode of operation.

One editing program - the problem-oriented matrix screen editor MULTED - is included in the package. By using the menus and different types of screens it enters and edits the parameters of the programs, the upper bounds and the lower bounds on the variables, the initial solution, the type of constraints, the value of the right-hand sides, the type of criteria, the coefficients of the criterion matrix, the coefficients of the constraint matrix and the symbol names of the variables and the constraints. The MULTED problem-oriented matrix editor has some specific features. For easier comprehension of its general description, three of them will be discussed below. (1.) MULTED employs chiefly mass storage. It is the reason for the slower execution of the separate operations, however, it allows entering and editing of input data of large size problems. The higher version of the editor will be able to operate on the main memory in case there is sufficient free space available. (2.) The problem matrices (the criterion matrix and the constraint matrix) are stored on a disk in a packed form i.e. what is stored are the column numbers, the row numbers and the values of the nonzero elements. This impedes the execution of the separate operations to a certain extent, but allows for more efficient utilization of the memory especially for problems of large size with a small number of nonzero elements. The reading of the constraint matrix of the optimization program in the latter case is considerably faster. (3.) MULTED can be called from any disk drive. The input data files can be created on any available disk drive, however, the temporary files are created on the default drive. On terminating the program the temporary files are deleted automatically. The screen editor MULTED implements structural and syntactic control on the input data. Its operating is very easy. The user has to enter only the non-zero elements of the selected sets of input data into their respective positions.

Included in MOLP-16 are three methods for solving the multiobjective linear programming problem - the STEP method (STEM) (Benayoun et al., 1971), the Satisficing trade-off method (STOM) of Nakayama (Nakayama and Furukawa, 1985; Nakayama and Sawaragi, 1984; Sawaragi et al., 1985), an original interactive method of reference directions representing a hybrid of reference point methods and those founded in the local trade-off technique (Vassilev et al., 1992), as well as the single objective optimization subroutine RESM. The first two multicriteria interactive procedures are well-known and their efficiency has been proven by a number of real-life applications. About the hybrid method, to the DM's mind the method is analogous to the reference point methods and, regarding to the computation procedure, to the trade-off methods. This is achieved by the introduction of a scalarizing parametric problem. The problem is based on a reference direction, which is determined by the DM's aspiration level and a solution is obtained in previous iterations. The feasible solution of the single criterion optimization problem at each iteration lies on or is in close proximity of the weak efficient surface. The parametrization allows, as in Korhonen's method, the visualization of the criterion function changes at different values of the used parameter. The interactive optimization procedures have common features, the most essential of which are: identical organization of the interaction with the user; an option of interrupting; an option of repeated output of the obtained results; dynamic allocation of main memory required for their operation. The optimization program may not find a solution to the problem during operation for the following reasons: the program constraints are incompatible; some of the constraints are unbounded; insufficient main memory; hardware failure or user's errors in running the product. In these four instances the user is assisted in making decisions by the information and the diagnostic messages generated by the respective program or operating system. The RESM single optimization subroutine implements a modified primal simplex method with multiplication form of storing the inverse basis matrix (Murtaph, 1981). This program stores and operates only with the non-zero values of the constraint matrix and objective function elements. It is less liable to approximation errors as the initial constraint matrix is not affected and the calculations of the inverse basis matrix are carried out addressing the initial matrix. This subroutine allocates additional main memory for the inverse basis matrix.

The interactive user interface is menu-oriented. The decision maker is asked to select the respective functions by pressing keys and to enter only the digital values of the criteria functions, the number of the significant digits, and alphanumeric information about the file name when storing the solution. For this purpose the screen is divided into three fields (fig. 1). The first field is designed for displaying results, additional information and selection of a criterion to be accepted or changed. The second field is intended for selection, implementing the necessary functions for solving and investigating the multiobjective linear programming problem, and messages output. The third field is designed for entering alphanumeric and digital information necessary for solving the problem. In the first - the uppermost and widest field - the problem solution is displayed. It includes the values of the criterion functions and, on request, the variable values. For the sake of greater convenience and comprehensiveness the values of the criterion functions are displayed simultaneously in two ways. The first way is pseudo-graphical (visual) - by green or star lines, with lengths representing the scaled values of the criterion functions. In the right-hand part of the first field the criterion function value is shown together with the respective element of the ideal vector. The other way is alphanumeric. Additionally, in front of each line, from the left-hand side, the criterion functions identifiers F1, F2, ..., FN are displayed. In this field also additional information is displayed that helps the decision maker when making intermediate or final decisions: the ideal vector, the set value of a criterion function, the pay-off table. This additional information is displayed analogously to the criterion functions values in the above described two ways. The full length of each line - the green line extended with a blue line or the star line extended with a continuous line represents pseudo-graphically the ideal vector elements. Alongside these lines the values of these elements are displayed. The set value of a criterion function is represented pseudo-graphically through a line - brown or continuous, placed immediately under the line that visualizes the current value of this function. The value is displayed alphanumerically from the right-hand side against this line. The pay-off table is displayed in columns and in rows. The elements of the pay-off table are displayed basically in the described above two ways. The obtained optimum values of the variables are used (substituted) for calculating the criteria functions represented by green or star lines. On displaying the elements of the table in rows from the right-hand side, the ideal vector (the main diagonal of the matrix) is visualized. In the first field, from the left-hand side, a cursor is displayed - a blinking yellow or white triangle, pointing out the selected criterion function of which the value will be changed, analyzed and so on. In the second field - the middle one, within the framework of two rows, the main menu for selection of interface functions is set out. The said functions include: setting the level, solving the problem, displaying the variables, displaying the pay-off table, setting the significant digits, storing the solution, switching the scales and the way of visualizing concerning the monitor - color or monochrome and exit the interface. The functions are implemented by a set of submenus. The third field - the lowest, consisting of one row is used for entering digital information - new values of the functions and a number for the significant digits, as well as alphanumeric information - a name of the file for storing the output data. The place

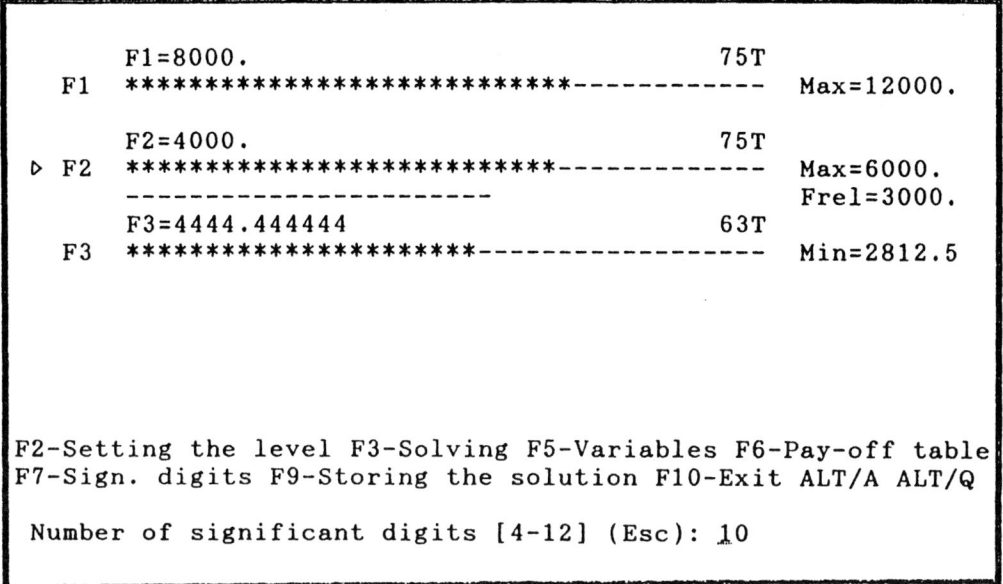

Figure 1: Sample screen layout of the user interfaces.

where this information should be entered is marked by a blinking cursor.

The MOLP-16 package includes two utility programs. The first one visualizes a text file containing basic information on: the purpose and the functional potentials of the product; components and structure of the product; editing programs operation; optimization programs operation. The text file can be called from the three menus of the control program. On calling from the main menu the text file is displayed from the beginning. If called from the menus for selection of the editing or optimization programs, the respective description of the editing or optimization program is displayed. The second utility program is a subroutine integrated in the optimization module. It provides an option to the user for execution of arbitrary system commands and other executable programs (including MOLP-16) not connected with the product. On terminating commands or program execution, control is returned to the control program.

4   MONP-16

The MONP-16 package is designed for solving the basic class of multiobjective nonlinear programming problems. It consists of the following program modules: an editing and visualizing module (screen editor); optimization programs - multiobjective interactive procedures and single objective optimization subroutines.

By means of the screen editor, data are entered and edited. The problem description in the form of mathematical expressions is stored in a standard ASCII file. The editing module functions as an ordinary text editor. Meanwhile it is actually an environment under which other functions, aside from the editing ones, are activated. The problem description editing is assumed to be a default mode of operation. It is automatically activated on starting MONP-16. The other functions specific for the screen editor are: storing of the problem description (and the results) into a file; creating a new file; string search; obtaining help information concerning MONP-16 operation; end of MONP-16 operation; specifying the problem parameters; solving of the problem from the initial point; solving of the problem from the current point.

Included in MONP-16 are two methods for solving the multiobjective nonlinear programming problem - the Satisficing trade-off method (STOM) of Nakayama and an original interactive method. The second method is a modification of the original interactive method of reference directions for the nonlinear case. It represents a hybrid of reference point methods and those founded in the local trade-off technique. At each iteration the feasible solutions of the single criterion optimization problem lie on or are in close proximity of the weak efficient surface. In some cases this allows approximate solutions of the single criterion optimization problem to be used, which is of particular importance when solving NP-hard nonlinear problems. The interactive optimization procedure has the following essential features: improved organization of the interaction with the user; an option for automatic storing of interim results into a file; an option of interrupting computation and examining the interim results and the alteration of parameters, e.g. relative precision, the number of digits during visualization of the results, exposure time of the interim results and the time for automatic storage into file; an option of repeated output of the obtained results; dynamic allocation of main memory, required for the operation of the single optimization program modules; double-precision operation; an option for obtaining general information about the package. Incorporated in MONP-16 are two single objective optimization subroutines LAGR and SQP. In the Lagrange multipliers method (Bertsekas, 1982) (the LAGR optimization subroutine) a modified function of Lagrange is used. At each iteration the unconstrained minimization is implemented. This scheme allows concurrent optimization for the straight variables (minimization) as well as for the dual variables (maximization) - multipliers corresponding to the equalities and multipliers corresponding to the inequalities until reaching the saddle point of the Lagrange function. For this manner of computation the multipliers, corresponding to inequalities which are not active at the optimum point, after a finite number of iterations become equal to 0. The method implements a scheme for asymptotically precise solution of the unconstrained minimization subproblems. The SQP optimization subroutine implements a method of sequential quadratic programming for solving the problem of constrained optimization (Schittkowski, 1983). This is one of the best algorithms for nonlinear programming which are checked in practice.

The interactive user interface is similar to that in the MOLP-16 package, described in the previous section.

On running the packages, two types of messages are displayed - information and diagnostic messages. The information messages affect the program modules operation to the extent that the user must react to them, observing the requirements for normal operation of the different modules and the product as a whole. Diagnostic messages can be generated by the operating system, the editing programs and the optimization programs.

5 Input/Output data structures

Considering the specific structure of multiobjective linear programming problems, a special system of input files has been developed in the MOLP-16 package, which is suitable for computer processing, and efficient with regard to memory allocation on the magnetic medium. The different types of MOLP- problem data i.e. the type of constraints and criteria, the right-hand sides of the constraints, the constraint matrix, the criterion matrix, the upper bounds on the variables, etc. are written to separate files with distinct filename extensions. The basic input file which contains information about the other files is the problem parametric vector file. All data files are a sequence of records of real numbers of 4-bytes length. The number of records in these files depends on the size of the problem. The constraint matrix and the criteria matrix are stored in the file in a packed form, which means that only the non-zero elements of the matrix are stored.

Considering the specific structure of multiobjective nonlinear programming problems a language has been developed for describing the problem model. The description itself is stored as a standard ASCII file. The description language specifies the form of writing the interim and the final results as part of the description of the problem - at the end of the text file containing the description. Thus, in operating MONP-16 we use only one type of files which serve as input files in one case and output ones in other cases. Another advantage of this approach is that the interim results obtained in solving the problem can be used as input results - initial point for the subsequent solving. This option is additional to the output data files, obtained without using the screen editor. The problem is described in the following order: - description of the constants (non-obligatory); - defining the names of the variables and eventually specifying their initial values; - description of the problem model - the description itself.

The output results can be stored in a file, displayed on screen or ported to a printer in a definite format easy to be interpreted. They are written to an ASCII file. They can be analyzed with the help of an arbitrary text or program editor, of external file processing programs, of DOS-commands, etc. The optimization programs implementing multiobjective linear and nonlinear programming methods use the same structure for the output data. They are arranged in three groups as follows: optimum values of the objective functions, an optimum solution, an ideal vector. The alphanumeric information about the three output data groups is complimented by pseudo-graphic display (by lines in text mode) of the objective functions values and the ideal vector. The variable values in the optimum solution are displayed in a separate window.

6 Conclusion

The presented software tools for multicriteria programming, including the two separate packages MOLP-16 and MONP-16 have been delivered for distribution by the Software Products and Systems Corporation. The future development of these software tools will stress on the following issues:

- design of new versions with an improved user interface for dynamic visualization of the interim and final results;

- including other prominent multicriteria interactive procedures like Korhonen's reference and also inclusion of new efficient single criterion optimization methods;

- incorporating the software tools in the Integrated Software System for Mathematical Optimization.

References

Benayoun, R., J. de Montgolfier, J. Tergny and O. Larichev (1971). Linear Programming with Multiple Objective Functions: Step Method (STEM), Mathematical Programming 1 pp. 366-375.

Bertsekas, D. P. (1982). Constrainted Optimization and Lagrange Multiplier Methods. Academic Press, New York.

Fishburn P. C., R. E. Steuer, J. Wallenius, and S. Zionts (1990). Multiple Criteria Decision Making/Multiatribute Utility Theory - The Next Ten Years. Working Paper 747, SUNY at Buffalo.

Korhonen, P. and J. Laakso (1986). A Visual Interactive Method for solving the Multiple Criteria Problem, European Journal of Operational Research 24, pp. 277-287.

Lewandowski, A. and A. P. Wierzbicki (1989). Decision Support Systems Using Reference Point Optimization. In: Lewandowski, A. and A. P. Wierzbicki, Eds., Aspiration Based Decision Support Systems, Theory, Software and Applications, Lecture Notes in Economics and Mathematical Systems, Vol. 331, Springer-Verlag, Berlin.

Murtaph, B. A. (1981). Advanced Linear Programming: Computation and Practice. McGraw-Hill, New York.

Nakayama, H. and K. Furukawa (1985). Satisficing Trade-off Method with an Application to Multiobjective Structural Design. Large Scale Systems 8, pp. 47-57.

Nakayama, H. and Y. Sawaragi (1984). Satisficing Trade-off Method for Multiobjective Programming. In: Grauer, M. and A. P. Wierzbicki, Eds., Interactive Decision Analysis, Lecture Notes in Economics and Mathematical Systems, Vol. 229, Springer-Verlag, Berlin.

Nakayama, H. and Y. Sawaragi (1984). Satisficing Trade-off Method for Multiobjective Programming and Its Applications. In: Preprints of the 9th World Congress of IFAC - A Bridge Between Control Science and Technology, Vol. V, Budapest, Hungary, pp. 247-252.

Roy, A and J. Wallenius (1992). Nonlinear and Unconstrained Multiple Objective Optimization: Algorithm, Computations, and Application (to appear).

Sawaragi, Y., H. Nakayama and T. Tanino (1985). Theory of Multiobjective Optimization. Mathematics in Science and Engineering, Vol. 176, Academic Press.

Schittkowski, K. (1983). On the Convergence of a Sequential Quadratic Programming Method with an Augmented Lagrangien Line Search Function, Math. Operationsforsch. u. Optimization, 14, 197-216.

Shin, W. S., A. Ravindran (1991). Interactive Multiple Objective Optimization: Survey I - Continuous Case, Computers & Operations Research, 18, pp. 97-114.

Steuer, R. E. (1986). Multiple Criteria Optimization: Theory, Computation, and Application, John Wiley & Sons, New York.

Steuer, R. E. and A. W. Whisman (1986). Towards the Consolidation of Interactive Multiple Objective Programming Procedures. In: Fandel, G., M. Grauer, A. Kurzhanski and A.P. Wierzbicki, Eds., Large-Scale Modelling and Interactive Decision Analysis. Preceedings, Lecture Notes in Economics and Mathematical Systems 273, Springer-Verlag, Berlin, pp. 232-241.

Vassilev, V., L. Kirilov and A. Atanassov (1992). Reference Direction Approach for Solving Multiple-Criterion Programming Problems (Forthcoming in Problems of Engineering Cybernetics and Robotics, BAS).

Wierzbicki, A. P. (1982). A Mathematical Basis for Satisficing Decision Making. Mathematical Modelling 3, pp. 391-405.

Zionts, S. (1988). Multiple Criteria Mathematical Programming: An Updated Overview and Several Approaches. In: Mitra G., Ed., Mathematical Models for Decision Support, NATO ASI Series, Vol F48, Springer-Verlag, Berlin, pp. 135-167.

On Engineering Applications of Interactive Multiobjective Programming Methods

Hirotaka Nakayama

Department of Applied Mathematics
Konan University
8-9-1 Okamoto, Higashinada, Kobe 658, Japan

Abstract

The methodology of interactive multiobjective programming has been developed remarkably over the past decade. Now it seems to be the time to apply it to real problems. The author has been trying to apply an interactive multiobjective programming method to various kinds of real problems. In this paper, some of these results will be reported, and gaps between theory and practice will be discussed.

1 Introduction

The major difficulties in DSS are how to deal with the uncertainty and the value judgment. Traditional modeling tries mainly to make models of (usually, complex) problems based on statistical data. In other words, it only treats the uncertainty in the decision making process. However, the decision making is originally subjective, that is, the decision strongly depends on the decision maker. Therefore, decision support systems must lead decision makers to their own decisions whatever they are. Now we have a problem with the multiplicity of value judgment. On the other hand, it is easily observed that a decision maker changes his/her attitude often even in the decision making process. This is very natural because the information available changes very often throughout the decision making process. Therefore, we also have a problem of the inconsistency of value judgment.

Multiobjective programming mainly treats these difficulties of value judgment rather than the uncertainty. Therefore, the type of information regarding the value judgement decision makers use and how it is used is very important. Above all, it should be noted that interactive programming methods for multiobjective decision problems have a role of a machine interface as a feature. Therefore, the following properties are imposed on desirable interactive multiobjective programming methods:

1. *(easy)* The way of trading-off is easy. In other words, decision makers can easily grasp the total balance among the objectives.

2. *(simple)* The judgment and operation required of decision makers is as simple as possible.

3. *(understandable)* The information shown to decision makers is as intuitive and understanable as possible.

4. *(quick response)* The treatment by computers is as quick as possible.

5. *(rapid convergence)* The convergence to the final solution is rapid.

6. *(explanatory)* Decision makers can easily accept the obtained solution. In other words, they can understand why it is so and where it came from.

7. *(learning effect)* Decision makers can learn through the interaction process.

The aspiration led methods for interactive multiobjective programming problems, e.g. the DIDAS family developed by Wierzbicki and his collaborators (Wierzbicki, 1981; Grauer et al., 1984; Lewandowski and Wierzbicki, 1989), a version of it with graphic interaction (Korhonen and Wallenius, 1988) and the satisficing trade-off method developed by the author (Nakayama, 1984), seem most promising from the above viewpoint. In Lewandowski and Wierzbicki (1989) and Eschenauer et al. (1990), we can see several examples of applications. In particular, in the latter reference many kinds of engineering applications are included. On the other hand, for the past several years, the author has been applying the satisficing trade-off method to various kinds of real problems. In the following, experiences of application in engineering fields are reported and discussed.

2 Some examples of engineering application

1. blending

   - feed
   - plastic materials (Nakayama et al., 1986)
   - cement production (Nakayama, 1991)
   - portfolio (Nakayama, 1989)

2. engineering design and management

   - camera lens
   - erection management of a cable-stayed bridge (Ishido et al., 1987)

3. planning

   - scheduling of string selection in steel manufacturing (Ueno et al., 1990)
   - long term planning of atomic power plants

These problems can be formulated as mathematical programming problems with plural objective functions. For example, stock farms in Japan have been modernized recently. Above all, the feeding systems in many farms is fully controled by a computer: each cow has its own place to eat which has a locked gate. And each cow has a key on her neck, which can only open the corresponding gate. Every day, on the basis of an ingredient analysis of the milk, the appropriate blending ratio for materials such as corn, cereals, fish meal, etc., from several viewpoints such as cost, nutrition, stock amount of materials, etc. This feeding problem is well known as the diet problem from the beginning of the history of mathematical programming, which can be formulated as the traditional linear programming problem. It is very easy to consider it as a multi-objective linear programming problem. For the application of the satisficing trade-off method to this problem, good results have been obtained in the experiments based on real data. Just recently, we completed an input-output interface which meets the requirements of actual decision makers (farmers, or consultants in some cases). This interface is in the stage of testing in actual feeding.

In cement production, raw materials such as lime stone, clay, iron, silica are crushed, mixed and burned. Usually, several kinds of cement are produced in a factory. Each kind of cement has to meet the standard imposed by the government. To this end, the decision of the blending ratio of raw materials is very important. The blending problem of raw material stones in cement production can be formulated as a linear fractional multiobjective programming problem (Nakayama, 1991). As yet, one has used the traditional goal programming method for an equivalently transformed linear multiobjective programming problem. As is well known, it is very difficult to decide upon the appropriate weight for each criterion in goal programming. In addition, since the objective functions are transformed into one without any practical meaning, it it more difficult to decide the corresponding weight. Therefore, one has been using a set of weights obtained after many trials through experience, and without modifying it even though the situation changes.

The author applied the satisficing trade-off method to the original linear fractional multiobjective programming problem (Nakayama, 1991). After the author's experiment, a company tested the satisficing trade-off method for the blending problem in one of its factories. As a result, they observed that the aspirational level is very easy to operate, and therefore they can get an appropriate blending ratio very easily by changing it depending on the situation. Throughout the test for half a year, they estimated that they can decrease the cost by six yen/ton by using the satisficing trade-off method rather than traditional goal programming.

However, this does not imply that they immediately replace their system for a calculation of the blending ratio with a new one. The replacement depends on the investment policy of the company. They do not replace a small part of the total production process at a large expense (e.g. for developing the computer software, and for stopping the production process and so on). Moreover, one of the major difficulties is that the chemical ingredient of raw materials changes very often. Of course, the company makes ingredient analyses for sampled materials in the production process. However, the raw material stones are already blended at the time that the result of the analysis is available. If they calculate an appropriate blending ratio on the basis of the result of the analysis, they have to stop the process. Therefore, they usually decide on the blending ratio on the basis of data by prediction. Their main interest is in the prediction system for the ingredient. The company is now engaged in developing a new prediction system for the ingredient, and considers the building of a decision system for the blending ratio based on the satisficing trade-off method after getting a newly developed prediction system.

A good example in engineering management is the erection management of a cable stayed bridge. In 1984, one of the biggest heavy industrial companies in Japan asked the author to collaborate in the development of a new method for adjusting the cable length in the construction of a cable stayed bridge, because a very big cable stayed bridge was planned at that time in Japan.

For the erection of cable stayed bridges, the following criteria are considered for accuracy control:

- residual error in each cable tension,

- residual error in camber at each node,

- amount of shim adjustment for each cable,

- number of cables to be adjusted.

Since the change of cable rigidity is small enough to be neglected with respect to shim adjustment, both the residual error in each cable tension and that in each camber are linear functions of the amount of shim adjustment.

Let us define n as the number of cables in use, ΔT_i $(i = 1, \ldots, n)$ as the difference between the designed tension values and the measured ones, and x_{ik} as the tension change of the i-th cable caused by the change of the k-th cable length by a unit. The residual error in the cable tension caused by the shim adjustments $\Delta l_1, \ldots, \Delta l_n$ is given by

$$p_i = |\Delta T_i - \sum x_{ik} \cdot \Delta l_k| \quad (i = 1, \ldots, n)$$

We also define m as the number of nodes, Δz_j $(j = 1, \ldots, m)$ as the difference between the designed camber values and the measured ones, and y_{jk} as the camber change of the j-th node caused by the change of the k-th cable length by a unit. Then the residual error in the camber caused by the shim adjustments of $\Delta l_1, \ldots, \Delta l_n$ is written by

$$q_j = |\Delta z_j - \sum y_{jk} \cdot \Delta l_k| \quad (j = 1, \ldots, m)$$

In addition, the amounts of shim adjustment can be treated as objective functions in the following form:

$$r_i = |\Delta l_i| \quad (i = 1, \ldots, n)$$

And the upper and lower bounds of shim adjustment inherent in the structure of the cable anchorage are as follows:

$$\Delta l_{L^i} \leq \Delta l_i \leq \Delta l_{U^i} \quad (i = 1, \ldots, n). \tag{2.1}$$

Our problem is to minimize p_i, r_i $(i = 1, \ldots, n)$ and q_j $(j = 1, \ldots, m)$ under the constraint (2.1). For this problem, we developed a method using the satisficing trade-off

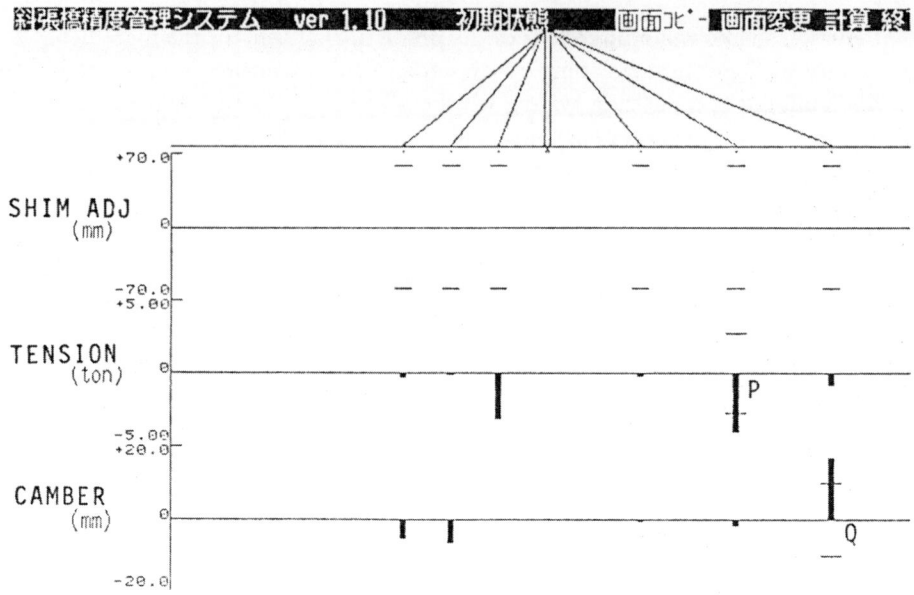

Figure 1: An example of erection management for a cable stayed bridge

method (Ishido et al., 1987). Unfortunately, however, the company did not get the job. In Japan, many companies are supposed to join a project of bridge construction to share jobs. It seems to the author that the reason why the company did not get the job is that it succeeded in getting another job with much more profit. Finally, another company got the job for the bridge, and therefore our method was not applied for it.

A few years later, another company wanted to know our method. At that time, the cable stayed bridge was very popular due to its beautiful shape. The company planned to join in some project constructing this kind of bridge, and the technology of adjusting the cable lengths during the construction is a necessary condition for it. The author gave the company the software with a graphic input/output interface which was developed in his laboratory. Since this problem usually has very many objective functions, it is very important to make a graphic interface so that the burden of decision makers can be decreased. The company tried to apply the software for constructing a relatively small sized cable stayed bridge in a golf course it owns. Unfortunately, the result was not successful. The reason was obvious: the operators were not familiar with the method and were confused as to whether the plus signed values correspond to shortening or lengthening of cables. The cable length adjustment is performed usually from 02:00 a.m to 08:00 a.m. so that the temperature is stable. They could not resolve their problem in time. A few months later, the author received the data from that experiment, and found a more important thing in it. Figure 2 shows the initial error caused during construction. The cable lengths should be adjusted in order to get these errors within some reasonable extent. After many trials, we found that both the camber error at Q and the tension error of the cable P cannot be improved at the same time. In particular, the sensitivity of the camber error at Q is very high, and therefore if we want to improve it even a little,

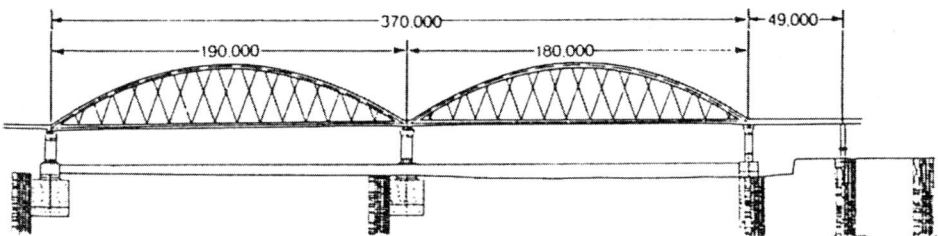

Figure 2: The Nielsen Bridge

we have to relax the other criteria including the tension error at P very much. Both errors are beyond the acceptable range, and in addition it should be noted that the end points of the bridge are fixed to the land. If the bridge was constructed in a normal way, these major errors might not have occurred. It seems that this happened due to the poor construction skills of the company as it was their first experience.

Meanwhile, the company which asked me to collaborate for the first time applied the method to the construction of another kind of bridge called the "Nielsen Bridge" is also gaining much popularity in Japan, and which also has the same problem of cable length adjustment in construction. One of the major differences is that this kind of bridge is constructed in a factory and then carried by ship to the place to be set up. Similarly, as in cable stayed bridges, the cable length adjustment is constructed in a factory as well as after setting up. The constructed bridge is of middle scale with 22 cables on one side. They implemented our method, and as a result they could keep all criteria within a reasonable range except for a camber error at one point. They could not bring the error within an allowable range by all means. Although the reason is not clear, they guessed that operators did not adjust correctly, because shims (iron pieces to adjust the cable length) were not used in this bridge, but operators adjusted it by the rotation angle of a screw. The rotation angle of a screw is very intuitive, and is very difficult to set correctly.

As was seen in the above examples, we cannot say that the actual implementation of our method was successful. However, very recently, the company informed me that they succeeded in getting sufficiently satisfactory results for the cable adjustment of a miniature bridge, which is of the reduced scale 1/50 of the real one to be to constructed in the near future. Even though it is a miniature, they made the cable adjustment under almost the same circumstances as with a real brigde. In addition, the bridge is of a large scale with 136 cables. Also, another company which constructed a small bridge in a golf course recently received an order to construct a new bridge from a local government. It is expected that our method will be applied successfully to some real bridge in the near future.

3   Some remarks on applications

The author has had good results for all experimental applications based on real data. As is readily seen in the above report, however, we cannot say that we got good results for practical implementation. This situation is quite similar to many prediction methods, which can fit to interpolation for past data, while they cannot predict (extrapolation) for the future in many practical cases. What causes this gap? One of the main reasons for its is that we have many uncertainties, in other words, unexpected happenings in real implementation. In the cases of erection management of a cable stayed bridge, we had many unexpected happenings such as misunderstanding of the method and misoperation. These are caused by the poverty of knowledge and skill due to lack of experience. Therefore, we can expect that this will be improved by stimulation of experiences.

On the other hand, there are inevitable uncertainties, e.g. errors in observed data. The cable tension is measured by the vibration of the cable. In this experiment, the wind and the change of temperature affect the measurement. The uncertainty included in the data is more severe in cement production. Although the measurement technology will be improved in the future, errors in data will be inevitable. For problems with measurement errors, a new methodology must be developed in the future.

Finally, it must be emphasized that the collaboration among the people, who are engaged in applications, not only between academic institutions and companies but also in companies themselves is very important for real implementation. It seems to me that the good human relationship such as good mutual communication and good mutual understanding can bring the success of real implementation.

4   Concluding remarks

Some people might not think that researchers need to make practical applications by themselves: researchers can leave them to practitioners. However, practitioners cannot apply methods to practical problems by themselves in general. If we would like to apply our methods to real problems, it is necessary for us to be involved in applications. Moreover, it is the author's belief that methods can be inspired in real life by real applications only.

References

Eschenauer, H.A., Koski, J. and Osyczka, A. (eds.) (1990). Multicriteria Design Optimization. Springer-Verlag, Berlin.

Grauer, M. Lewandowski, A. and Wierzbicki A.P. (1984). DIDAS Theory, Implementation and Experiences, in M. Grauer and A.P. Wierzbicki (eds.) Interactive Decision Analysis. Proceedings of the International Workshop on Interactive Decision Analysis and Interpretative Computer Intelligence, Lecture Notes in Economics and Mathematical Systems, vol. 229, Springer-Verlag, Berlin, Heidelberg, pp. 22-30.

Ishido, K., Nakayama, H., Furukawa, K., Inoue, K. and Tanikawa, K. (1987). Management of Erection for Cable Stayed Bridge Using Satisficing Trade-off Method. In Y. Sawaragi, K. Inoue and H. Nakayama (eds.) *Toward Interactive and Intelligent Decision Support Systems, Volume 1*, Lecture Notes in Economics and Mathematical Systems, vol. 285, Springer-Verlag, Berlin, Heidelberg, pp 304-312.

Korhonen, P. and Wallenius, J. (1988). A Pareto Race. Naval Research Logistics Quarterly 35, pp. 615-623.

Lewandowski A. and Wierzbicki, A.P. (eds.) (1989). Aspiration Based Decision Support Systems. Lecture Notes in Economics and Mathematical Systems, vol. 331, Springer-Verlag, Berlin, Heidelberg.

Nakayama, H. (1984). Proposal of Satisficing Trade-Off Method for Multi-objective Programming. Transact. SICE, 20, pp, 29-35 (in Japanese).

Nakayama, H. and Sawaragi, Y. (1984). Satisficing Trade-off Method for Interactive Multiobjective Programming Methods. In M. Grauer and A.P. Wierzbicki (eds.) *Interactive Decision Analysis*, Proceedings of the International Workshop on Interactive Decision Analysis and Interpretative Computer Intelligence, Lecture Notes in Economics and Mathematical Systems, vol. 229, Springer-Verlag, Berlin, Heidelberg, pp. 113-122.

Nakayama, H., Nomura, J., Sawada, K. and Nakajima, R. (1986). An Application of Satisficing Trade-off Method to a Blending Problem of Industrial Materials. In G. Fandel et al. (eds), *Large Scale Modelling and Interactive Decision Analysis*, Lecture Notes in Economics and Mathematical Systems, vol. 273, Springer-Verlag, Berlin, Heidelberg, pp. 303-313.

Nakayama, H. (1989). An Interactive Support System for Bond Trading. In A. G. Lockett and G. Islei (eds.) *Improving Decision Making in Organizations*, Lecture Notes in Economics and Mathematical Systems, vol. 335, Springer-Verlag, Berlin, Heidelberg, pp. 325-333.

Nakayama, H. (1991). Satisficing Trade-off Method for Problems with Multiple Linear Fractional Objectives and its Applications. In A. Lewandowski and V. Volkovich (eds.) *Multiobjective Problems of Mathematical Programming*, Lecture Notes in Economics and Mathematical Systems, vol. 351, Springer-Verlag, Berlin, Heidelberg, pp. 42-50.

Sawaragi, Y., Nakayama, H. and Tanino, T. (1985). Theory of Multiobjective Optimization. Academic Press, Orlando.

Ueno, N., Nakagawa, Y., Tokuyama, H., Nakayama, H. and Tamura, H. (1990). A Multiobjective Planning for String Selection in Steel Manufacturing, Communications of Operations Research Society of Japan, Vol. 35, No. 12, pp. 656-661.

Wierzbicki, A. P. (1981). A Mathematical Basis for Satisficing Decision Making, in J. Morse (ed.) *Organizations: Multiple Agents with Multiple Criteria*, Lecture Notes in Economics and Mathematical Systems, vol. 190, Springer-Verlag, Berlin, Heidelberg, pp. 466-485.

Modelling of Allocation of Social Resources and Decision Support

Roman Kulikowski

Systems Research Institute

Polish Academy of Sciences, Poland

1   Introduction

The notion of social resources as used here denotes the stock of qualified labour, expressed in working time (e.g. man-year) units, as well as capital or financial resources, which are used to produce public goods, such as: education, science, health and social care, environment quality etc.

In numerous social-economic systems market mechanisms do not work satisfactorily and the decisions regarding allocation of government expenditures among the producers of public goods, such as universities, hospitals, research institutes etc. are left to bureaucracy. Such practices are often opposed by public opinion and it occurs frequently that before the final decision is taken the programs of activities are formulated in the form of proposals, followed by negotiation processes. The negotiations take place at the committee composed of representatives of public-goods producers and consumers elected or nominated (e.g. by the government). In such a system the group or "social choice" within the committee should support taking of the optimum decision. However, in real situations due to imperfect information on the supply and demand side, as well as the large number of institutions involved, there is no guarantee that the decision proposed is optimal. There is also a feeling that computerized support should be used to improve the system's efficiency.

In the present paper a general concept of such a system, meant for allocation of research funds, is studied.

The main objective of the systems considered are:

a. Efficiency – the system should employ all the available resources efficiently, i.e., for instance when there is a surplus of labour the "less efficient" individuals should be transferred to an alternative activity.

b. Effectiveness – it is required in addition that allocation strategies be Pareto optimal and stable.

In other words, in an effective system all program or project leaders, who compete for government grants, have equal rights of access and they allocate labour resources in such a way that their utilities be maximum. A similar requirement can be formulated with respect to people responsible for capital when they compete for labour resources. Stability is required here to obtain a converging negotiation and decision processes.

Effectiveness is understood here as an ultimate standard against which the management of organization, dealing with allocation of social resources, can be evaluated. Formal introduction of the efficiency and effectiveness measure makes it possible to compare different organizations, institutions and management systems. Using such measures one can also analyse the deficiencies and improve the performance of organizations and institutions, and also design organizations and institutions which are best in the sense of obtaining effective performance.

As a concrete example a typical research fund allocation system (RFAS), shown in Fig. 1, will be studied.

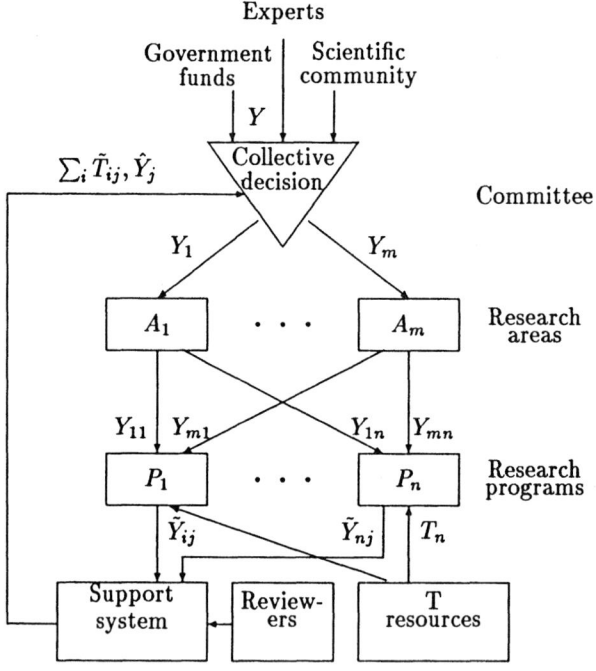

Fig. 1. RFAS system

The allocation process consists of two stages.

I. Allocation of the total fund Y among m research areas (A_j, $j = 1, 2, \ldots, m$) in such a way that

$$\sum_{j=1}^{m} Y_j \leq Y \quad Y_j \geq 0, \quad \forall j \tag{1}$$

II. Allocation of the area funds Y_j among n programs $(P_i, \ i = 1, 2, \ldots, n)$ in such a way that

$$\sum_{j=1}^{n} Y_{ji} \leq Y_j \quad Y_{ji} \geq 0, \quad \forall j, i \tag{2}$$

The function of the support system is to

a. collect information regarding demands $\hat{Y}_{ij}$, $\forall i, j$, claimed by research organizations,

b. organize the reviewing process for evaluation of programs P_i and the negotiations for allocation of Y among A_j,

c. propose the effective strategies $\hat{Y}_j, \hat{Y}_{ij}, \ \forall i, j$.

A simple version of the proposed decision support system has been already tested experimentally and applied to allocate research funds at the Systems Research Institute of Polish Academy of Sciences.

In the above mentioned work, described in (Kulikowski, R. et al., 1986), all the area funds (Y_j) were regarded as given and one did not take into account the general form of individual and and collective utilities, as postulated by modern decision and choice theory.

In the present paper the theory of effective organizations and methodology based on a number of assumptions (axioms), developed in (Kulikowski, R., 1991), has been used. An interesting feature of that theory is the explicit form of effective strategies, this form not depending on the analytic form of particular utilities.

2 Models of particular research activities

In the modern decision and utility theory no concrete form of utility is assigned to an individual to describe his activity but rather it is proven, under a number of assumptions, that such a utility function exists.

Let us consider as an example the case of risky actions or lotteries when the outcomes $Y_j, j = 1, 2 \ldots, m$ occur with certain, known probabilities $q_j \geq 0, \sum q_j = 1$. Following the von Neumann – Morgenstern axiomatic approach one can show (see e.g. Luce and Raiffa, 1959) that for an ordered set of outcomes

$$Y_1 \succsim Y_2 \succsim \cdots \succsim Y_m$$

a real–valued utility function U, such that

$$U(Y_1, \ldots, Y_m) = \sum_{j=1}^{m} q_j u_j, \quad u_j = U(Y_j), \quad \forall j, \tag{3}$$

exists.

This function is invariant under positive linear transformations. Thus, whenever the axioms hold there exists a utility function U preserving order and satisfying the expectation principle (3): the utility of the lottery equals the expected utility of its outcomes.

Dealing with research activities, an individual who proposes his program (P_i) to the research areas (A_j) is aware of the risk that his proposal can be rejected due to the competition from other proponents $(P_\nu, \nu \neq i)$. In order words, competition reduces his access to research funds and his expected gain becomes

$$U_i = \sum_{j=1}^{m} x_{ij} \frac{c_j Y}{\sum_\nu x_{\nu j}}. \tag{4}$$

where

$x_{\nu j} = $ demand for research time at A_j, claimed by $P_\nu, \forall j$,

$c_j = $ part of total fund Y assigned to area A_j, $\sum c_j = 1$.

Since the individual's resources of research time are limited

$$\sum_{j=1}^{m} x_{ij} \leq T_i, \quad T_i = \text{given} \tag{5}$$

one can formulate the following optimization problem:

Find the strategies $x_i \overset{\Delta}{=} (x_{i1}, \ldots, x_{im}) = \hat{x}_i$, $\forall i$, such that

$$U_i(\hat{x}_i) = \max_{x_i \in \Omega_i} U_i(x_i) \quad \forall i \tag{6}$$

where

$$\Omega_i = \{x_{ij} \mid \sum_{j=1}^{m} x_{ij} \leq T_i, \ x_{ij} \geq 0, \ \forall i, j\} \tag{7}$$

One can observe that in the model (4) the gain is expressed by working time and it may be viewed as too simplified. Indeed, research has also monetary value and the experimental evidence indicates that utility increases (with decreasing rate) along with money. Besides, utility is usually a function, say $F_i(Y_j, x_{ij})$, of outcome and effort, expressed by working time x_{ij}. To analyse the impact of these two factors let us ignore for a moment the competition.

The explicit form of F_i is, of course, unknown. However, one can restrict the class of possible functions F_i using the dimensional analysis and assuming that under constant cost of labour F_i is "constant returns to scale" (otherwise one could generate utility by simple changing units, e.g. changing 1\$ to 100 cents) i.e.:

$$F_i(Y_j, x_{ij}) = Y_j f_j \left(\frac{x_{ij}}{Y_j} \right)$$

where $f_i(\cdot)$ is strictly concave, differentiable and $f'_i(\cdot) > 0$, (it is called also "risk averse").

It should be also observed that the expected outcome depends on the "production function" (assumed to be Cobb-Douglas) i.e.:

$$Y_j = p_j T_{ij} \beta_j \left(\frac{K_j}{T_{ij}} \right)^{\alpha_j}, \quad 0 < \alpha_j < 1, \tag{8}$$

where

p_j = price attached to the outcome,

T_{ij} = working time at A_j with program P_i,

K_j = capital used for program at A_j,

α_j, β_j = given positive coefficients.

One can also assume that capital is used in optimum proportions, i.e. $K_j = \hat{u}_j T_{ij}$, see Appendix 1, and the expected outcomes:

$$Y_j = a_j b_j p_j T_i, \quad \forall j$$

where

$b_j = \beta_j \left(\frac{\alpha}{1-\alpha} \cdot \frac{w_k}{w_l} \right)^{\alpha_j}$ is the productivity of working time,

$a_j = T_{ij}/T_i$ is the individual willingness to spend the part $a_j T_i$ of total time T_i on the alternative A_j (when $p_j b_j$ are ignored or constant).

The choice coefficients (a_j) can be also regarded as attractiveness measure of A_j. Generally, modelling of choice coefficients is not easy. A possible approach (see Intriligator, M. D., 1982), is to evaluate separately each alternative A_j from the point of view of a given criterion. When there is a given set of K criteria one can gets a table A of numbers $a_{kj} \geq 0$, $\forall k, j$, and $\sum_j a_{kj} = 1$, $\forall k$.

For evaluation of the relative importance of each criterion the weight vector $w = (w_1, \ldots, w_k), w_k > 0, \forall k, \sum_k w_k = 1$, is introduced. Then, the preference vector becomes:

$$a = wA, \tag{9}$$

where

$$a_j = \sum_{k=1}^{K} w_k a_{kj}, \quad \forall j \tag{10}$$

Relation (9),(10) can be also used in the probabilistic choice model where numbers a_j should be regarded as probabilities (see Intriligator, M. D., 1982), while $Y_j = a_j p_j b_j T$, $\forall j$ as expected outcomes.

Coming back to the model with competition one can define a measure of access M_i, say $\Psi_i[C_j, (\sum_\nu x_{\nu j})^{-1}]$, which is an increasing function of $C_j = c_j Y$, decreasing along with demand $\sum_\nu x_{\nu j}$. Assuming that measure to be "constant returns to scale" one can write

$$M_i = C_j \Psi_i \left(\frac{C_j}{\sum_\nu x_{\nu j}} \right),$$

where $\sum_j c_j = 1$.

Now one is able to formulate the utility function for an individual researcher in the presence of competition (Kulikowski, R., 1991) in the product form of f_i, M_i:

$$U_i(x_i) = Y T_i \sum_{j=1}^{m} B_j f_i \left(\frac{x_{ij}}{B_j T_i} \right) \Psi_i \left(\frac{Y B_j}{\sum_\nu x_{\nu j}} \right), \quad \forall i, \tag{11}$$

where $B_j = a_j b_j p_j c_j$, $f_i(\cdot) and \Psi_i(\cdot)$ are strictly concave, differentiable and $\|\text{grad } U_i(x_i)\| > 0$, $\forall i$.

An assertion, proved in (Kulikowski, R., 1991), stipulates that for (11) the unique set of optimum strategies, maximizing $U_i(x_i)$ in Ω_i:

$$\hat{x}_{ij} = \frac{B_j}{B} T_i, \quad B = \sum_{j=1}^{m} B_j, \quad \forall i, j \tag{12}$$

exists, and

$$U_i(\hat{x}_i) = BYT_i f_i(1/B)\Psi_i\left(\frac{BY}{\sum_\nu T_\nu}\right), \quad \forall i. \tag{13}$$

It should be noted that $\hat{x}_{ij}$, $\forall i, j$, do not depend on the particular forms of individual utilities f_i, Ψ_i $\forall i$. These strategies are cooperative (Pareto optimal) which means that none of the competing individuals can increase his utility by departing from the optimal strategy. In that sense cooperativeness contributes to a consensus in case the access–induced conflicts arise.

It should be also noted that condition $\|\text{grad } U_i(x_i)\| > 0$, $\forall i$, is essential here. When it is neglected, U_i loses concavity and unstable strategies (bifurcations of strategies) follow (see Kulikowski, R., 1990; 1991).

3 Model of effective research organizations

An organization is understood here to be a voluntary collective (team) of individuals with chosen (or nominated) leader (director of research institute) who is making decisions in the name of the team. The leader is also organizing (directing implementation of decisions) and awarding individuals out of the organization income. Suppose that there are n organizations given, with N_i, $i = 1, \ldots, n$ individuals in each. The utilities of the individuals are assumed in the form (11) i.e.

$$U_{il}(x_{il}) = YT_{il} \sum_{j=1}^{m} B_{jl} f_{il}\left(\frac{x_{ilj}}{B_{jl}T_{il}}\right) \Psi_{il}\left(\frac{YB_{jl}}{\sum_\nu x_{\nu lj}}\right), \tag{14}$$

where

$$B_{jl} = a_{jl} b_j c_j p_j, \quad \forall l, j, \quad B_l = \sum_j B_{jl}.$$

Each individual has an admissible set of activities

$$\Omega_{il} = \{x_{ilj} \mid \sum_j x_{ilj} \le T_{il}, \ x_{ilj} \ge 0, \ \forall i, l, j\}$$

It is assumed that the leader possesses (as an individual) the utility of the form (11) but instead of using his personal resources he uses the collective, aggregate resources

$$T_i = \sum_{l=1}^{N_i} T_{il}, \quad \forall i \tag{15}$$

and averaged preferences, which for $T_{il} = \tilde{T}_i, \forall l, i$, become

$$\frac{\tilde{B}_j}{\tilde{B}} = \frac{1}{N_i} \sum_{l=1}^{N_i} \frac{B_{jl}}{B_l} T_{il}, \quad \tilde{B} = \sum_j \tilde{B}_j \tag{16}$$

Assumption (16) means that leaders are equally sensitive to individual preferences (B_{jl}/B_l) within the organization. This requires a democratic relationship between the leaders and the members of collectives.

It is also assumed that an individual will join an organization when it ensures for him the accomplishment of a program of activities with an income not less than he could earn by acting elsewhere. On the other hand a leader may remunerate the members of the collective according to individual efficiencies (b_{jl}), by setting the salaries

$$\omega_l = \sum_j (p_j b_{jl} - \omega_k \hat{u}_{jl}),$$

where $\omega_k \hat{u}_{jl}$ is the cost of capital used by l-th individual in the j-th program.

One can say that the leaders or organizations are efficient when they employ all the aggregate resources i.e.

$$x_{ij} = \sum_{l=1}^{N_i} x_{ilj}, \quad \forall i, j \tag{17}$$

and they are effective if, in addition, their strategies are cooperative (Pareto optimal) and stable.

The main assertion, describing the existence of effective strategies, (proved in Kulikowski, R., 1991) says that under assumptions (15) and (16) the unique set of optimum, effective strategies

$$\hat{x}_{ij} = \frac{\tilde{B}_j}{\tilde{B}} \tilde{T}_i, \quad \forall i, j, \quad \tilde{B} = \sum_j \tilde{B}_j, \tag{18}$$

$$\hat{x}_{ilj} = \frac{B_{jl}}{B_l} T_{il}, \quad \forall i, l, j, \quad B_l = \sum_j B_{jl}, \tag{19}$$

exists, and

$$U_{il}(\hat{x}_{il}) = B_l Y T_{il} f_i (1/B_l) \Psi_i \left(\frac{B_l Y}{\sum_{\nu,l} T_{\nu l}} \right), \quad \forall i, l, \tag{20}$$

$$U_i(\hat{x}_i) = \tilde{B} Y \tilde{T}_i f_i (1/\tilde{B}) \Psi_i \left(\frac{\tilde{B} Y}{\sum_\nu \tilde{T}_\nu} \right), \quad \forall i. \tag{21}$$

It should be noted that in the model studied here the attractiveness of A_j (expressed through coefficients a_i) is the same for all the organizations. The case when $a_{ij} \neq a_j$, for different i, was considered in (Kulikowski, R., 1990).

4 Towards an effective system of allocation of research funds

The present section is devoted to the problem of effectiveness of a RFAS, shown in Fig.1, using the methodology proposed and the computer–based support.

Obviously, there are two sides of the funds allocation problem. On the demand side one has research institutes (or individuals representing institutes) who have preferences as to program A_j, represented by motivation coefficients a_j, $j = 1, \ldots, m$, and demand functions:

$$\hat{x}_{ij} = \frac{B_j^r}{B^r}, \quad B_j^r = a_j b_j \bar{c}_j p_j, \quad B^r = \sum_j B_j^r, \quad \sum_i T_i = T.$$

where $\bar{c}_j$ is a coefficient of fund supply, as expected by the demanding bodies.

On the other side one has the fund suppliers $(S_\nu, \nu = 1, \ldots, M)$ who, however, can also be regarded as demanders of research work resources $(a_j T_i, i = 1, \ldots, n)$, represented by the expected $\bar{a}_j$, $\forall j$ coefficient values. these demand functions can in a way analogous to $\hat{x}_{ij}$ be written in the following form

$$\hat{y}_{\nu j} = \frac{B_j^s}{B^s} Y_\nu \tag{22}$$

where

$$B_j^s = \bar{a}_j \bar{b}_j c_j p_j, \quad B^s = \sum_j B_j^s, \quad \sum_{\nu=1}^M Y_\nu = Y.$$

Indeed, as shown in Appendix 1, the output $Y_j = p_j \bar{b}_j \bar{Y}_j$, $\forall j$ while $\bar{Y}_j = c_j \bar{Y}$ with $\bar{b}_j$ being productivity of input costs. When the capital endowment $(\hat{u}_j)$ is constant for all j, the $\bar{b}_j$ values of differ from b_j by constant multiplier γ only and

$$B_j^r = a_j b_j \bar{c}_j p_j, \quad B_j^s = \bar{a}_j b_j c_j p_j \gamma, \quad \forall j$$

Now one can derive the so called supply–demand structure:

$$s_j = \frac{\sum_\nu y_{\nu j}}{\sum_i x_{ij}} : \frac{Y}{T} = \frac{\bar{a}_j c_j}{a_j \bar{c}_j}, \quad \forall j, \tag{23}$$

When the expected coefficients $\bar{a}_j, \bar{c}_j$ are equal a_j, c_j $\forall j$, respectively, $s_j = 1$, $\forall j$, and the amounts of funds "per capita" are the same and equal Y/T for each research area A_j. Such a situation may be regarded as satisfactory from the efficiency point of view. However, there are two main reasons to give financial preferences to certain research areas (even at the expense of some loss of utility in other areas). First – some areas require higher costs since appropriate capital endowments are higher. Second – it is necessary to stimulate the development of the new, promising research areas.

In the case of our RFAS the fund allocation strategy $(\hat{Y}_j, \forall j)$ is deliberated and decided by a committee composed of representatives of government, universities and research institutes.

Assuming that all members of the committee, say L in number, have "equal rights and weights" (i.e. social choice probabilities are strictly and equally sensitive to individual choice probabilities and when all members of the committee reject an alternative so does the committee (one can use the probabilistic social choice model (see Intriligator, M. D., 1982). In that model the social choice probabilities $\alpha \overset{\triangle}{=} a_j c_j$, $\forall j$, are obtained by averaging the individual choice probabilities α_{jl}, $\forall jl$:

$$\alpha_j = \frac{1}{L} \sum_{l=1}^L \alpha_{jl}, \quad \forall j. \tag{24}$$

In order to describe the individual choice probabilities α_{jl} one can use (dropping for convenience the index l) the probabilistic individual choice model (9),(10) (see also Intriligator, M. D., 1982), where the probability of choosing alternative A_j depends on given (adopted) K criteria with weights $w_i > 0$, $\forall i, \sum_i w_i = 1$:

$$\alpha_j = \sum_{i=1}^{K} w_i \alpha_{ij}, \quad \forall j. \tag{25}$$

A numerical example will explain how the models proposed can be used to arrive at the social choice solution.

EXAMPLE 1 A committee is allocating research funds among Mechanical (M), Electrical & Electronics (E^2) and Computer (C) sciences. The committee has decided to find α_j coefficients by (25), (using the questionnaires filled by all committee members) and 3 criteria: development of basic science, applications and education.

Consider a typical questionnaire filled by an individual

Criterion	Weights & areas			
	w	M	E^2	C
Basic science	0.3	0.2	0.3	0.5
Application	0.4	0.3	0.3	0.4
Education	0.3	0.25	0.25	0.5

On the basis of data from the table, i.e. w_i and α_{ij} one gets from (25) $\alpha_1 = 0.255, \alpha_2 = 0.285, \alpha_3 = 0.460$.

When all questionnaires are collected one can easily derive the committee choice structure by (24). Then, taking into account the information regarding $p_j b_j$, i.e. the expected values of research productivity and outputs (which can be negotiated or supplied by independent experts) one can derive the final strategy

$$\hat{Y}_j = \frac{B_j}{B} Y, \quad \forall j, \quad B_j = \alpha_j p_j b_j$$

When using models (24) and (25) one should be aware of the possibility of appearance of the exaggeration tactic, which can be used by certain committee members who want to increase preferences given to a particular research area.

For purposes of detecting if there is a bias towards a research area one can derive the numbers

$$\delta_{jl} = \alpha_{jl} - \frac{1}{L-1} \sum_{i \neq l}^{L} \alpha_{ji}, \quad \forall j, l.$$

When there are numbers $l = l_0, j = j_0$, such that $\delta_{j_0 l_0} \gg 0$, the committee may punish the l_0-th individual by deleting his participation in social choice, i.e. in the averaging formula (24).

It should be noted that the idea of punishment can be also realized in the so called "double payment" originated by Groves and Ledyard, (1977), though generally no good (i.e. strategy – proof) algorithms have been devised as yet, especially for cases when coalitions among committee members are possible.

For that reason negotiations may prove to be more effective than punishment in arriving at consensus among the members of the committee. The negotiation process can be accelerated when each committee member knows the criteria (in particular - the government policy), the weights attached to the criteria and his opponents', preference structure over alternatives, as postulated by model (10).

When all the relevant information is supplied in the form of questionnaires, or – on the computer monitor, one can find easily:

    a. points or areas of common interest as well as the discrepancies in interests, according to the structure of criteria,

    b. weights attached to criteria (policies) indicated by opponents,

    c. preferences of opponents given to the particular criteria.

The chairman of the committee can find the areas where the rapprochement of points of view an arbitration as well as partial or complete consensus is possible.

At the lower level of RFAS concrete decisions of acceptance or rejection of a grand proposal should be taken. In the case of research programs the collective decision is concerned with acceptance or reduction of demands $(\check{Y}_{ji}, \forall i, j)$. All these decisions should respect the financial constrain $\sum_i Y_{ij} \leq Y_j$, where Y_j is postulated by higher level of RFAS.

When, due to imperfect information, regarding e.g. coefficients c_j, the demand for funds at a particular A_j is higher then the supply of funds the $s_j < 1$. In order to increase the fund per capita index in that case the decision makers have two possibilities: to reject more proposals than necessary, or to ask the higher level for larger Y_j. Both these options lower the system's effectiveness.

In other word, for an effective system of research funds allocation a fast, computer supported information exchange system between supply and demand side is necessary.

References

Groves T., Ledyard J. (1977). Optimal Allocation of Public Goods: A Solution to the Free-Rider Problem, *Econometrica* 45, 783–809.

Intriligator M. D. (1973). A probabilistic model of social choice, *Review of Economic Studies* 40, 553–560.

Intriligator M. D. (1982). Probabilistic models of choice, *Math. Social Sciences* 2, 157–166.

Kulikowski R. (1990). Equilibrium and Disequilibrium in Negotiation of Public Joint Venture, in: Systems Analysis a. Computer Science. Proc. of II Polish–Spanish Conf., Sept. 1990. R. Kulikowski, J.Rudnicki, eds. OMNITECH Press, Warszawa.

Kulikowski R. (1991). Model for Decision Support in Allocation of Social Resources, *Control and Cybernetics* 2.

Kulikowski R., Jakubowski A., Wagner D. (1986). Interactive system for collective decision making, *System Analysis, Modelling and Simulation* 3, 13.

Luce R. D., Raiffa H. (1959). Games and Decisions, New York, Wiley.

Appendix 1

Assume the value of output product (in monetary term) be given by Cobb–Douglas production function

$$Y = p\beta K^\alpha T^{1-\alpha} = p\beta T \left(\frac{K}{T}\right)^\alpha, \tag{A.1}$$

where p = output price, K = capital, T = working time, α, β = given positive coefficients, $0 < \alpha < 1$.

Let the input costs be limited, i.e.:

$$\omega_k K + \omega_l T = \bar{Y} \tag{A.2}$$

where ω_k, ω_l = the input costs of capital and labour, $\bar{Y}$ = given, and it is necessary to find $\hat{K}, \hat{L}$ such that Y attains maximum.

From the necessary conditions of optimality

$$\phi'_k = 0, \quad \phi'_T = 0, \quad \text{where} \quad \psi - Y + \lambda(\omega_k K + \omega_l T),$$

λ = Lagrange multiplier, one gets

$$\hat{u} = \frac{\hat{K}}{\hat{T}} = \frac{\alpha}{1-\alpha}\frac{\omega_l}{\omega_k}, \tag{A.3}$$

or

$$\hat{K} = \frac{\alpha}{\omega_k}\bar{Y}, \quad \hat{T} = \frac{1-\alpha^{1-\alpha}}{\omega_l}\bar{Y}. \tag{A.4}$$

Then (A.1) can be written as

$$Y = p\beta \left(\frac{\alpha}{1-\alpha}\frac{\omega_l}{\omega_k}\right)^\alpha T = p\beta \left(\frac{\alpha}{\omega_k}\right)^{1-\alpha} \left(\frac{1-\alpha}{\omega_l}\right)\bar{Y}$$

or

$$Y = pbT = pb\bar{Y}, \quad b = \beta\left(\frac{\alpha}{1-\alpha}\frac{\omega_l}{\omega_k}\right)^\alpha, \quad \bar{b} = b\left(\frac{\omega_l}{1-\alpha}\right)^{1-2\alpha} \tag{A.5}$$

The coefficient b can be called "time productivity", while $\bar{b}$ = input cost productivity, $\bar{b} = \gamma b$, $\gamma = \left(\frac{\omega_L}{1-\alpha}\right)^{1-2\alpha}$.

It should be noted that the general form of production function (A.1) can be applied to different A_j activities with, generally, different $p_j\beta_j\alpha_j$ coefficients. When the costs ω_l, ω_k do not change and neither does technology, i.e. $\alpha_j = \alpha$, $\forall j$, then $u_j = u, \gamma_j = \gamma$ are constant for $\forall j$.

Application of Processing Meat Production Optimization System Operating as a Decision Support System

Anna Bogucka, Albin Rydzewski
Meat and Fat Research Institute in Warsaw, Poland
Waclaw Szymanowski
Agricultural University of Warsaw, Poland

1   The development of the system

The development of the system started in 1987 and about 20 specialists were engaged in works at different periods during this time. The main works and groups doing this work are:

– Determining users requests and coordination of works - 4 persons from Meat and Fat Research Institute (leaders: A. Bogucka, A. Rydzewski).

– Development of *Data Base Generating System* which allowed for creation of the *Processing Meat Production Control System* - 6 persons from theFaculty of Management of University of Warsaw (leaders: J.Kisielnicki, W.Radzikowski).

– Development of *Optimization System* - 5 persons mainly from the Mathematical Faculty of University of Warsaw (leader: W. Ogryczak).

– Parallel development of *Processing Meat Production Control System* and especially on *Optimization System* - 6 persons from Wroclaw Academy of Economics (leader: E. Konarzewska-Gubala).

– Testing of *Optimization System* - 3 persons from Agricultural University of Warsaw (leader: W. Szymanowski).

The Meat and Fat Research Institute intends to develope the *Optimization System in future*, adding an additional phase of sensitivity analysis.

2   The purposes and the properties of system

The *Processing Meat Production Control System* is developed as a *Decision Support System* for the management of meat processing production in the meat plants. *The Optimization System* is developed as an integral part of the *Processing Meat Production Control System* which includes the Data Base with the data that partially can be used by the *Optimization System*. The main goal of the Optimization System is to determine the assortment structure of the processing meat production and the necessary raw materials for this production in the operating plant management. The *Optimization System* can be exploited also in meat plants, where the *Processing Meat Production Control System* is not used. Therefore the *Optimization System* has its own *Data Subbase* which gives possibility to operate it together with the Data Base of the *Processing Meat Production Control System* or without. The system is working on IBM PC microcomputers in a basisconfiguration. It can be used on computers working separately as well as in computer networks.

3   The structure of the Processing Meat Production Control System

The microcomputer *Processing Meat Production Control System* realises the following base functions:
   - the planning and record of raw materials supply,
   - the optimization of production,
   - the settlement of accounts and analysis of production phasis,
   - the planning and record of products selling,
   - the settlement of accounts of products selling.
The general structure of the *Processing Meat Production Control System* is following:

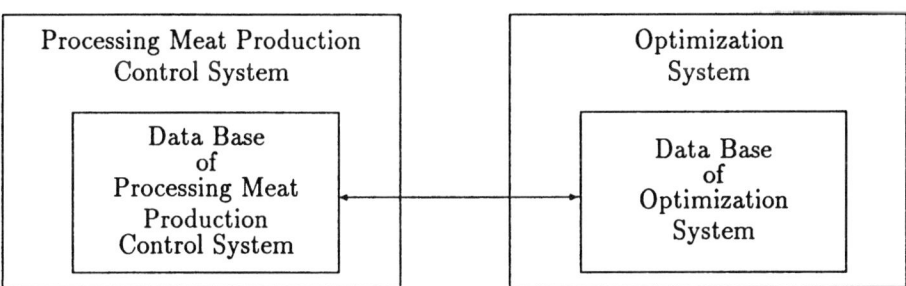

The *Optimization System* can use some data from Data Base of *Processing Meat Production Control System* as follows:

   – names and prices of final products, raw meat materials, spices, auxiliary raw materials and names of market product groups,

   – stock levels and circulations of raw meat materials,

   – informations about market orders for goods and informations about expedition stock levels.

4 The structure of Optimization System

The *Optimization System* consists of three modules:

- the *Feeding File Module* to create and modify all data used by the Optimization System;

- the *Models Generation Module* to build the optimization models for the planning of the processing meat production;

- the *Solver*, which can prepare the solutions of the optimization models.

The general structure of the *Optimization System* is following:

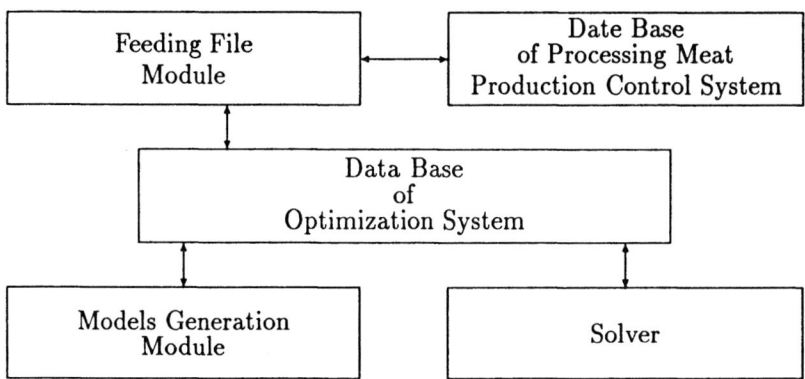

As was mentioned earlier, the *Optimization System* can exist without *Data Base of Processing Meat Production Control System*.

4.1 The Feeding File Module

The main goal of the *Feeding File Module* is to create and to modify the following data files (see Figure 1):
- the file of final products;
- the file of raw meat materials;
- the file of market product groups;
- the file of spices;
- the file of auxiliary raw materials;
- the file of recipes;
- the file of raw material circulations.

The possibility of the ingredients substitution in the recipes is one of the advantages of the *Optimization System*. In case, when *Optimization System* is connected with *Control System*, a part of data is taken from Data Base of Control System.

```
┌─────────────────────────────────────────────────────────┐
│  PROCESSING  MEAT  PRODUCTION  OPTIMIZATION              │
│      SYSTEM      -       FEEDING  FILE  MODULE           │
└─────────────────────────────────────────────────────────┘

   Enter index file for actualization:
  ╔══════════════════════════════════════════════════════╗
  ║File of products              -  ASOR                  ║
  ║File of raw meat materials    -  SUROWCE               ║
  ║File of spices                -  PRZYPRAW              ║
  ║File of auxiliary materials   -  MATPOM                ║
  ║File of market product groups-  GH, AG                 ║
  ║Raw materials circulations    -  OBROTY               ║
  ║Reindexing of index files                              ║
  ║Global actualizations with new prices                  ║
  ╚══════════════════════════════════════════════════════╝
```

F1-help   Enter - choice of activity   Esc-exit

Figure 1: The Feeding File Module: Choice the sphere of activity at beginning of work with the module

4.2   Models Generation Module

The Models Generation Module, connecting the Feeding File Module and The Solver, can prepare the input data files.

This module gives possibilities of:

- generating and modifying the optimization models,

- automatically taking necessary information from the data base for preparing the optimization models,

- modifying the data obtained from the data base.

The generation of the optimization model is realized in the following steps:

1. Choosing the kind of generated model. The program gives the possibility of generating two types of models: the production on the basis of the pickled or fresh meat.

2. Determining whether the quantitative data should be taken from data base, from input model files or determined by the user.

3. Determining the parameters of the model. This step includes the list of products, product groups and raw materials which are included to the optimization calculations and the production constraints. Constraints for decision variables can be given as three values: lower and upper bound and user's aspiration level. For product groups this constraints can be also given in the form of production structure percentage constraints.

4. Controling the data coincidence in the Optimization Model.

5. Preparing the data files for the Solver which includes:

```
╔══════════════════════════════════════════════════════════════╗
║  ENTERING  THE  PARAMETERS  FOR  MODEL      TEST               ║
╚══════════════════════════════════════════════════════════════╝
```

Choose from the following possibilities:

```
┌──────────────────────────────────────────────────────────────┐
│ Determining  product  groups  in  optimization  model         │
│ Determining  products  in  optimization  model                │
│ Determining  raw  meat  materials                             │
│ Determining  percentage  structure  of  product  groups       │
│ Change  of  input  data  source                               │
│ Save  model  in  form  for  calculation                       │
│ Print  production  plan                                        │
│ Exit  without  saving  model  in  form  for  calculation      │
└──────────────────────────────────────────────────────────────┘
```

F1-Help   Enter-choice of activities   Esc-Exit

Figure 2: Models Generation Module: Choice of activities during generation of model

- the specifications of different types of input data in rows and columns,
- the list of rows and columns of the simplex matrix,
- the simplex matrix i.e.

  number of rows and columns,

  the constraint types of rows and columns,

  the constraint values for rows and columns,

  the factor values of matrix.

6. The analysis and printing of the solution calculated by the *Solver*.

4.3   The Solver

The *Solver* uses the files generated by the *Models Generation Module*. The following specific operations are possible :

1. To provide the solution of the optimization model with one or all five following objective functions (see Figure 3):

   - to maximize the profit,
   - to minimize the deviation between computed processing meat programme and market items requirements; the market items requirements are introduced in model as aspiration levels for products and groups of product,
   - to minimize the deviation between the computed using up raw materials and the appropriate resources (it means values introduced in model as an aspiration levels for raw materials),
   - to maximize the global quantity of processing meat production,

```
┌──────────────────────────────────────────────────┐ ┌─────────────┐
│ PROCESSING MEAT PRODUCTION OPTIMIZATION          │ │ Model:      │
│       SYSTEM     -      SOLVER                    │ │ TEST        │
└──────────────────────────────────────────────────┘ └─────────────┘
            ┌═ Enter objective function: ═══════════┐
            │ PROFIT                           (max) │
            │ DEVIATIONS FROM EXPECTED PROD.   (min) │
            │ DEVIAT.OF RAW MAT.FROM ASPIR.    (min) │
            │ QUANTITY OF MEAT PRODUCT PROD.   (max) │
            │ LABOUR CONSUMPTION               (min) │
            │ Multi objective optimization           │
            └────────────────────────────────────────┘
```

Figure 3: Solver: Choice of objective function

```
┌──────────────────────────────────────────────────┐ ┌─────────────┐
│ THE PROCESSING MEAT PRODUCTION OPTIMIZATION       │ │ Model:      │
│       SYSTEM     -      SOLVER                     │ │ TEST        │
└──────────────────────────────────────────────────┘ └─────────────┘
┌══════════════ Current solution Otimal ═════════════════════════════┐
│                               Value:      actual     preceding      │
│ PROFIT                        (max)  28806549.6  28830766.2         │
│ DEVIATIONS FROM EXPECTED PROD.(min)      1546.3      2919.2         │
│ DEVIAT.OF RAW MAT.FROM ASPIR. (min)        58.7       600.0         │
│ QUANTITY OF MEAT PRODUCT PROD.(max)     13833.3     14596.6         │
│ LABOUR CONSUMPTION            (min)     15100.0     20561.1         │
└────────────────────────────────────────────────────────────────────┘

┌═ What next? ═══════════════════┐
│ New objective function         │
│ Analyzing and modifying of model│
│ Seeing the solution            │
│ Write current solution         │
│ Delete current solution        │
│ Read one of previous solution  │
│ Comparison of solutions        │
│ End of optimization            │
└────────────────────────────────┘
```

Figure 4: Solver: Choice of operations

- to minimize the labour consumption.

The multiple objective optimization problem utilizes an extention of classical reference point approach (see Ogryczak, W. et al., 1987; Lewandowski, A. et al., 1987). In this approach the Decision Maker forms his requirements in terms of aspiration and reservation levels i.e. he specifies in an interactive way the acceptable and required values for given objectives according to the earlier determined types of objective functions.

After calculations with any of the objective functions the *Solver* returns to the begining, to allow for a choice of the next operation (see Figure 4).

2. To analyze more deeply actual level of the model formulation *Solver* gives possibility to come back to the *Models Generation Module*;

```
┌──────────────────────────────────────────────────────┐ ┌──────────────┐
│  PROCESSING MEAT PRODUCTION OPTIMIZATION               │ │ Model:       │
│        SYSTEM      -       SOLVER                       │ │ TEST         │
└──────────────────────────────────────────────────────┘ └──────────────┘
┌═══ Enter levels of Aspiration/Rezarvation for objectivs: ═══┐
│                    Values:  Aspiration  Rezervation Solution│
│PROFIT                       (max) 30000000   10000.0  28806549.6│
│DEVIATION FROM EXPECTED PRO (min)   1000.0    2000.0      1546.3│
│DEVIAT.OF RAW MAT.FROM ASP. (min)      0.0       1.0        58.7│
│AMOUNT OF MEAT PRODUCT PRO. (max)  15000.0   10000.0     13833.3│
│LABOUR CONSUMPTION          (min)  15000.0   20000.0     15100.0│
└─────────────────────────────────────────────────────────────┘
Exit: <Esc>
```

Figure 5: Solver: Determining the parameters for multicriterial optimization

3. To visualize new efficient (Pareto-optimal) solution computed by the Solver and presented to the Decision Maker as the current solution;

4. To registrate current solution;

5. To delete current solution;

6. To read one of previous solutions;

7. to compare diferent efficient solutions by simultaneous vizualization;

8. The end of the interactive procedure determines the current solution as a base for the future analysis.

5 The applications of the Optimization System

In this part of the paper we would like to present the results of some applications of the *Optimization System* in one of selected meat plants. To obtain a satisfying solution, the multicriteria method was used (see Figure 5).

The solution is printed in following form:

SOLUTION OF OPTIMIZATION   IN PERIOD: 04/27/91 – 04/27/91

NAME OF PRODUCT GROUP	Lower B.	Upper B.	Aspirat.	Solution
KIELB. POLTRWALE	0.0			838.8
KIELB. TRWALE	0.0			937.2

NAME OF FINAL PRODUCT	Lower B.	Upper B.	Aspirat.	Solution
KIELB. JALOWCOWA	0.0	200.0	150.0	0.0
KIELB. PIWNA	100.0	400.0	400.0	263.8
KIELB. BIALOSTOCKA	50.0	300.0	200.0	50.0
KIELB. POLSKA WEDZ.	0.0	2500.0		375.0
KIELB. TORUNSKA	100.0	100.0		100.0
SERWOLATKA	50.0	50.0		50.0
KIELB. KRAKOWSKA SUCHA	200.0	3000.0		437.2
KIELB. MYSLIWSKA SUCHA	500.0	500.0		500.0
KIELB.PAROWKOWA	0.0	1000.0	500.0	516.6
PAROWKI	50.0	500.0		50.0
KIELB. ZWYCZAJNA	0.0	1000.0		328.4
BALERON GOT.	100.0	2000.0		291.0
POLEDWICA SOPOCKA	300.0	700.0	450.0	475.0
SCHAB WEDZ.	50.0	250.0		250.0
SZYNKA WP GOT.	0.0	200.0	100.0	95.0
SZYNKA WP WEDZ.	100.03	200.0		100.0

NAME OF RAW MATERIAL	Lower B.	Upper B.	Aspirat.	Solution
BOCZEK SKOR.	0.0	645.0		22.4
KARCZEK B/K	0.0	300.0	200.0	300.0
PODGARDLE SKOR.	0.0	100.0		100.0
POLEDWICA WP	200.0	500.0	300.0	500.0
SCHAB	200.0	7699.0		260.4
SLONINA	0.0	3744.0		64.3
SZYNKA WP B/K	0.0	200.0		200.0
TLUSZCZ DROBNY	0.0	205.0		205.0
WOL B/K I KL.	100.0	200.0	150.0	163.1
WOL B/K II KL.	100.0	400.0	300.0	400.0
WOL B/K III KL.	0.0	100.0	50.0	20.5
WOL B/K IV KL.	0.0	100.0		1.3
WP B/K I KL.	0.0	800.0	600.0	800.0
WP B/K II KL.	500.0	1200.0	700.0	1200.0
WP B/K III KL.	0.0	100.0		34.7
WP B/K IV KL.	0.0	50.0		21.7
KREW	0.0	152.0		0.0
KAZEINIAN SODU	0.0	100.0		75.6
MIESO Z MECH. ODMIES. KOSCI	0.0	500.0		21.7
SKORKI	0.0	100.0		21.7

RECIPES FOR FINAL PRODUCTS   IN PERIOD: 04/27/91 - 04/27/91

NAME OF FINAL PRODUCT	AMOUNT	NAME OF RAW MATERIAL	AMOUNT	%
KIELB. PIWNA	263.8	TLUSZCZ DROBNY	91.0	30.0
		WOL B/K II KL.	60.6	20.0
		WP B/K I KL.	60.6	20.0
		WP B/K II KL.	60.6	20.0
		WP B/K III KL.	30.3	10.0
KIELB. BIALOSTOCKA	50.0	TLUSZCZ DROBNY	11.1	20.0
		WOL B/K II KL.	27.8	50.0
		WOL B/K III KL.	16.7	30.0
KIELB. POLSKA WEDZ.	375.0	WP B/K I KL.	174.4	40.0
		WP B/K II KL.	261.6	60.0
KIELB. TORUNSKA	100.0	WOL B/K II KL.	23.0	20.0
		WP B/K II KL.	92.0	80.0
SERWOLATKA	50.0	BOCZEK SKOR.	22.4	35.0
		WOL B/K I KL.	16.0	25.0
		WOL B/K II KL.	20.5	32.0
		WOL B/K III KL.	3.8	5.9
		WOL B/K IV KL.	1.3	2.0
KIELB. KRAKOWSKA SUCHA	437.2	SLONINA PRZEM.	64.3	10.0
		WOL B/K II KL.	64.3	10.0
		WP B/K I KL.	417.9	65.0
		WP B/K II KL.	96.4	15.0
KIELB. MYSLIWSKA SUCHA	500.0	WOL B/K I KL.	147.1	20.0
		WOL B/K II KL.	73.5	10.0
		WP B/K I KL.	147.1	20.0
		WP B/K II KL.	367.6	50.0
KIELB. PAROWKOWA	516.6	PODGARDLE SKOR.	86.8	20.0
		TLUSZCZ DROBNY	86.8	20.0
		WP B/K II KL.	151.9	35.0
		WP B/K IV KL.	21.7	5.0
		KAZEINIAN SODU	43.4	10.0
		MIESO Z MECH. ODMIES. K.	21.7	5.0
		SKORKI	21.7	5.0
PAROWKI	50.0	PODGARDLE SKOR.	13.2	30.0
		WOL B/K II KL.	17.5	40.0
		WP B/K II KL.	8.8	20.0
		WP B/K III KL.	4.4	10.0
KIELB. ZWYCZAJNA	328.4	TLUSZCZ DROBNY	16.1	5.0
		WOL B/K II KL.	112.7	35.0
		WP B/K II KL.	161.0	50.0
		KAZEINIAN SODU	32.2	10.0
BALERON GOT.	291.0	KARCZEK B/K	300.0	100.0
POLEDWICA SOPOCKA	475.0	POLEDWICA WP	500.0	100.0
SCHAB WEDZ.	250.0	SCHAB	260.4	100.0
SZYNKA WP GOT.	95.0	SZYNKA WP B/K	99.0	100.0
SZYNKA WP WEDZ.	100.0	SZYNKA WP B/K	101.0	100.0

One of the interesting features of the system is that the recipes for products, where substitution of the ingredients is allowed, are calculated during the optimization.

The applications made in this year in one of selected meat plants show the possibility to improve the rentability of processing meat production up to 5%. To solve a problem with the matrix of dimension 100 x 200 on microcomputer IBM PC/AT one needs about two minutes. It takes about two hours for the Decision Maker to find a satisfying efficient solution. The interactive dialog between Decision Maker and *Optimization System* gives possibility for an easy adaptation of the model to the specific Decision Maker requirements, such as:

– calculating the most profitable production structure on the basis of the given structure of raw materials,

– calculating the needed structure of raw materials on the basis of the given structure of products,

– optimization with given restrictions on final products as well as on raw materials.

The application shows that this *System* can be a good instrument for a better planning of the meat processing production in meat plants.

6   Problems for further research

In market economy, the marketed items requirements are strictly defined and it is dificult to change the needed structure of production without changing the prices. Fortunately, meat plants in Poland have now possibility to change prices of their products (which was impossible in the state-controlled economy). The main question now is: which changes of prices for products may ameliorate the structure of products demanded by market while increasing the profit of a meat plant? Generally, the Meat and Fat Research Institute intends in the future to develop further the *Optimization System*, adding an additional phase of sensitivity analysis in order to make the management in meat plants more efficient.

References

Ogryczak, W., Studzinski, K and K. Zorychta (1987). General concepts of the Dynamic Interactive Network Analysis System (DINAS). *Archiwum Automatyki i Telemechaniki*, Vol. 32, pp. 277-287.

Lewandowski, A., Kreglewski, T., Rogowski, T. and A. P. Wierzbicki (1987). Decision Support System of DIDAS Family (Dynamic Interactive Decision Analysis and Support). *Archiwum Automatyki i Telemechaniki*, Vol. 32, pp. 221-245.

Bogucka, A., Pasternak, K. et al. (1986). The Application of Optimization in Meat Industry. *Gospodarka Miesna*, 8'86, 9-10'86.

Radzikowski, W., Dabrowski, P. (1988). Processing Meat Production Control System, Part II.2: Data Base Management, works of: Meat and Fat Research Institute, 8'88.

Radzikowski, W., Chmielarz, W., and B.&W. Ogryczak (1988). Processing Meat Production Control System. Part II.3: Optimization Subsystem - Integration of programming tools, works of: Meat and Fat Research Institute, 11'88.

Competitive Selection of R & D Projects by a Decision Support System

Alexy B. Petrovsky and Gennadiy I. Shepelyov

Institute for Systems Studies
Russian Academy of Sciences
Moscow, Russia

Abstract

The paper describes a methodological approach and techniques concerning multicriterial selection of basic R&D projects on the competitive basis in the USSR Academy of Sciences. The main features of the Decision Support System (DSS) that was developmed to support the selection are given. Experts' estimations and factographical data related to R&D projects form the DSS information base.

The DSS to be used by the competition committee aims at solving problems of multicriterial choice and group decision making. The system provides a comfortable interface with users, a simple interpretation of a choice model (decision rule) in an accepted terminology, and a solution of choice problem within the framework of any model in a real time. The methodological approach to multicriterial alternatives selection and the DSS were verified in the course of a competition of basic research projects.

1 Problems of competitive selection of R&D projects

The elaboration of scientific and technological research programs and R&D projects on the competitive basis is rather widespread in ogranization of R&D activities. The system of R&D grants in the U.S.A. of a competition of basic research projects in Hungary should be mentioned as an example. Also the competitive system has recently gained some acceptance in the Soviet Union for elaboration of the governmental science and technology programs as well as basic and applied research programs in the U.S.S.R. Academy of Sciences.

The foundations of competitive selection of R&D alternatives include the following tasks:

- an assessment and choice of projects meeting certain criteria connected with requirements of science policy;

- an allocation of resources to selected projects (budgeting individual scientists or research teams engaged in concrete projects, rather than organizations conducting various studies).

With a view to conducting a competition special organizational structures are set up which ussually includes experts estimating the bids. Thus as in the US National Science Foundation it is the well-known peer review system. The USSR Academy of Sciences set up special scientific boards and competition committees which include prominent scientists of this country representing different scientific fields.

The problem of competitive selection of basic research projects has the following peculiarities. Fundamental science is characterized by a high level of uncertainty and risk connected with elicitation of new knowledge. So, planning basic research, in general, and its assessment and comparision, in particular, are referred to ill-structured problems of unique choice.

There are both quantitative and qualitative factors essential for projects assessment usually dominated by the former. The key factors may be evaluated only by experts working in the same field of science.

Another peculiarity of the considered problem is the availability of various, often conflicting interests among scientists participating in the competition, and scientists estimating R&D projects. At the same time basic science differs from applied science in a more democratic character of problem discussion and decision making. It means that during the competition of basic research projects each member of the competition committee may have a considerable influence on the final results, though sometimes the opinion of the program director or the chiarman of scientific board may be decisive.

Finally, the competition is not usually to take much time. The discussion of projects presented to the competition lasts from one to several days. In the course of discussion members of the competition committee are to have every opportunity to analyse different variants of scientific policy and quickly receive all project-related information required for decision making.

2 Methodological approach

A lot of problems arising in the assessment of R&D activities may be analysed by methods of multicriterial alternative estimation and comparison. The suggested approach to evaluate R&D activities has the following methodological features:

- the considered problems are described in a language which is used by decision makers and management bodies in real-life situations, is habitual for managers and understandable for experts;

- the information elicited from decision makers and experts is consistent with a certain scientific and technological policy;

- the information proccesing techniques enable the manager to monitor all the stages of decision preparing and making, to assess the implications of the choice made;

- the reliability of the information must be secured with the collection and processing proedures accounting for the specifics of the handled problem.

The description of a problem situation by making use of qualitative criteria with verbal definitions on the scales constitutes a decision model which can expose comprehensive qualitative notions, take into account an uncertainty associated with incomplete knowledge of all factors. Also, the set of criteria reflects all essential aspects of the problem situation or in our case the scientific policy pursued by the management body or the decision maker. The completeness of this set of criteria is verified by a logical analysis with the decision maker's participation.

The estimate scales of each criterion are formualted in cooperation with the decision maker. The estimates' descriptions contain precisely those quality grades that are usually accounted for by the management body or the decision maker in preparing and making decisions. Criterion estimates are a routine and understandable language of manager-expert communications. In choosing the estimates by each criterion the expert evaluates the objects making use of the management body preferences.

The rules for the alternatives comparison (decision rules) are developed on the basis of the manager preferences with regard to his/her skills and intuition, therefore the manager trusts more such a description of the problem and the results obtained.

The necessity to improve decision making has led to new information technologies – Decision Support Systems (DSS). There are a variety of DSS concepts. According to the decision-based approach suggested by Larichev and Petrovsky (1988), DSSs are man-machine systems that help users to formulate and analyze decision alternatives in a number of ways, to solve complex ill-structured problems by making use of objective and subjective data, models, knowledge.

The "use-system" interface of DSS included tools for generating and controlling the dialogue, data and model management. It can also contain facilities for data and knowledge elicitation, model construction.

The blocks of problem analysis and decision making incorporate procedures and techniques with help formulate the problem, analyse approaches to its solution, and generate the result. In order to perform its functions the decision making block must contain a library of decision methods including those for solution of multicriteria and single criterion problems on objective and subjective models.

3 Computer support of R&D projects competition

Research teams and individual scientists willing to take part in competitions organized by the USSR Academy of Sciences presented their applications for R&D projects. An application included the following information: general factorgraphical data concerning R&D project (title, leader, organization, schedule, resource requirements); objectives of the project; a brief characteristic of the present day situation in the problem area; backlogs of results and their comparision with foreign ones; substance of the proposal and expected results. Every application for the R&D project was evaluated by one or two experts. The experts' conclusions were used as a basis for project selection by the competition committee.

Special information-analytical tools were developed to support the competitions. These are methods of multicriterial expert assessment and DSS "COMPETITION". Experts' estimations and factorgraphical data related to R&D projects formed the DSS information base. A questionnaire for experts' estimations of applications contained the following qualitative criteria; scientific importance; expectations of the project; novelty of operators; available scienfitic backlog; available resources; detailed verbal definitions of quality divisions. For example, the scale criterion "Qualtification level of project operators" looked like this:

a – project operators are one of the best research teams by their experience and qualification level;

b – project operators have experience and qualification level sufficient for project realization;

c – project operators have insufficient experience and qualification level;

d – qualification level of project operators is unknown.

The techniques to be used by the competition committee conducting the competition should be aimed at solving problems of multicriterial choice, and at group decision making. The system must provide a comfortable interface with users, a simple interpretation of choice model (decision rule) in an accepted terminology and a real time solution of choice problem within the framework of any model. From our point of view, satisfaction meeting this demand is essential for an effective application of DSS when the competition committee is making agreed colletive decisions related to R&D project selection (Petrovsky, et al., 1987).

The designed DSS meeting the above requirements integrates technology of data base management systems and original techniques of multicriterial alternatives' selection developed by Larichev et al. (1979, 1987). Gnedenko et al. (1986), Petrovsky and Shepelyov (1989). The method of ordinal classification CLASS is effective in drawing a dividing surface between objects in multicriterial space and classify a set of objects into several groups. The ZAPROS method is helpful for ordering objects in accordance with their estimates including objects classified by method CLASS into one class. To realise quickly any model of subjective choice during sessions of the competition committee a simple, easily interpretable and rather flexible method using the technique of dividing planes was developed. These techniques provide partitioning a set of objects into groups on the basis of decision makers' preferences, identification and elimination of inconsistencies in user's responses.

4 DSS Application

The methodological principles and techniques of multicriterial alternatives selection and the DSS were verified in the course of competition of basic research projects in the USSR Academy of Sciences (Larichev et al., 1989). The user formulated various decision rules (science policy alternatives) while interacting with the DSS by imposing some or other constraints on criteria values. On obtaining the lists of projects meeting the constraints the user quickly found a satisfactory solution. To build more accurate boundaries of classes in a multicriterial space the methods CLASS or ZAPROS can be used but these methods are more time consuming.

"User-sytem" interface in the DSS prototype was realized with the help of system designers. In the course of further DSS development and with a view of facilitating a dialogue between lay users and the system the interface was developed in the form of hierarchical menu providing the answers to the most typical quetions arising during competitions.

Making use of the system the user can do the following:

- selecting the projects which have the given set of expert estimates on the criteria and meets certain resource limits;

- searching the projects on their titles, names of leaders, experts and/or organizations, and so on;

- receiving information about the experts, their estimates of projects, and the results of the expertise.

Information required by the user can be received in detail for an individual project and/or it can be aggregated for a set of projects. This information is presented in textual and graphic forms which is convenient for an analysis.

The experience gained in DSS application showed that the system was effective when it is built in the traditional decision making procedure in the following way. At the first stage members of the competition committee study the experts' conclusions, discuss competetive R&D projects and make a preliminary selection of the best projects. The typical ranges of task: the number of applications for R&D projects – from several scores to hundreds, the number of preliminarily selected projects accounts for 50 – 70% of the total quantity.

On the basis of preliminarily selected projects the Chairman of the competition committee assisted by DSS designers build a model of subjective choice which was the most accordant on this opinion with preferences of the competition committee and leading to the preliminary selection of projects. This led to identification of projects that had been rejected by the committee at the initial stage but satisfying the available decision rule, and projects not meeting it but earlier accepted by the committee. Along with solving the problem of choice in the framework of built model the system generated aggregate data on the total resource requirements for the selected projects. It took 0.5 – 1 minutes on an IBM PC to solve a choice problem with the fixed model of choice. The building of a decsion rule satisfying the user lasted for half an hour.

The subsequenct sessions of the competition committee largely focused on the discovered inconsistencies and final decision concerning the project approval and resource allocation was made. The DSS was used to supply members of the committee with a complete set of materials required for the competition and final decisions. The DSS-generated decision rule reflecting the scientific policy of the competition committee made is possible to convinsingly substantiate the competition results and to explain to competition participants the motives behind the acceptance or rejection of projects.

5 Prospectives

Consider some problems to be solved. First, completeness of a set of factors criteria adequately describing the considered alternatives and accounted for by the user during decision making. In our cases, in building choice models users often applied simple choice strategies and used only fragments of criteria. The problem of using statistic analysis methods to analyse initial and intermediate samples of projects is connected with the problem of completeness. In particular, we mean methods of comparison of experimental distribution to select the criteria not used by members of the competition committee.

The second problem is an aggregation of multicriterial estimates if several experts evaluate the same project. We used various heuristic procedures, for instance, taking into account only the best or the worst estimates followed by an analysis of the decision developed by the user.

The third problem is designing a model of resource allocation that meets various strategies of resource application given different variants of scientific policy. All these issues will be dealt with in the process of designing the next generation of Decision Support Systems.

References

Gnedenko, L.S., O.I. Larichev, H.M. Moshkovich and E.M. Furems (1986). Quasiordering of a Set of Multiattribute Alternatives on the Basis of Reliable Information on Decision-Maker's Preferences. *Avutomatika i Telemkhanika* 9; 104-113.

Larichev, O.I., Yu.A. Zujev and L.S. Gnedenko (1979). ZAPROS (Closed Procedures by Reference Situation) Technique for Solution of Ill-structured Multiple Criteria Choice Problems. Preprint. Moscow, VNIISI.

Larichev, O.I., A.I. Mechitov, H.M. Moshkovich and E.M. Furems (1987). Expert Knowledge Elicitation Systems in Classification Tasks. *Technical Cybernetics* 2; 74-84.

Larichev, O.I. and A.B. Petrovsky (1988). Decision Support Sytems for Ill-structured Problems: Requirements and Constraints. In: Organizational Decision Support Systems. R.M. Lee, A.M. McCosh and P. Migliarese (Eds.) North-Holland, 247-257.

Larichev, O.I., A.S. Prokhorov, A.B. Petrovsky, M.Yu. Sternin and G.I. Shepelyov (1989). The Experience of Fundamental Research Planning on the Competition Base. Vestnik AN SSSR (Papers of the USSR Academy of Sciences), 7; 51-61.

Petrovsky, A.B., M.Yu. Sternin and V.K. Morgoev (1987). Decision Support Systems. Preprint. Moscow, VNIISI.

Petrovksy, A.B. and G.I. Shepelyov (1989). Decision Support System for Competitive Selection of R&D Projects. Proceedings of the International Workshop *Multiple Criteria Decision Support*, Helsinki, Finland, August 7-11, 1989.

THE INTERNATIONAL INSTITUTE FOR APPLIED SYSTEMS ANALYSIS

is a nongovernmental research institution, bringing together scientists from around the world to work on problems of common concern. Situated in Laxenburg, Austria, IIASA was founded in October 1972 by the academies of science and equivalent organizations of twelve countries. Its founders gave IIASA a unique position outside national, disciplinary, and institutional boundaries so that it might take the broadest possible view in pursuing its objectives:

To promote international cooperation in solving problems arising from social, economic, technological, and environmental change

To create a network of institutions in the national member organization countries and elsewhere for joint scientific research

To develop and formalize systems analysis and the sciences contributing to it, and promote the use of analytical techniques needed to evaluate and address complex problems

To inform policy advisors and decision makers about the potential application of the Institute's work to such problems

The Institute now has national member organizations in the following countries:

Austria
The Austrian Academy of Sciences

Bulgaria
The National Committee for Applied Systems Analysis and Management

Canada
The Canadian Committee for IIASA

Czech and Slovak Federal Republic
The Committee for IIASA of the Czech and Slovak Federal Republic

Finland
The Finnish Committee for IIASA

France
The French Association for the Development of Systems Analysis

Germany
The Association for the Advancement of IIASA

Hungary
The Hungarian Committee for Applied Systems Analysis

Italy
The National Research Council (CNR) and the National Commission for Nuclear and Alternative Energy Sources (ENEA)

Japan
The Japan Committee for IIASA

Netherlands
The Netherlands Organization for Scientific Research (NWO)

Poland
The Polish Academy of Sciences

Russia
The Russian Academy of Sciences

Sweden
The Swedish Council for Planning and Coordination of Research (FRN)

United States of America
The American Academy of Arts and Sciences

Lecture Notes in Economics and Mathematical Systems

For information about Vols. 1–210
please contact your bookseller or Springer-Verlag

Vol. 253: C. Withagen, Economic Theory and International Trade in Natural Exhaustible Resources. VI, 172 pages. 1985.

Vol. 254: S. Müller, Arbitrage Pricing of Contingent Claims. VIII, 151 pages. 1985.

Vol. 255: Nondifferentiable Optimization: Motivations and Applications. Proceedings, 1984. Edited by V.F. Demyanov and D. Pallaschke. VI, 350 pages. 1985.

Vol. 256: Convexity and Duality in Optimization. Proceedings, 1984. Edited by J. Ponstein. V, 142 pages. 1985.

Vol. 257: Dynamics of Macrosystems. Proceedings, 1984. Edited by J.-P. Aubin, D. Saari and K. Sigmund. VI, 280 pages. 1985.

Vol. 258: H. Funke, Eine allgemeine Theorie der Polypol- und Oligopolpreisbildung. III, 237 pages. 1985.

Vol. 259: Infinite Programming. Proceedings, 1984. Edited by E.J. Anderson and A.B. Philpott. XIV, 244 pages. 1985.

Vol. 260: H.-J. Kruse, Degeneracy Graphs and the Neighbourhood Problem. VIII, 128 pages. 1986.

Vol. 261: Th.R. Gulledge, Jr., N.K. Womer, The Economics of Made-to-Order Production. VI, 134 pages. 1986.

Vol. 262: H.U. Buhl, A Neo-Classical Theory of Distribution and Wealth. V, 146 pages. 1986.

Vol. 263: M. Schäfer, Resource Extraction and Market Struucture. XI, 154 pages. 1986.

Vol. 264: Models of Economic Dynamics. Proceedings, 1983. Edited by H.F. Sonnenschein. VII, 212 pages. 1986.

Vol. 265: Dynamic Games and Applications in Economics. Edited by T. Basar. IX, 288 pages. 1986.

Vol. 266: Multi-Stage Production Planning and Inventory Control. Edited by S. Axsäter, Ch. Schneeweiss and E. Silver. V, 264 pages.1986.

Vol. 267: R. Bemelmans, The Capacity Aspect of Inventories. IX, 165 pages. 1986.

Vol. 268: V. Firchau, Information Evaluation in Capital Markets. VII, 103 pages. 1986.

Vol. 269: A. Borglin, H. Keiding, Optimality in Infinite Horizon Economies. VI, 180 pages. 1986.

Vol. 270: Technological Change, Employment and Spatial Dynamics. Proceedings, 1985. Edited by P. Nijkamp. VII, 466 pages. 1986.

Vol. 271: C. Hildreth, The Cowles Commission in Chicago, 1939–1955. V, 176 pages. 1986.

Vol. 272: G. Clemenz, Credit Markets with Asymmetric Information. VIII,212 pages. 1986.

Vol. 273: Large-Scale Modelling and Interactive Decision Analysis. Proceedings, 1985. Edited by G. Fandel, M. Grauer, A. Kurzhanski and A.P. Wierzbicki. VII, 363 pages. 1986.

Vol. 274: W.K. Klein Haneveld, Duality in Stochastic Linear and Dynamic Programming. VII, 295 pages. 1986.

Vol. 275: Competition, Instability, and Nonlinear Cycles. Proceedings, 1985. Edited by W. Semmler. XII, 340 pages. 1986.

Vol. 276: M.R. Baye, D.A. Black, Consumer Behavior, Cost of Living Measures, and the Income Tax. VII, 119 pages. 1986.

Vol. 277: Studies in Austrian Capital Theory, Investment and Time. Edited by M. Faber. VI, 317 pages. 1986.

Vol. 278: W.E. Diewert, The Measurement of the Economic Benefits of Infrastructure Services. V, 202 pages. 1986.

Vol. 279: H.-J. Büttler, G. Frei and B. Schips, Estimation of Disequilibrium Modes. VI, 114 pages. 1986.

Vol. 280: H.T. Lau, Combinatorial Heuristic Algorithms with FORTRAN. VII, 126 pages. 1986.

Vol. 281: Ch.-L. Hwang, M.-J. Lin, Group Decision Making under Multiple Criteria. XI, 400 pages. 1987.

Vol. 282: K. Schittkowski, More Test Examples for Nonlinear Programming Codes. V, 261 pages. 1987.

Vol. 283: G. Gabisch, H.-W. Lorenz, Business Cycle Theory. VII, 229 pages. 1987.

Vol. 284: H. Lütkepohl, Forecasting Aggregated Vector ARMA Processes. X, 323 pages. 1987.

Vol. 285: Toward Interactive and Intelligent Decision Support Systems. Volume 1. Proceedings, 1986. Edited by Y. Sawaragi, K. Inoue and H. Nakayama. XII, 445 pages. 1987.

Vol. 286: Toward Interactive and Intelligent Decision Support Systems. Volume 2. Proceedings, 1986. Edited by Y. Sawaragi, K. Inoue and H. Nakayama. XII, 450 pages. 1987.

Vol. 287: Dynamical Systems. Proceedings, 1985. Edited by A.B. Kurzhanski and K. Sigmund. VI, 215 pages. 1987.

Vol. 288: G.D. Rudebusch, The Estimation of Macroeconomic Disequilibrium Models with Regime Classification Information. VII,128 pages. 1987.

Vol. 289: B.R. Meijboom, Planning in Decentralized Firms. X, 168 pages. 1987.

Vol. 290: D.A. Carlson, A. Haurie, Infinite Horizon Optimal Control. XI, 254 pages. 1987.

Vol. 291: N. Takahashi, Design of Adaptive Organizations. VI, 140 pages. 1987.

Vol. 292: I. Tchijov, L. Tomaszewicz (Eds.), Input-Output Modeling. Proceedings, 1985. VI, 195 pages. 1987.

Vol. 293: D. Batten, J. Casti, B. Johansson (Eds.), Economic Evolution and Structural Adjustment. Proceedings, 1985. VI, 382 pages.

Vol. 294: J. Jahn, W. Knabs (Eds.), Recent Advances and Historical Development of Vector Optimization. VII, 405 pages. 1987.

Vol. 295. H. Meister, The Purification Problem for Constrained Games with Incomplete Information. X, 127 pages. 1987.

Vol. 296: A. Börsch-Supan, Econometric Analysis of Discrete Choice. VIII, 211 pages. 1987.

Vol. 297: V. Fedorov, H. Läuter (Eds.), Model-Oriented Data Analysis. Proceedings, 1987. VI, 239 pages. 1988.

Vol. 298: S.H. Chew, Q. Zheng, Integral Global Optimization. VII, 179 pages. 1988.

Vol. 299: K. Marti, Descent Directions and Efficient Solutions in Discretely Distributed Stochastic Programs. XIV, 178 pages. 1988.

Vol. 300: U. Derigs, Programming in Networks and Graphs. XI, 315 pages. 1988.

Vol. 301: J. Kacprzyk, M. Roubens (Eds.), Non-Conventional Preference Relations in Decision Making. VII, 155 pages. 1988.

Vol. 302: H.A. Eiselt, G. Pederzoli (Eds.), Advances in Optimization and Control. Proceedings, 1986. VIII, 372 pages. 1988.

Vol. 303: F.X. Diebold, Empirical Modeling of Exchange Rate Dynamics. VII, 143 pages. 1988.

Vol. 304: A. Kurzhanski, K. Neumann, D. Pallaschke (Eds.), Optimization, Parallel Processing and Applications. Proceedings, 1987. VI, 292 pages. 1988.

Vol. 305: G.-J.C.Th. van Schijndel, Dynamic Firm and Investor Behaviour under Progressive Personal Taxation. X, 215 pages.1988.

Vol. 306: Ch. Klein, A Static Microeconomic Model of Pure Competition. VIII, 139 pages. 1988.